TOPICS IN
NUMERICAL ANALYSIS II

Sponsors

Allied Irish Banks Ltd.
Bank of Ireland Ltd.
Bord Fáilte Éireann
United States Army Research and Development Group (Europe)
IBM Ireland Ltd.
Trinity College Dublin
University College Dublin

TOPICS IN NUMERICAL ANALYSIS II

Proceedings of the Royal Irish Academy
Conference on Numerical Analysis, 1974

Edited by

JOHN J. H. MILLER

School of Mathematics,
Trinity College, Dublin,
Ireland

Published for

THE ROYAL IRISH ACADEMY
by ACADEMIC PRESS · LONDON AND
NEW YORK

ACADEMIC PRESS INC. (LONDON) LTD
24–28 Oval Road
London NW1 7DX

U.S. Edition published by
ACADEMIC PRESS INC.
111 Fifth Avenue
New York, New York 10003

Library of Congress Catalog Card Number: 73–1474

ISBN: 0-12-496952-6

Printed by William Clowes & Sons Limited
London, Colchester and Beccles

Preface

A Conference on Numerical Analysis was held under the auspices of the National Committee for Mathematics of the Royal Irish Academy at University College, Dublin from 29th July to 2nd August, 1974. It brought together approximately 150 numerical analysts and 40 associates from 26 countries. The technical sessions of the conference included 20 forty-five minute invited papers and 34 fifteen minute contributed papers. This was the second of a series of conferences on this topic, the first of which was held in 1972; its proceedings were published in 1973 with the same title, publisher and editor as the present companion volume. The third of these conferences is planned for 16th to 20th August 1976.

This volume contains, in complete form, the papers given by the invited speakers, with the exception of the lecture "Recent results and open problems in analytic computational complexity" by J. F. Traub, which was a summary of work published elsewhere. Also it contains a paper by Ioan A. Rus, who had accepted an invitation to deliver a paper but who, at the last minute, was unable to attend the conference in person. In addition the titles of the contributed papers are listed together with the names and addresses of the authors who presented them at the conference.

The final event of the conference was a memorial to the late Professor Cornelius Lanczos M.R.I.A., who died in Budapest on 24th June, 1974. The session was chaired by Professor David Greene, President of the Royal Irish Academy, and tributes were paid by the Chairman, Professor J. R. McConnell (Dublin Institute for Advanced Studies), Professor M. Urabe (Kyushu University), Professor Géza Freud (Hungarian Academy of Sciences) and Mr. Geoffrey Phillips (B'nai B'rith). In addition to the participants at the conference, those who attended the memorial included Mrs. Lanczos, Dr. Isaac Cohen (Chief Rabbi of Ireland), Dr. Eamon de Valera (former President of Ireland) and Dr. A. J. McConnell (Provost of Trinity College, Dublin).

Cornelius Lanczos was to have delivered an invited paper to the conference. No manuscript of the paper has been found, but just before leaving for Budapest he wrote the following abstract:

"Fourier analysis of random sequences"

Random sequences generated by the machine were submitted to a Fourier analysis. Against expectation, the distribution of the phase angle was not uniform over the circle, but curious "bald patches" remained. The Fourier analysis itself allows the construction of ideal random sequences. For an "ideal white noise" the amplitudes were chosen as one, whereas the phase angles were uniformly distributed. For an "ideal random noise" both phase angles and amplitudes were uniformly distributed. In both cases the sequence of the identification index k was chosen in Monte Carlo fashion, thus obtaining ideal noise patterns, which were tabulated and analysed.

A full account of this work is being written by his co-worker, B. Gellai, which is due to appear this year in a memorial issue of the new journal "Computers and Mathematics with Applications".

The success of the conference was due principally to the participants, who came from so many different countries, and to the high quality of the invited and contributed papers. In addition the generous financial support of the sponsors was vital in meeting the greatly increased travel costs of the invited speakers. Finally, the hard work and varied contributions of many people "behind the scenes" in Dublin was essential to the success of this event. To all of the above I extend my sincere thanks and appreciation.

April, 1975.   John J. H. Miller.

Names and Addresses of Invited Speakers

E. L. ALLGOWER. *Department of Mathematics, Colorado State University, Fort Collins, Colorado 80523, U.S.A.*

R. ANSORGE. *Institut für Angewandte Mathematik, Universität Hamburg, 2, Hamburg 13, Rothenbaumchaussée 41, Germany.*

J. DESCLOUX. *Département de Mathématiques, École Polytechnique Fédérale de Lausanne, 61 Av. de Cour, CH-1007 Lausanne, Switzerland.*

GÉZA FREUD. *Department of Mathematics, Ohio State University, Columbus, Ohio 43210, U.S.A.*

SIN HITOTUMATU. *Research Institute for Mathematical Sciences, Kyoto University, Kitashirakawa, Sakyoku, Kyoto 606, Japan.*

K. R. KELLY. *Research Center, Amoco Production Company, P.O. Box 591, Tulsa, Oklahoma 74102, U.S.A.*

J. D. LAMBERT. *Department of Mathematics, The University, Dundee DD1 4HN, Scotland.*

JEAN MEINGUET. *Institut de Mathématiques, Université de Louvain, Chemin du Cyclotron 2, B-1348 Louvain-La-Neuve, Belgium.*

A. R. MITCHELL. *Department of Mathematics, The University, Dundee DD1 4HN, Scotland.*

M. R. OSBORNE. *Computer Centre, Australian National University, P.O. Box 4, Canberra A.C.T., Australia 2600.*

ALLADI RAMAKRISHNAN. *Matscience, Institute of Mathematical Sciences, Madras 600020, India.*

P. A. RAVIART. *Université Paris VI, Analyse Numérique, Tour 55/65, 5E 4 Place Jussieu, 75230 Paris Cedex 05, France.*

IOAN A. RUS. *Faculty of Mathematics, University of Cluj, Str. Kogalniceanu 1, Cluj, Rumania.*

E. SCHECHTER. *Faculty of Mathematics, University of Cluj, Str. Kogalniceanu 1, Cluj, Rumania.*

Bl. SENDOV. *Faculty of Mathematics, University of Sofia, bul. Ruski 15, Sofia, Bulgaria.*

G. W. STEWART. *Department of Computer Science, Carnegie-Mellon University, Schenley Park, Pittsburgh, Pennsylvania 15213, U.S.A.*

FRIEDRICH STUMMEL. *Mathematisches Seminar, Universität Frankfurt, D6000 Frankfurt am Main, Robert-Meyer-Strasse 10, Germany.*

J. F. TRAUB. *Department of Computer Science, Carnegie-Mellon University, Schenley Park, Pittsburgh, 15213 Pennsylvania, U.S.A.*

M. URABE. *Department of Mathematics, Faculty of Science, Kyushu University, Fukuoka 812, Japan.*

EMIL VITÁSEK. *Institute of Mathematics, Academy of Sciences, Žitná ulice 25, Prague 1, Czechoslovakia.*

MILOŠ ZLÁMAL. *Technical University, Obráncŭ Míru 21, 60200 Brno, Czechoslovakia.*

Titles and Authors of Contributed Papers

* indicates the presenter.

J. ABAFFY. A unified iteration scheme and a new nonsymmetric quasi-Newton method.

J. F. G. AUCHMUTY. Numerical models of rotating stars.

OLE CAPRANI. On the minimal upper bound for the accumulated round-off error in the floating-point summation of positive addends.

M. G. COX. Numerical computations associated with Chebyshev polynomials.

MICHEL CROUZEIX. On the stability of one-step methods for time dependent problems.

Z. CSENDES. General conditions for least-squares solutions of matrix equations.

E. P. CUNNINGHAM* and G. MAHON. Numerical simulation of the effect of experimental design on the efficiency of cattle selection schemes.

ALAIN DIGUGLIELMO. Method of splitting for unilateral problems of parabolic type.

M. E. A. EL TOM. High order spline function approximations for solutions of Volterra integral equations.

A. M. ERISMAN. On computing certain elements of the inverse of a sparse matrix.

FOLORUNSO F. FAMBO. ADI solution of the two dimensional triharmonic equation.

W. FORSTER. Remarks on the use of certain abstract notions in the context of computational methods.

PAUL O. FREDERICKSON. Fast approximate inversion of large elliptic systems.

B. GABUTTI, P. LEPORA* AND G. MERLO. Numerical solution of a large deflection problem.

JÓZSEF GERGELY. Solution of a system of equations by dimension expansion.

V. GIRAULT. An extension of the M.A.C. method for solving Navier-Stokes equations on polyhedral domains.

L. HAYES AND E. WASSERSTROM*. Solution of nonlinear eigenvalue problems by the continuation method.

Names and Addresses of Presenters of Contributed Papers

J. ABAFFY. *Computer and Automation Institute, Hungarian Academy of Sciences, H 1502 Budapest XI, Kende Utca 13-17, Hungary.*

J. F. G. AUCHMUTY. *Department of Mathematics, Indiana University, Bloomington, Indiana 47401, U.S.A.*

OLE CAPRANI. *DIKU Datalogisk Institut, Kobenhavns Universitet, Sigurdsgade 41, DK-2200 Kobenhavn, Denmark.*

M. G. COX. *Division of Numerical Analysis and Computing, National Physical Laboratory, Teddington, Middlesex TW11 OLW, England.*

MICHEL CROUZEIX. *Département de Mathématiques, Université de Rennes, B.P. 25A, Rennes Cedex 35, France.*

Z. CSENDES. *Department of Electrical Engineering, McGill University, Montreal 101, Quebec, Canada.*

E. P. CUNNINGHAM. *Department of Animal Breeding and Genetics, An Foras Talúntais, Dunsinea Research Centre, Castleknock, Co. Dublin, Ireland.*

ALAIN DIGUGLIELMO. *Faculté des Sciences et des Techniques, 25030 Besançon Cedex, France.*

M. E. A. EL TOM. *DD Division, CERN, 1211 Geneva 23, Switzerland.*

A. M. ERISMAN. *Boeing Computer Services, Seattle, Washington 98124, U.S.A.*

FORLORUNSO F. FAMBO. *Institute of Computer Sciences, University of Lagos, Lagos, Nigeria.*

W. FORSTER. *Department of Mathematics, The University, Southampton, England.*

PAUL O. FREDERICKSON. *Department of Mathematical Sciences, Lakehead University, Thunder Bay, Ontario, Canada.*

JÓSZEF GERGELY. *Computer and Automation Institute, Hungarian Academy of Sciences, H 1502 Budapest XI, Kende Utca 13-17, Hungary.*

V. GIRAULT. *Université Paris VI, Analyse Numérique, Tour 55/65 5E, 4 Place Jussieu, 75230 Paris Cedex 05, France.*

F. J. JACOBS. *Department of Mathematics, Technological University Eindhoven, P.O. Box 513, Eindhoven, Netherlands.*

PIERRE JAMET. *Centre d'Etudes de Limeil, B.P. 27, 94190 Villeneuve St. Georges, France.*

P. LEPORA. *Istituto Matematico, Politecnico di Torino, Turin, Italy.*

MATTI MÄKELÄ. *Institute of Mathematics, Helsinki University of Technology, SF-02150 Otaniemi, Finland.*

NABIL R. NASSIF. *Department of Mathematics, American University, Beirut, Lebanon.*

OLE ØSTERBY. *Matematisk Institut, Ny Munkegade, 8000 Aarhus, Denmark.*

M. H. C. PAARDEKOOPER. *Tilburg School for Economics, Social Sciences and Law, Hooeschoollaan 225, Tilburg, Netherlands.*

M. J. D. POWELL. *Building 8.9, A.E.R.E., Harwell, Didcot, Berkshire, England.*

REIMUND RAUTMANN. *Institut für Angewandte Mathematik, Universität Hamburg, 2 Hamburg 13, Rothenbaumchaussée 41, Germany.*

V. RULOFF. *Oxford University Computing Laboratory, 19 Parks Road, Oxford OX1 3PL England.*

R. A. SACK. *Department of Mathematics, University of Salford, Salford M5 4WT, Lancashire, England.*

A. SHARMA. *Department of Mathematics, University of Alberta, Edmonton, Alberta, Canada.*

M. VAN VELDHUIZEN. *Wiskundig Seminarium, Vrije Universiteit, De Boelelaan 1081, Amsterdam, Netherlands.*

J. G. VERWER. *Mathematical Centre, Boerhaavestraat 49, Amsterdam, Netherlands.*

H. A. VAN DER VORST. *Academisch Computer Centrum Utrecht, Budapestlaan 6, de Uithof-Utrecht, Netherlands.*

S. J. WADE. *Department of Mathematics, University of Salford, Salford M5 4WT, Lancashire, England.*

E. WASSERSTROM. *Department of Aeronautical Engineering, Technion, Haifa, Israel.*

D. G. WILSON. *Union Carbide Corporation, Nuclear Division, P.O. Box Y, Oak Ridge, Tennessee 37830, U.S.A.*

LUC WUYTACK. *Department of Mathematics, University of Antwerp, Universiteitsplein 1, B-2610, Wilrijk, Belgium.*

Contents

On a Discretization of $y'' + \lambda y^k = 0$

E. L. Allgower

Section 1

The nonlinear two-point boundary value problem

$$y'' + \lambda y^k = 0 \quad (\lambda > 0, k \text{ a non-negative rational number} \neq 1)$$
$$y(0) = 0 = y(L) \tag{1.1}$$

and several other closely related problems have recently been treated using primarily two distinct techniques:

(i) variational methods and classical analysis, e.g., [1], [8], [9];
(ii) topological methods such as Lyusternik–Schnirelman theory or concepts such as genus or transversality, e.g., [2], [5], [6], [10].

The nonlinear boundary value problem (1.1) relates to certain physical problems connected with oscillations. For example, if $k = 3$, (1.1) may describe the motion of a mass suspended between a pair of springs (see, e.g., [3]).

The approach which is taken here is to study the discrete analogue of (1.1):

$$\delta_h^2 y_i + \lambda y_i^k = 0, \quad i = 1, 2, \ldots, n$$
$$y_0 = 0 = y_{n+1} \tag{1.1_n}$$

where $h = L/(n + 1)$, $x_i = ih$, $y_i = y(x_i)$ for $y \in C^2[0, 1]$ and δ_h^2 represents the standard difference approximation to y''. It will be seen that by a simple trick the problem of solving (1.1_n) may be reduced to that of finding the real zeros of an algebraic function which is recursively defined. This approach may be used both to obtain numerical approximations to the solutions to (1.1) and to deduce qualitative results concerning the solutions to (1.1). Approximate solutions to (1.1_n) are not found by standard techniques, but in fact by utilizing certain properties of a recursive relation arising from (1.1_n). The method of finding numerical approximations is analyzed and seen to be reasonably efficient.

1

The reason for undertaking this study, however, is not its surprising tractability, but rather to obtain insights into discretizations of nonlinear boundary value problems which may have a countable infinitude of solutions, e.g., if k is an odd integer.

Since k may be a non-negative rational, it is necessary to define what is to be understood by y^k in this context. If y^k is real only for non-negative values of y, y^k will denote the non-negative root of $y(x)^k$ for each $x \in [0, L]$. This case will be referred to as the *even case*. If y^k is real both for positive and negative values of y, y^k will denote the principal root of $y(x)^k$ for each $x \in [0, L]$. This case will be referred to as the *odd case*. Thus

$y^k \geqslant 0$ on $[0, L]$ in the even case,

y^k may be positive or negative in the odd case.

For example, $k = 3/2$ is an even case and $k = 1/3$ is an odd case.

Section 2

Since it is intended to make a comparison of the qualitative features of solutions of (1.1) with solutions of (1.1$_n$), several results concerning solutions of (1.1) will be quoted here. Most of these results may be found in the articles cited in Section 1. The results in the articles cited in Section 1 pertain to the case that k is an odd integer, however, it is seen below that they carry over to rational k under the interpretation made in Section 1.

There always exist two solutions to (1.1) viz. the trivial solution $y^{[0]} \equiv 0$ and $y^{[1]}$ which is concave downward on $[0, L]$.   (2.1)

The solutions $y^{[1]}$ attain their maxima at $x = L/2$ and are symmetric about $x = L/2$ ([7]).   (2.2)

For the odd case, to each positive integer m, there corresponds a solution $y^{[m]}$ having $m - 1$ oscillations in $(0, L)$. (If y satisfies (1.1), then so does $-y$ and hence only solutions with $y'(0) > 0$ will hereafter be considered.)   (2.3)

For the even case, there are only two solutions to (1.1), $y^{[0]}$ and $y^{[1]}$.   (2.4)

A proof of (2.4) will now be given since the author has not seen this result specifically stated elsewhere and in the process formulae for $\max\limits_{[0,L]} y^{[1]}$ and $y^{[1]'}(0)$ for any k will be obtained. For the proof the superscript will be omitted, i.e., $y = y^{[1]}$. Multiplication of (1.1) by y' and subsequent integration yields

$$y'(x)^2 - y'(0)^2 = \frac{-2\lambda}{k+1} y^{k+1}.$$

By (2.1) and (2.2)

$$y'(x) = \left(y'(0)^2 - \frac{2\lambda}{k+1} y^{k+1}\right)^{1/2} = \left(\frac{2\lambda}{k+1}\right)^{1/2} (t - y^{k+1})^{1/2} \qquad (2.5)$$

where

$$t = \left(\frac{k+1}{2\lambda}\right) y'(0)^2.$$

Thus

$$x = \left(\frac{k+1}{2\lambda}\right)^{1/2} \int_0^y (t - y^{k+1})^{-1/2} \, dy \quad \text{for } x \in [0, L]. \qquad (2.6)$$

Since by (2.2), $y'(x) = 0$ if and only if $x = L/2$ and by (2.5), $y'(x) = 0$ if and only if $y = t^{1/(k+1)}$, (2.6) yields

$$L/2 = \left(\frac{k+1}{2\lambda}\right)^{1/2} \int_0^{t^{1/(k+1)}} (t - y^{k+1})^{-1/2} \, dy$$

$$= \left(\frac{k+1}{2\lambda}\right)^{1/2} (t^{-(k-1)/(2k+2)}) \left[1 + \sum_{j=1}^{\infty} (1/2)(3/4) \cdots \left(\frac{2j-1}{2j}\right)\right.$$

$$\left. \times \left(\frac{1}{jk+j+1}\right)\right].$$

Thus

$$t = \left[\frac{2C_k^2(k+1)}{L^2\lambda}\right]^{(k+1)/(k-1)}$$

where

$$C_k = 1 + \sum_{j=1}^{\infty} (1/2)(3/4) \cdots \left(\frac{2j-1}{2j}\right)\left(\frac{1}{jk+j+1}\right).$$

Now

$$\max y = t^{1/(k+1)} = \left[\frac{2C_k^2(k+1)}{L^2\lambda}\right]^{1/(k-1)} \qquad (2.7)$$

and

$$y'(0) = \left[\left(\frac{k+1}{2\lambda}\right)\left(\frac{2C_k}{L}\right)^{k+1}\right]^{1/(k-1)}. \qquad (2.8)$$

Table 1 lists $y'(0)$ and $y_{\max}$ for several values of k.

The uniqueness of the solution $y^{[1]}$ now follows from the uniqueness of its value at $x = L/2$. Problem (1.1) may now be recast as an initial value problem by using (2.8) and numerical approximations to the solution $y^{[1]}$

TABLE 1

k	C_k	y_{max}	$y'(0)$
0	2	$\lambda L^2(1.25 \times 10^{-1})$	$\lambda L(0.5)$
0.4	1.764	$\lambda^{5/3}L^{10/3}(2.71069 \times 10^{-2})$	$\lambda^{5/3}L^{7/3}(9.56315 \times 10^{-2})$
0.6	1.68568	$\lambda^{5/2}L^5(4.01097 \times 10^{-3})$	$\lambda^{5/2}L^4(1.37934 \times 10^{-2})$
2	1.40102	$\lambda^{-1}L^{-2}(11.7773)$	$\lambda^{-1}L^{-3}(3.30000 \times 10)$
3	1.31007	$\lambda^{-1/2}L^{-1}(3.7054)$	$\lambda^{-1/2}L^{-2}(9.7085)$
4	1.25290	$\lambda^{-1/3}L^{-2/3}(2.50386)$	$\lambda^{-1/3}L^{-5/3}(6.27418)$

may be obtained by techniques designed for such problems, e.g., Runge–Kutta methods. However, the method which will be presented below is at least as efficient for moderate values of k and does not require the calculation of C_k.

Now suppose that the odd case obtains. A solution $y^{[m]}$ having $m - 1$ oscillations in $(0, L)$ is formed as follows. First replace L by L/m in (1.1). Then there is a unique solution $u^{[1]}$ which is concave downward on $[0, L/m]$ and

$$\max_{[0,L/m]} u^{[1]} = u^{[1]}(L/2m) = \left[\frac{2m^2 C_k^2(k + 1)}{L^2\lambda} \right]^{1/(k-1)}. \qquad (2.9)$$

Now the function

$$y^{[m]}(x) = (-1)^j u^{[1]}(x) \quad \text{if } x \in \left[\frac{j}{m}, \frac{j+1}{m} \right], \quad j = 0, 1, \ldots, m - 1 \qquad (2.10)$$

satisfies (1.1) and has zeros at $x = j/m, j = 1, \ldots, m - 1$.

By the preceding argument $u^{[1]}$ is unique. Any solution y to (1.1) having its first zero at $x \neq 1/m$ will either be incompatible with the boundary condition $y(L) = 0$ or with the condition that y has exactly $m - 1$ zeros in $(0, L)$. Thus $y^{[m]}$ is uniquely determined; its norm is given by (2.9), and $y^{[m]'}(0)$ may be computed via (2.5) and (2.7).

Owing to (2.10), to find the solutions to (1.1), it suffices to obtain the *fundamental solution* $y^{[1]}$. To conclude this section it is noted that (2.7) reflects the behavior of the solutions to (1.1) corresponding to the super- and sub-linear cases discussed in [6].

Section 3

The system of nonlinear equations (1.1_n) may be written

$$y_{i-1} - 2y_i + y_{i+1} + \lambda h^2 y_i^k = 0, \quad i = 1, \ldots, n,$$
$$y_0 = 0 = y_{n+1}.$$

In particular, for $i = 1$,

$$-2y_1 + y_2 + \lambda h^2 y_1^k = 0$$

or

$$y_2 = y_1(2 - z)$$

where

$$z = \lambda h^2 y_1^{k-1}.$$

Now for $i = 2$,

$$y_3 = -y_1 + 2y_2 - \lambda h^2 y_2^k = y_1(-1 + 2(2 - z) - z(2 - z)^k).$$

Proceeding recursively,

$$\begin{cases} y_i = y_1 f_i(z), \quad i = 1, 2, \ldots, n + 1 \quad \text{where} \\ f_1(z) \equiv 1, \quad f_2(z) = 2 - z \quad \text{and} \\ f_{i+1}(z) = -f_{i-1}(z) + 2f_i(z) - zf_i(z)^k, \quad i \geqslant 2. \end{cases} \tag{3.1}$$

The boundary condition $y_{n+1} = 0$ yields
$$y_1 f_{n+1}(z) = 0. \tag{3.2}$$

Thus to each solution to (3.2) there corresponds a solution to (1.1_n),

$$\bar{y} = y_1(1, 2, -z, \ldots, f_i(z), \ldots, f_n(z))$$

where $y_1 f_{n+1}(z) = 0$. Hence either $\bar{y} = \bar{0}$ or

$$\bar{y}(z) = \left[\left(\frac{n+1}{L} \right)^2 \left(\frac{z}{\lambda} \right) \right]^{1/(k-1)} (1, 2 - z, \ldots, f_n(z)) \tag{3.3}$$

where z is a real root of the algebraic function f_{n+1}. (If k is an integer, $f_{n+1}(z)$ is a polynomial of degree $(k^n - 1)/(k - 1)$.)

It will, however, not be necessary to calculate the roots of f_{n+1} for the purpose of obtaining numerical approximations to the solutions to (1.1). A study of the properties of the real roots of the recursively defined functions f_n will be undertaken rather in order to obtain the analogous results to those in Section 2 for the solutions to (1.1_n).

For each integer $j \geqslant 2$, the distinct real roots of f_j shall be denoted by $z_1(j) < z_2(j) < \ldots < z_{m_j}(j)$. Thus $z_1(j)$, $z_{m_j}(j)$ denote the minimum and maximum real roots of $f_j(z)$, respectively.

It is elementary to establish that $f_j(0) = j$ for every j and for the odd case,

$$f_{3j}(1) = 0 \qquad f_{2j}(2) = 0 \qquad f_{3j}(3) = 0$$
$$f_{3j+s}(1) = (-1)^j, \quad f_{2j+1}(2) = (-1)^j, \quad f_{3j+s}(3) = (-1)^{s+1}, \quad s = 1, 2;$$

$$(3.4)$$

while for the even case, $f_2(2) = 0$ and $f_3(1) = 0$. Since in either case, f_j is an algebraic function, it has only a finite number of real roots.

Theorem 1
For the even case, the real roots of $f_j(z)$ lie in $(0, 2)$ and the interlacement relation $0 < z_1(j+1) < z_1(j)$ holds for every integer $j \geq 2$.

Proof
Suppose that for $z \leq 0$, the condition $0 < f_{j-1}(z) < f_j(z)$ holds. In fact this holds for $j = 2$ regardless of whether the even or odd case obtains. Then

$$f_{j+1}(z) = [-f_{j-1}(z) + f_j(z)] + f_j(z) - zf_j(z)^k \geq f_j(z) \quad \text{for} \quad z \leq 0$$

since in the even case, $f_j(z)^k \geq 0$ for all j, z.

Suppose that for $z \geq 2$ and $j \geq 2$, the condition $f_j(z) < f_{j-1}(z) \leq 0$ holds. This certainly holds for $j = 3$ since

$$f_3(z) = -1 + 2(2 - z) - z(2 - z)^k < 2 - z \quad \text{for} \quad z \geq 2.$$

From the assumption $f_j(z) < f_{j-1}(z) \leq 0$ for $z \geq 2$, it follows that

$$f_{j+1}(z) = [-f_{j-1}(z) + f_j(z)] + f_j(z) - zf_j(z)^k < f_j(z) \leq 0 \quad \text{for} \quad z \geq 2, j \geq 2.$$

Suppose that $0 < z_1(j) < z_1(j-1)$. This holds for $j = 3$ since $z_1(3) \leq 1$ and $z_1(2) = 2$. Since $f_{j-1}(0) = j - 1$ and $z_1(j-1)$ is the first real root of f_{j-1}, $f_{j-1}(z) > 0$ for $z \in [0, z_1(j-1))$. Hence

$$f_{j+1}(z_1(j)) = -f_{j-1}(z_1(j)) < 0$$

and consequently $0 < z_1(j+1) < z_1(j)$.

Theorem 2
For the odd case, the real roots of $f_j(z)$ lie in $(0, 4)$ and the interlacement relations $0 < z_1(j+1) < z_1(j)$ and $z_{mj}(j) < z_{mj+1}(j+1) < 4$ hold for all $j \geq 2$.

Proof
That $f_j(z) > 0$ for $z \leq 0$ follows from the proof of Theorem 1. Define $g_j(z) = (-1)^{j+1}f_j(z)$. Then $g_j(z)$ has the same roots as $f_j(z)$ and in the odd case,

$$g_{j+1}(z) = (-1)^{j+2}f_{j+1}(z) = (-1)^{j+2}(-f_{j-1}(z) + 2f_j(z) - zf_j(z)^k)$$
$$= (-1)^{j+2}((-1)^{-j-1}g_{j-1}(z) + 2(-1)^{-j-1}g_j(z) + z(-1)^{-(j+1)k-1}g_j(z)^k)$$
$$= zg_j(z)^k - 2g_j(z) - g_{j-1}(z).$$

Also $g_1 \equiv 1$, $g_2(z) = z - 2$. Suppose that for $z \geq 4$, the condition $g_j(z) > g_{j-1}(z) \geq 1$ holds. This certainly holds for $j = 2$. Now

$$g_{j+1}(z) = (z - 4)g_j(z)^k + 2(g_j(z)^k - g_j(z)) + (g_j(z)^k - g_{j-1}(z)) +$$
$$+ g_j(z)^k > g_j(z) \geq 1$$

holds for $z \geq 4$ and $j \geq 1$.

The proof of the first interlacement assertion is the same as that in Theorem 1. If $z \in (z_{m_{j-1}}(j-1), 4)$, then $g_{j-1}(z) > 0$ and sgn $f_{j-1}(z) = (-1)^j$. Suppose that the second interlacement assertion holds for $j - 1$, i.e., $z_{m_{j-1}}(j-1) < z_{m_j}(j) < 4$. In fact it holds for $j = 3$ since $z_{m_2}(2) = 2 < 3 \leq z_{m_3}(3) < 4$. Now $f_{j+1}(z_{m_j}(j)) = -f_{j-1}(z_{m_j}(j))$ and hence sgn $f_{j+1}(z_{m_j}(j)) = -\text{sgn} f_{j-1}(z_{m_j}(j)) = (-1)^{j+1}$. But from the preceding, sgn $f_{j+1}(4) = (-1)^{j+2}$. Hence the second interlacement assertion holds.

For the odd case a further interlacement relation generally holds. If sgn $f_{j-1}(z) = (-1)^i$ for $z \in (z_i(j-1), z_{i+1}(j-1))$, $i = 0, 1, \ldots, m_{j-1}$ (here $z_0(j) = 0$, $z_{m_j+1}(j) = 4$) and if

$$z_{i-2}(j-1) < z_s(j) < z_{i-1}(j-1) < z_{s'}(j) < z_i(j-1),$$

then

$$\text{sgn} f_{j+1}(z_s(j)) = -\text{sgn} f_{j-1}(z_s(j)) = (-1)^{i-1}$$

and

$$\text{sgn} f_{j+1}(z_{s'}(j)) = -\text{sgn} f_{j-1}(z_{s'}(j)) = (-1)^i.$$

Thus in general f_{j+1} has at least j real roots and hence (1.1_n) in general has at least n solutions in the odd case.

Example 3.1
For the purpose of studying the sequence of algebraic functions $f_j(z)$, $k = 1$ is a permissible value. Then $f_1 \equiv 1$, $f_2(z) = 2 - z$ and $f_{n+1}(z) = -f_{n-1}(z) + 2f_n(z) - zf_n(z)$ for $n \geq 2$. This sequence of polynomials may be recognized as the sequence of characteristic polynomials of the $n \times n$ matrices A_n with $a_{ii} = 2$, $a_{ij} = -1$ if $|i - j| = 1$ and $a_{ij} = 0$ if $|i - j| \geq 2$. The roots of f_{n+1} are $z_i(n + 1) = 4 \sin^2 [i\pi/2(n + 1)]$, $i = 1, \ldots, n$ and the roots of f_n are interlaced between those of f_{n+1}.

This example also suggests the matrix vector formulation of (3.1), (3.2) as a nonlinear algebraic eigenvalue problem

$$\begin{bmatrix} 2 - zf_1^{k-1} & -1 & & 0 \\ -1 & \ddots & \ddots & \\ & \ddots & \ddots & -1 \\ 0 & & -1 & 2 - zf_n(z)^{k-1} \end{bmatrix} \begin{bmatrix} f_1(z) \\ \vdots \\ \vdots \\ f_n(z) \end{bmatrix} = \begin{bmatrix} 0 \\ \vdots \\ 0 \\ f_{n+1}(z) \end{bmatrix} = \begin{bmatrix} 0 \\ \vdots \\ \vdots \\ 0 \end{bmatrix}.$$

Example 3.2
If $k = 0$, then $f_{n+1}(z) = -f_{n-1}(z) + 2f_n(z) - z$, $f_1 \equiv 1$, $f_2(z) = 2 - z$. Thus $f_n(z) = n - [n(n - 1)/2]z$, $n = 0, 1, \ldots$ and f_{n+1} has only one real root, $z_1(n + 1) = 2/n$ for $n \geqslant 1$. This example suggests the following generalization. In the even case, if $f_{j-1}(z) > 0$ and $f_j(z) \leqslant 0$, then $f_{n+j}(z) < 0$ for $n \geqslant 1$ since $f_{j+1}(z) = -f_{j-1}(z) + 2f_j(z) - zf_j(z)^k < 2f_j(z) \leqslant 0$ and if $f_{j+q}(z) < f_{j+q-1}(z)$, then

$$f_{j+q+1}(z) = [-f_{j+q-1}(z) + f_{j+q}(z)] + f_{j+q}(z) - zf_{j+q}(z)^k < f_{j+q}(z) < 0.$$

Example 3.3
If $k = 4$, then $f_3(z) = (1 - z)(3 - 15z + 17z^2 - 7z^3 + z^4)$ which has three real roots, $0 < z_1(3) < 1 = z_2(3) < z_3(3) < 2$. Hence for $n = 2$, the solutions to (1.1_n) are $\bar{y} = \bar{0}$,

$$\bar{y}(z_1) = \left[\left(\frac{3}{L}\right)^2 \left(\frac{z_1}{\lambda}\right)\right]^{1/3} \begin{pmatrix} 1 \\ 2 - z_1 \end{pmatrix}, \quad \bar{y}(z_2) = \left[\left(\frac{3}{L}\right)^2 \left(\frac{1}{\lambda}\right)\right]^{1/3} \begin{pmatrix} 1 \\ 1 \end{pmatrix},$$

$$\bar{y}(z_3) = \left[\left(\frac{3}{L}\right)^2 \left(\frac{z_3}{\lambda}\right)\right]^{1/3} \begin{pmatrix} 1 \\ 2 - z_3 \end{pmatrix}.$$

Note that $\bar{y}(z_1)$ and $\bar{y}(z_3)$ are not symmetric vectors and hence properties (2.2) and (2.4) do not carry over directly to (1.1_n).

    Example 3.3 indicates that if it is desired that positive solutions to (1.1_n) be symmetric, it is necessary to impose the condition

$$f_{n+1-j}(z) = f_j(z), \quad j = 1, \ldots, n. \tag{3.5}$$

The following theorem shows that (3.5) may be replaced by a single equation.

Theorem 3.3
The system (1.1_n) *has at least one solution satisfying* (3.5). *In particular, any real root of*

$$\begin{aligned} f_{n/2}(z) &= f_{(n/2)+1}(z) & \text{if } n \text{ is even} \\ f_{(n-1)/2}(z) &= f_{(n+3)/2}(z) & \text{if } n \text{ is odd} \end{aligned} \tag{3.6}$$

yields a solution to (1.1_n) *satisfying* (3.5).

Proof
Since the remaining case may be similarly handled, it suffices to consider just one of the cases (3.6), say $n = 2j$. Consider the sequence of functions $g_{j+1}(z) = f_{j+1}(z) - f_j(z)$. Then $g_{j+1}(0) = 1$ and $g_{j+1}(z_1(j + 1)) = -f_j(z_1(j + 1)) < 0$

Hence g_{j+1} has at least one root ζ and $\zeta < z_1(j+1)$. Suppose that $f_{j-r}(\zeta) = f_{j+1+r}(\zeta)$ for $r = 0, 1, \ldots, p$. Since

$$f_{j-p+1}(\zeta) = -f_{j-p-1}(\zeta) + 2f_{j-p}(\zeta) - \zeta f_{j-p}(\zeta)^k,$$
$$f_{j-p-1}(\zeta) = -f_{j-p+1}(\zeta) + 2f_{j-p}(\zeta) - \zeta f_{j-p}(\zeta)^k$$
$$= -f_{j+p}(\zeta) + 2f_{j+p+1}(\zeta) - \zeta f_{j+p+1}(\zeta)^k = f_{j+p+2}(\zeta).$$

Hence (3.5) is satisfied.

Let $\zeta_1(n+1)$ be the least root satisfying (3.5). Then $f_j(\zeta_1(n+1)) > 0$ for $j = 1, 2, \ldots, n$ and $\bar{y}(\zeta_1(n+1))$ in (3.3) corresponds to $y^{[1]}$ in Section 2. Moreover, in the odd case,

$$f_{n+2}(\zeta_1) = -f_n(\zeta_1) + 2f_{n+1}(\zeta_1) - \zeta_1 f_n(\zeta_1)^k = -f_n(\zeta_1) = -f_1(\zeta_1) = -1$$

and if

$$f_{n+1+j}(\zeta_1) = -f_j(\zeta_1) \quad \text{for} \quad j = 1, \ldots, q < n,$$

then

$$f_{n+1+q+1}(\zeta_1) = -f_{n+q}(\zeta_1) + 2f_{n+1+q}(\zeta_1) - \zeta_1 f_{n+1+q}(\zeta_1)^k$$
$$= -f_{q-1}(\zeta_1) - 2f_q(\zeta_1) + \zeta_1 f_q(\zeta_1)^k = -f_{q+1}(\zeta_1).$$

Thus for the odd case $f_{n+1+j}(\zeta_1) = -f_j(\zeta_1)$ for $j = 0, 1, \ldots, n+1$. More generally, $f_{m(n+1)+j}(\zeta_1) = (-1)^m f_j(\zeta_1), j = 0, 1, \ldots, n$. Hence $\bar{y}(\zeta_1(n+1))$ is the discrete analogue of the fundamental solution $y^{[1]}$ in Section 2. $\bar{y}(\zeta_1(n+1))$ may be used in the same way as $y^{[1]}$ was used to obtain $\bar{u}(\zeta_1(n+1), L/m)$. This in turn is used to construct a solution to $(1.1_{m(n+1)+1})$ which oscillates $m-1$ times in $(0, L)$.

Consider the even case and suppose that (3.6) has a root $\zeta_2 > \zeta_1 \geqslant z_1$. The sequence $f_j(\zeta_1)$ initially dominates the sequence $f_j(\zeta_2)$ and hence $f_p(\zeta_2) < 0$ for some $p \leqslant n+1$. But then by the remark following Example 3.2, $f_{n+1}(\zeta_2) < 0$ which is a contradiction. It is similarly seen that in the odd case, there is only one root $(\zeta_1(n+1))$ which yields a positive symmetric solution to (1.1_n).

Example 3.3 and the subsequent discussion show that the discrete analogue (1.1_n) of (1.1) may have "irrelevant" solutions. This phenomenon was observed in a similar context in [4]. The behavior of the irrelevant solutions observed in [4] was that they diverged to infinity as $n \to \infty$. The behavior of the irrelevant solutions in the present problem is that for n sufficiently large, they are simply no longer present. This results from the fact that for n sufficiently large, $z_1(n+1)$ and $\zeta_1(n+1)$ eventually coincide. The lesson which this problem instructs is that the occurrence of irrelevant solutions to the discretization might be prevented by forcing the discrete analogue to fulfill the properties which the solutions of the boundary value problem are known to have.

Ordinarily, if a boundary value problem has more than one solution, the convergence of solutions of the discretization to a solution of the boundary value problem must be proved by applying compactness considerations such as the Arzela–Ascoli theorem. However, owing to the fact that the oscillating solutions to (1.1) and (1.1_n) may be expressed in terms of the fundamental solutions $y^{[1]}$ and $\bar{y}(\zeta_1(n+1))$, respectively, there is no difficulty in extracting sequences of solutions such that $\{\bar{y}_n^{[m]}\}_{n=1}^{\infty} \to y^{[m]}$ as $n \to \infty$. In the odd case, if the roots of $f_{n+1}(z)$ were calculated for $n = 1, 2, \ldots$, the sequence extraction would generally be $\bar{y}(z_m(n+1)) \to y^{[m]}(x_i)$, $i = 1, \ldots, n$.

From (3.3), it follows that in order for a sequence of respective roots $\{z(n+1)\}_{n=1}^{\infty}$ of the sequence $\{f_{n+1}\}_{n=1}^{\infty}$ to yield a sequence of solutions $\{\bar{y}(z(n+1))\}_{n=1}^{\infty}$ which converges to a solution y to (1.1), it is necessary that

$$y_1 = \left[\left(\frac{n+1}{L} \right)^2 \left(\frac{z(n+1)}{\lambda} \right) \right]^{1/(k-1)} \to 0 \text{ as } n \to \infty,$$

i.e., $z(n+1) = o(n^{-2})$ for $k > 1$ or $1/[z(n+1)] = o(n^2)$ for $k < 1$, as $n \to \infty$. But then $f_{\lceil n/2 \rceil}(z(n+1)) = 0(n)$ as $n \to \infty$. Since $y_1 f_{\lceil n/j \rceil}(z(n+1)) \to y(x_j) \neq 0$ for some $j = 1, \ldots, n$ as $n \to \infty$, it follows that

$$z_m(n+1) = 0(n^{-(k+1)}) \quad \text{as} \quad n \to \infty. \tag{3.7}$$

Thus it is plain that it is also possible to construct irrelevant sequences of solutions $\bar{y}_n(z)$ to (1.1_n), $n = 1, 2, \ldots$. For example, if $n = 3j - 1, j = 1, 2, \ldots$, then $z = 1, 3$ are roots of $f_{n+1} = f_{3j}$ and

$$\bar{y}_{3j-1}(1) = \left[\left(\frac{3j}{L} \right)^2 \left(\frac{1}{\lambda} \right) \right]^{1/(k-1)} \quad (1, 1, 0, -1, -1, 0, \ldots, 0, (-1)^j, (-1)^j),$$

$$\bar{y}_{3j-1}(1) = \left[\left(\frac{3j}{L} \right)^2 \left(\frac{3}{\lambda} \right) \right]^{1/(k-1)} \quad (1, -1, 0, 1, -1, \ldots, 1, -1)$$

are solutions to (1.1_{3j-1}) for $j \geqslant 1$ which clearly do not converge to any solution to (1.1) as $j \to \infty$. These solutions should rather be considered as (initial) terms in sequences of solutions to (1.1_{3j-1}) which converge to solutions of (1.1) having $j - 1$ and $j + 1$ oscillations, respectively.

Section 4

Theorem 4.1
Let the sequence $\{f_n(z)\}$ be defined by the recursion relation (3.1). Then for any $z \in (0, 4)$, there is an integer N_z such that $f_{N_z}(z) \leqslant 0$.

Proof
Suppose that there is a z such that $f_n(z) > 0$ for all n. Then the sequence $\{f_n(z)\}$ is dominated by the sequence $g_{n+1}(z) = -g_{n-1}(z) + 2g_n(z)$, $g_1(z) = 1$,

$g_2(z) = 2 - z$. But $g_n(z) = n - (n-1)z$ and $g_n(z) \leqslant 0$ for $n \geqslant 1/(z-1)$, i.e., for $z \geqslant n/(n-1)$. Since $f_3(1) = 0$, only $z \in (0, 1)$ remains possible.

For any $z \in (0, 1)$ and $\alpha \in (0, 1]$, the sequence $f_n(z)$ initially satisfies the condition $f_n(z)^k \geqslant \alpha$. Hence the sequence

$$h_{n+1}(z) = -h_{n-1}(z) + 2h_n(z) - \alpha z, \quad h_1(z) = 1, \quad h_2(z) = 2 - z$$

dominates the sequence $\{f_n(z)\}$ for all initial n such that $f_n(z)^k \geqslant \alpha$. But $h_n(z) = n - [(n-1)/2](2 + (n-2)\alpha)z$, $n = 1, 2, \ldots$ and hence $h_n(z) \leqslant 0$ for n sufficiently large. Hence either $f_n(z) \leqslant 0$ or $f_n(z)^k < \alpha$ for some n. In the latter case, there is a first n such that $f_{n-1}(z)^k \geqslant \alpha$ and $f_n(z)^k < \alpha$. Then $f_{n+1}(z) < -\alpha^{1/k} + 2\alpha^{1/k} - z\alpha < 0$ if α is chosen so that $z \geqslant \alpha^{(1/k)-1}$.

Theorem 4.1 suggests an extremely simple procedure to obtain approximations to solutions to (1.1_n). It consists merely of selecting a small value for z and computing the values of the sequence $\{f_n(z)\}$ via the recursion formula (3.1) until $f_{N_z}(z) \leqslant 0$. Then by setting $n + 1 = N_z$ in (3.3), an approximation to a $\bar{y}(z)$ is completely determined. The particular solution to (1.1) which is approximated is $y^{[1]}$ and hence in fact it suffices to compute the sequence $\{f_n(z)\}$ only until $f_{j+1}(z) < f_j(z)$ and extend the remainder of the vector symmetrically with $n + 1 = 2j - 1$, $2j$ or $2j + 1$ according as the increment $f_j(z) - f_{j-1}(z)$, $|f_{j-1}(z) - f_{j+1}(z)|$ or $f_{j+1}(z) - f_{j-1}(z)$ is smallest. The numerical approximation to the oscillating solutions of (1.1_n) are immediately obtained by using this solution in the same way that $y^{[1]}$ was used to construct $y^{[m]}$ in Section 2.

Extensive computational experience with the procedure outlined above shows that it performs more efficiently than either reducing (1.1) to an initial value problem as outlined in Section 2 or actually solving for the $\zeta_1(n + 1)$ mentioned in Section 3. The reason for this lies in the fact that very few arithmetical operations are required to obtain numerical approximations. Table 2 furnishes a few sample results.

TABLE 2

k	z	$n + 1$	max $\bar{f}(z)$	$\bar{y}_{max}(L^2\lambda)^{k-1}$	$y_{max}(L^2\lambda)^{k-1}$†	r_k
10^{-4}	10^{-4}	19,986	4.996742×10^3	1.2522×10^{-1}		10
0.2	10^{-5}	25,809	9.58948×10^3	1.59355×10^{-1}		6.8
0.4	10^{-7}	273,839	7.750968×10^4	2.697977×10^{-2}	2.71069×10^{-2}	5.2
0.6	10^{-7}	69,613	2.062677×10^4	3.99006×10^{-3}	4.01097×10^{-3}	4.2
0.8	10^{-7}	23,712	7.302412×10^3	1.29952×10^{-5}		3.6
2	10^{-7}	691	2.466213×10^2	1.17757×10^1	1.17773×10^1	2.15
3	10^{-7}	175	6.687366×10^1	3.70078	3.7054	1.8
4	10^{-7}	75	3.017299×10^1	2.49073	2.50386	1.6

† The figures in this column are taken from Table 1.

The term max $\bar{f}(z)$ denotes the value of $f_j(z)$ when $f_j(z) > f_{j+1}(z)$ occurs for the first time, $\bar{y}_{max}$ is the maximum component in the vector $\bar{y}(z)$. The term r_k denotes the approximate ratio of the number of iterations required to complete the calculation of the vector $\bar{f}(z)$ for $z = 10^{-i}$ over $z = 10^{-(i-1)}$. (Computational evidence strongly indicates that $r_k \to 10^{1/(1+k)}$ as $i \to \infty$.) The ratio r_k may be used to forecast the approximate number of iterations and hence also the total number of arithmetical operations which will be required to complete the calculation of $\bar{f}(z)$ for $z = 10^{-i}$. For example, if $k = 2$, the number of iterations which will be required to complete the calculation of $\bar{f}(10^{-i})$ is approximately $3 \times 10^{i/3}$ and the number of arithmetical operations will be approximately $6(3 \times 10^{i/3})$.

The ratio r_k may also be utilized in the formulation of truncation error estimates for the numerical procedure outlined above. Let $y_1(\zeta_1)\bar{f}(\zeta_1)$ denote the fundamental solution to (1.1_n). Let N_z be the least integer for which $f_{N_z}(z) \leqslant 0$. Then by the preceding remarks for $z = 10^{-i}$, $N_z \simeq B_k(1)r_k^{i-1}$ where $3 \leqslant B_k(1) \leqslant 21$. (Since N_z will be largest for $k = 0$.) Now for $n + 1 = N_z$ familiar considerations involving Taylor's formula yield

$$|y^{[1]}(x_j) - y_1(\zeta_1)f_j(\zeta_1)| \leqslant \frac{1}{2}\left(\frac{L}{N_z}\right)^2 y_{\max}^{[1]} = \frac{1}{2}\left(\frac{L}{N_z}\right)^2\left[\frac{2C_k^2(k+1)}{L^2\lambda}\right]^{1/(k-1)},$$

$$j = 1, \ldots, N_z - 1$$

or for $z = 10^{-i}$,

$$|y^{[1]}(x_j) - y_1(\zeta_1)f_j(\zeta_1)| \leqslant \frac{1}{2}\left[\frac{2C_k^2(k+1)}{L^2\lambda}\right]^{1/(k-1)}\left(\frac{L}{B_k(1)}r_k^{i-1}\right)$$

for $j = 1, 2, \ldots, N_z - 1$. $\hspace{3cm}$ (4.1)

For the procedure outlined above, the integer N_z minimizes $|z - \zeta_1(N_z)|$; in particular for $z = 10^{-i}$, it minimizes $|10^{-i} - \zeta_1(N_z)|$. It has been previously noted that $f_n(0) = n$. It may be further demonstrated that

$$f_n(z) = n + \frac{(n^{k+2} - n^k)}{(k+1)(k+2)}z + \text{(higher powers of } z\text{)}. \hspace{2cm} (4.2)$$

Since N_z minimizes $|z - \zeta_1(N_z)|$, either $\zeta_1(N_z + 1) \leqslant z \leqslant \zeta_1(N_z)$ or $\zeta_1(N_z) \leqslant z \leqslant \zeta_1(N_z - 1)$. In either case it follows from (3.7) that

$$|\zeta_1(N_z) - z| = 0(N_z^{-(k+2)}) \quad \text{as} \quad z \to 0, \text{ i.e., as } N_z \to \infty. \hspace{1cm} (4.3)$$

For the purpose of simplicity, the subscript on N_z will be suppressed and it will be assumed that $N = N_z$ is even. It will be obvious the latter assumption imposes no actual restriction. Now

$$|y_1(\zeta_1)f_j(\zeta_1) - y_1(z)f_j(z)|$$

$$= \left[\left(\frac{N}{L}\right)^2\left(\frac{1}{\lambda}\right)\right]^{1/(k-1)}|\zeta_1^{1/(k-1)}f_j(\zeta_1) - z^{1/(k-1)}f_j(z)|$$

$$\leqslant \left[\left(\frac{N}{L}\right)^2 \left(\frac{1}{\lambda}\right)\right]^{1/(k-1)} (\zeta_1^{1/(k-1)}|f_j(\zeta_1) - f_j(z)| + |\zeta_1^{1/(k-1)} - z^{1/(k-1)}|f_j(z))$$

$$\leqslant \left[\left(\frac{N}{L}\right)^2 \left(\frac{1}{\lambda}\right)\right]^{1/(k-1)} (\zeta_1^{1/(k-1)}|f_{N/2}(\zeta_1) - f_{N/2}(z)| + |\zeta_1^{1/(k-1)} - z^{1/(k-1)}|f_{N/2}(z))$$

$$\text{for } j = 1, \ldots, N \tag{4.4}$$

by (4.2) and the fact that $\max \bar{f}(z) = f_{N/2}(z)$. From (3.7), (4.2), (4.3), it follows that

$$\zeta_1^{1/(k-1)}(N) = 0(N^{-(k+1)/(k-1)}),$$
$$|f_{N/2}(\zeta_1) - f_{N/2}(z)| = 0(N^{k+2})z = 0(1),$$
$$|\zeta_1^{1/(k-1)} - z^{1/(k-1)}| = 0(N^{-2k/(k-1)})$$
$$f_{N/2}(z) = 0(N) \quad \text{as} \quad N \to \infty.$$

Substituting these relations into 4.4 yields

$$|y_1(\zeta_1)f_j(\zeta_1) - y_1(z)f_j(z)| = 0(N_z^{-1}) \quad \text{as} \quad z \to 0 \text{ for } j = 1, \ldots, N_z \tag{4.5}$$

and $\zeta_1 = \zeta_1(N_z + 1)$. In particular for $z = 10^{-i}$, combining (4.1) and (4.5) yields

$$|y^{[1]}(x_j) - y_1(z)f_j(z)| = 0(N_z^{-1}) = 0(10^{-i/(k+1)}) \quad \text{as} \quad i \to \infty. \tag{4.6}$$

The constant which has been suppressed in the relation (4.6) may be expressed in terms of k, C_k, $B_k(1)$ and r_k. Corresponding truncation error estimates on the numerical approximations to the oscillating solutions to (1.1) may also be obtained.

Section 5

The procedure outlined in Section 3 may be applied to a wider class of non-linear two-point boundary value problems than (1.1). A few of the possible generalizations are suggested below. The problem

$$y'' + p(x)y^k = 0, \quad y(0) = 0 = y(L) \quad (k \text{ non-negative rational}) \tag{5.1}$$

where $p(x)$ is a non-negative continuous function, may be treated by the same technique as that in Section 3. Under the convention that $x_i = iL/(n+1)$, $p_i = p(x_i)$, the analogue of (3.1) assumes the form

$$y_{i+1} = y_1(-f_{i-1}(y_1) + 2f_i(y_1) - h^2 p_i y_1^{k-1} f_i(y_1)^k) = y_1 f_{i+1}(y_1),$$
$$i = 1, \ldots, n \tag{5.2}$$

with $y_0 = 0 = y_{n+1}$ and $f_1(y_1) = 1$, $f_2(y_1) = 2 - h^2 p_2 y_1^{k-1}$. A similar analysis to that in Section 3 may also be carried through for (5.2).

Certain higher order boundary value problems may be treated by the method of Section 3. For example, the analogue of (3.1) for the problem

$$y''' + y^k = 0, \quad y(0) = 0 = y(L) = y'(0) \tag{5.3}$$

under the approximation $y_i''' = (y_{i+1} - 3y_i + 3y_{i-1} - y_{i-2})/h^3$ assumes the form

$$y_{i+1} = y_1 f_{i+1}(z), \quad i = 1, \ldots, n$$

where

$$y_0 = 0 = y_{n+1}, \quad y_{-1} = y_1, \quad z = \lambda h^3 y_1^{k-1} \tag{5.4}$$

and

$$f_{i+1}(z) = f_{i-2}(z) - 3f_{i-1}(z) + 3f_i(z) - z f_i(z)^k, \quad f_0(z) = 0, \quad f_1(z) = 1,$$
$$f_2(z) = 4 - z.$$

Emden's boundary value problem,

$$y'' + \left(\frac{\mu}{x}\right) y' + \lambda y^k = 0 \quad (\lambda, \mu \geqslant 0, k \text{ a non-negative rational} \neq 1) \tag{5.5}$$

$$y'(0) = 0 = y(L)$$

has the analogous discrete formulation to (3.1):

$$y_{i+1} = \left(\frac{y_1}{2i+\mu}\right) f_{i+1}(z) = \left(\frac{y_1}{2i+\mu}\right) [-(2i-\mu)f_{i-1}(z) + 4if_i(z) - 2iz f_i(z)^k],$$

$$i = 1, 2, \ldots, n \text{ with } y_{n+1} = 0, \quad z = h^2 \lambda y_1^{k-1}, \quad f_1(z) \equiv 1, \tag{5.6}$$

$$f_2(z) = \left(1 - \frac{2z}{2+\mu}\right).$$

An analysis of the recursion relation (5.6) similar to that carried out in Section 3 shows that (5.5) has only the trivial solution $y \equiv 0$ and a fundamental solution $y^{[1]}$ which is concave downward on $[0, L]$.

The method outlined above seems best suited for the particular class of problems (1.1), (5.1), (5.3), (5.5); however, it may also be applied to certain other nonlinear boundary value problems. The stability of the corresponding recursion formulae can also be established.

References

[1] Atkinson, F. V. (1955). On second-order nonlinear oscillations, *Pacific J. Math.* 5, 643–647.
[2] Coffman, D. (1973). Lyusternik–Schnirelman theory and eigenvalue problems for monotone operators, *J. Functional Analysis*, 14, 237–252.
[3] Collatz, L. (1966). *Differentialgleichungen.* Teubner, Stuttgart.

[4] Gaines, R. (1974). Difference equations associated with boundary value problems for second order nonlinear ordinary differential equations, *SIAM J. Num. Anal.* 11, 411–434.

[5] Krasnosellski, M. A. (1964). *Topological Methods in the Theory of Nonlinear Integral Equations*. MacMillan, New York.

[6] Kuiper, H. J. and Turner, R. E. L. Sturm-Liouville problems with prescribed nonlinearities (to appear).

[7] Laetsch, T. (1970). The number of solutions of a nonlinear two point boundary value problem, *Indiana Univ. Math. J.* 20, 1–13.

[8] Moore, R. A. and Nehari, Z. (1959). Nonoscillation theorems for a class of nonlinear differential equations, *Trans. Amer. Math. Soc.* 93, 30–52.

[9] Nehari, Z. (1960). On a class of nonlinear second order differential equations, *Trans. Amer. Math. Soc.* 95, 101–123.

[10] Turner, R. E. L. (1973). Superlinear Sturm–Liouville problems, *J. Diff. Eqs.* 13, 157–171.

Some Applications of Approximation Theory to Numerical Analysis

R. Ansorge

Introduction

Within the framework of numerical analysis, approximation theory is considered from several standpoints: In the first place approximation theory itself is usually understood to be a part of numerical analysis. From this standpoint, the main interest is, roughly speaking, in approximating an explicitly given but complicated function as well as possible with respect to a given metric by an element of a given set of functions having simpler structures. However, this aspect will not be considered in this paper.

During the last ten years, another circle of problems has been taken up, namely, data other than an explicit representation of the function that one wants to approximate have been taken into account, e.g. the function may be characterized as a solution to an initial or boundary value problem for ordinary or partial differential equations.

The first effort in this direction is due to Meinardus and Strauer [1], 1963. The approximation procedure is carried out in two steps. In the first step of this procedure, they replace the required solution of the problem (which characterizes the function) by a certain continuous function (not necessarily belonging to the given function set mentioned above) by means of some numerical method. This yields an intermediate solution. The second step consists in approximating this intermediate solution by approximation theoretic methods using elements from the given function set, and it is hoped that this approximation to the intermediate solution is at the same time a good approximation to the desired solution with respect to that given function set.

In the first section of this lecture the main problems of the Meinardus-Strauer-method are described without going into detail. There are classes of examples for which the method works well. But in many other linear cases and many more nonlinear cases the assumptions are hard to fulfill.

Therefore, I want to show in a second section how these difficulties may be avoided if one replaces the continuous method for finding the intermediate

17

solution by a discrete method: One finds then by the first step a discrete intermediate solution which is then approximated at the second step by a continuous function with respect to the given set of functions using methods of discrete Chebyshev-approximation [2].

It may certainly seem to be inefficient to compute in a first step an intermediate solution instead of directly seeking an approximating function with respect to the given set.

Indeed, there are many successful attempts in direct approximation (mainly due to Collatz [3], [4]) dealing with concrete examples of boundary and initial value problems. Although it is not yet possible to present in this context a general theory, I want to give a short description of some of these attempts because of the surprisingly good numerical results.

Finally, I want to introduce another field of application for some theorems of approximation theory to numerical analysis: Using discretization algorithms for solving initial or boundary value problems for ordinary or partial differential equations, or for computing integrals, it is usually possible to find error estimates if it is assumed that the given data (initial values, boundary values, the given integrand) are sufficiently smooth. On the other hand it is well known that such procedures are often convergent (for decreasing step sizes) in the case of less smooth data as well (e.g. in the case of generalized solutions of initial value problems for semilinear partial differential equations). The question of error estimates therefore arises (especially of the order of convergence) which depend on the degree of smoothness of the given data. Answers were given by Walsh and Young [5] 1953 for a special boundary value problem, by Peetre and Thomée [6] 1967 for linear initial value problems, by Davis and Rabinowitz [7] 1967 for quadrature formulas, by Bramble, Hubbard and Thomée [8] 1969 for linear boundary value problems with smooth boundaries, by Ansorge, Geiger and Hass [9] 1972 for semilinear initial value problems (and by others).

In Section 4 of this lecture I will give a short presentation of a rather transparent and general method for finding such error estimates by means of approximation theory. Of course there remain many difficulties regarding the realization of this general concept in concrete classes of problems. However, all the results just mentioned are obtainable by the method to be presented here and some of the results (Walsh and Young) can even be improved.

1. Indirect Continuous Approximation to Solutions of Linear Functional Equations with Respect to a given Set of Functions

We seek the solution $y \in \mathfrak{M}$ (normed space) of a given functional equation. We assume that the existence and uniqueness of y is guaranteed and we want to find a good approximation to y with respect to $\mathfrak{B} \subset \mathfrak{M}$.

Assume that we have found an intermediate approximating solution $u \in \mathfrak{B} \subset \mathfrak{M}$ by some suitable numerical method (first difficulty: which method is "suitable"?)

Assume: $u \notin \mathfrak{M} \cap \mathfrak{B}$.

We suppose that there is an error estimate

$$\| y - u \| \leq M$$

(second difficulty).

Now we approximate u with respect to $\mathfrak{B}$, that is: We want to find a function $v \in \mathfrak{B}$ with the property

$$\| u - v \| = \inf_{\tilde{v} \in \mathfrak{B}} \| u - \tilde{v} \| =: \rho_{\mathfrak{B}}(u).$$

For convenience we assume that there exist a "best approximation" v to u and also a best approximation $g \in \mathfrak{B}$ to y.

Obviously one has

$$\rho_{\mathfrak{B}}(u) - M \leq \| u - g \| - \| y - u \| \leq \| y - g \| = \rho_{\mathfrak{B}}(y) \leq \| y - v \|$$

$$\leq \| y - u \| + \| u - v \| \leq M + \rho_{\mathfrak{B}}(u).$$

Thus, we find the following estimate for $\| y - v \|$:

$$\rho_{\mathfrak{B}}(u) - M \leq \rho_{\mathfrak{B}}(y) \leq \| y - v \| \leq \rho_{\mathfrak{B}}(u) + M. \tag{1}$$

Therefore, the smallness of the ratio $M/\rho_{\mathfrak{B}}(u)$ is a measure of the quality of the approximation to y by v.

So the main problem to be faced in applying the ideas of Meinardus and Strauer is the realization of the assumption

$$\frac{M}{\rho_{\mathfrak{B}}(u)} \ll 1 \tag{2}$$

at a tolerable computational expense.

For certain classes of linear ordinary differential equations and linear integral equations, Meinardus and Strauer gave a method for finding a continuous intermediate solution u so that (using a suitable linear subspace $\mathfrak{B}$) (2) holds in a satisfactory manner. However, for partial differential equations, for nonlinear problems, or nonlinear sets $\mathfrak{B}$ there are, unfortunately, no results concerning the applicability of the method.

2. Indirect Discrete Approximation to the Solution of a Functional Equation

We content ourselves now with a discrete intermediate solution u only, instead of a continuous function u.

For this purpose we cover the domain of the elements of the function space $\mathfrak{M}$ by a grid G_{h_1}, where h_1 describes the mesh widths.

On this grid we replace the originally given (not necessarily linear) functional equation (or the given functional) by a discrete equation (e.g. difference equation), the discrete solution of which is denoted by u_{h_1}.

Such discretization procedures often allow error estimates of the form:

$$\| y - u_{h_1} \|_1 \leqq M_1 = M(h_1), \tag{3}$$

where y is the (unknown) solution of the given problem and $\| \ldots \|_1$ is the maximum norm on the space of (real-valued) functions with domain G_{h_1}.

We now reduce the step width to a new step size h_2 (defining a new grid G_{h_2}) so that

$$G_{h_1} \subset G_{h_2} \quad (h_1 > h_2).$$

Continuing in this way we obtain a set of mesh widths h_r $(r = 1, 2, \ldots)$ with

$$(1) \quad \lim_{r \to \infty} h_r = 0 \tag{4}$$

$$(2) \quad G_{h_p} \subset G_{h_q} \quad \text{for} \quad p < q. \tag{5}$$

The corresponding intermediate solutions are called u_{h_r} and the corresponding norms $\| \ldots \|_r$.

Suppose that the discretization algorithm leads to a convergent process:

$$(3) \quad \lim_{r \to \infty} M_r = 0 \ (M_r = M(h_r)). \tag{6}$$

We now approximate the discrete intermediate solution u_{h_r} in the sense of the discrete Chebyshev approximation by (continuous) functions $v_r \in \mathfrak{B}$ $(r = 1, 2, \ldots)$.

The assumption corresponding to the assumption (2) of Meinardus and Strauer is

$$\frac{M_r}{\rho_{\mathfrak{B}}^{[r]}(u_{h_r})} \ll 1 \quad \text{for suitable } r \tag{7}$$

$(\rho_{\mathfrak{B}}^{[r]}(u_{h_r}) := \inf_{\tilde{v} \in \mathfrak{B}} \| u_{h_r} - \tilde{v} \|_r$; if we compare a discrete function with a continuous function v, then the symbol v will denote the set of values of v on the grid).

From (5), it follows that the set $\{ \rho_{\mathfrak{B}}^{[r]}(y) \}$ is monotonically increasing, and it is also bounded because of the trivial inequality

$$\forall r : \rho_{\mathfrak{B}}^{[r]}(y) \leqq \rho_{\mathfrak{B}}(y).$$

Thus this set is convergent and from

$$\rho_{\mathfrak{B}}^{[r]}(y) - M_r \leqq \rho_{\mathfrak{B}}^{[r]}(u_{h_r}) \leqq \rho_{\mathfrak{B}}^{[r]}(y) + M_r$$

(which follows from an inequality analogous to (1)) we get

$$\lim_{r \to \infty} \rho_{\mathfrak{B}}^{[r]}(u_{h_r}) = \lim_{r \to \infty} \rho_{\mathfrak{B}}^{[r]}(y). \tag{8}$$

For $y \notin \mathfrak{B}$ (the case $y \in \mathfrak{B}$ is not under consideration) the right-hand side of (8) is almost always different from zero so that we get (from (6));

$$\lim_{r \to \infty} \frac{M_r}{\rho_{\mathfrak{B}}^{[r]}(u_{h_r})} = 0. \tag{9}$$

Thus for sufficiently small step sizes it is almost always possible to satisfy the inequality (7) in a satisfactory manner (also for nonlinear problems).

3. Discrete Approximation to Solutions of Boundary Value Problems

In this Section I want to consider only the approximation to solutions of boundary value problems with respect to given function spaces although there are similar attempts also for initial value problems and integral equations.

Without going into detail I shall touch upon the applicability of some interesting aspects of approximation theory which arise in the numerical implementation of certain principles of monotonicity to various boundary value problems.

For purposes of illustration, we restrict ourselves to the problem

$$Tz = -\Delta z + a(x, z) = 0 \quad \text{on} \quad B = \{(x_1, x_2, x_3) \mid \Sigma x_i^2 = r^2 < 1\} \tag{10}$$

$$Rz = \dot{z} - 1 \qquad = 0 \quad \text{on} \quad \Gamma = \{(x_i) \mid \Sigma x_i^2 = 1\}. \tag{11}$$

As a consequence of a theorem of Redheffer [10], we have: If

1) u, v are two elements of $C^2(B) \cap C^0(\bar{B}) =: U$,

2) $a(x, v + k) - a(x, v) \geqq 0 \quad$ for all $k > 0 \quad (k \in \mathbb{R})$

3) $Tu \leqq 0 \leqq Tv \quad$ on $\quad B$

4) $u \leqq v \quad$ on $\quad \Gamma$

it follows that $u \leqq v$ on $\bar{B}$ (inverse monotonicity of T with respect to u, v).
Let V be the set

$$V = \{v \in U \mid 2) \text{ holds}\}.$$

We see that if $u \in U, v \in V, u \leqq z \leqq v$ on Γ for any solution of the problem (10), then it follows that:

$$u \leqq z \quad \text{on} \quad B, \quad \text{if } z \in V$$

$$z \leqq v \quad \text{on} \quad B,$$

i.e.

$$u \leqq z \leqq v \quad \text{on} \quad B \text{ for } z \in V.$$

We want to find numerically such inclusion functions u and v for the solution z of the problem (10), (11).

For this purpose, choose now a function space

$$W := \{w \in U \mid w = w(x, \alpha), \alpha \in A_n \subset \mathbb{R}^n, Rw = 0 \text{ on } \Gamma\}$$

and try to solve the "one-sided" approximation problems

$$-\delta_1 \leqq [Tu](x) \leqq 0, \quad \delta_1 \overset{!}{=} \underset{u \in W}{\text{Min}} \tag{12}$$

$$0 \leqq [Tv](x) \leqq \delta_2, \delta_2 \overset{!}{=} \underset{v \in W}{\text{Min}} \tag{13}$$

(i.e. one wants to approximate the zero-function from above and from below respectively as well as possible with respect to the function set TW).

Depending on the choice of W and on the structure of $a(x, .)$, (12) and (13) lead to various (partly rather new) questions within approximation theory, which, in the sense that results concerning the existence of solutions $\hat{u}$ and $\hat{v}$ of (12) and (13) (respectively) and algorithms for their computation are available, are by and large unsolved.

But also if $\hat{u}$ and $\hat{v}$ are computable and if $z \in V$ is guaranteed, we have still to verify that $\hat{v} \in V$. The idea of verifying $\hat{v} \in V$ as part of a second step instead of making an *a priori* choice $W \subset V$ sometimes leads to a less complicated numerical effort for solving (12), (13), because the assumption $W \subset V$ involves additional restrictions, which sometimes complicate matters.

4. Error Estimates for Discretization Methods for Data with Less Restrictive Structure

Consider the following problem:

Given: Two normed spaces $\mathfrak{B}$, $\mathfrak{W}$, an operator $E: \mathfrak{B} \rightarrow \mathfrak{W}$ and certain "data" $v_0 \in \mathfrak{B}$.

Let

$$w := Ev_0.$$

Assume that v_0 is explicitly given. If E is explicitly given also, then (14) leads to a problem of "calculation".

If E is only implicitly characterized by a functional equation, then we call E the "solution operator" for the given functional equation and the problem is to solve this equation (at least numerically) with respect to the given data.

Let $\{P_h\}$ $(h \in (0, h_0), h_0 > 0)$ be a set of operators with domain $\mathfrak{B}$ and range $\mathfrak{W}$. We look on these operators P_h as approximations to the operator E (e.g. P_h is given by a quadrature formula for an integral operator E, or P_h is a difference operator given by a difference method for solving a given differential equation numerically, where h stands for a step size, etc.).

We now suppose:

1) There is an error estimate for sufficiently smooth data, i.e.:

$$\| P_h v_0 - E v_0 \| \leqq h^p \varphi(v_0), \quad \forall v_0 \in \mathfrak{G} \subset \mathfrak{B} \quad (p > 0), \qquad (15)$$

where it is possible to calculate the functional φ.

2) The operators P_h are Hölder-continuous on $\mathfrak{B}$:

$$\| P_h u - P_h v \| \leqq C \| u - v \|^\delta \quad (0 < \delta \leqq 1), \quad \forall u, v \in \mathfrak{B}. \qquad (16)$$

3) The operator E (possibly not explicitly known) is Hölder-continuous with the same constants C, δ as in (16), i.e.

$$\| Eu - Ev \| \leqq C \| u - v \|^\delta, \quad \forall u, v \in \mathfrak{B}. \qquad (17)$$

It may be mentioned that the assumption (17) is not very incisive because (15) guarantees the pointwise convergence of $\{P_h\}$ to E (for $h \to 0$) on $\mathfrak{G}$, and $\mathfrak{G}$ often is dense in $\mathfrak{B}$ (if $\mathfrak{G}$ is dense in $\mathfrak{B}$ and E is continuous, (17) follows from (16)).

Let $\mathfrak{M}$ be a subset of $\mathfrak{B}$ so that

$$\mathfrak{G} \subset \mathfrak{M} \subset \mathfrak{B}.$$

We ask for estimates of the error $\| P_h v - E v \|$ for $v \in \mathfrak{M} - \mathfrak{G}$; in particular we are interested in the question: How does the order p (valid for $v \in \mathfrak{G}$) decrease, if one replaces an element $v \in \mathfrak{G}$ by an element $v \in \mathfrak{M} - \mathfrak{G}$?

So let now $v \in \mathfrak{M} - \mathfrak{G}$. With elements $v_r \in \mathfrak{G}$ $(r = 1, 2, \ldots)$ which are available, we get as a consequence of (15), (16), (17):

$$\| P_h v - E v \| \leqq \| P_h v - P_h v_r \| + \| P_h v_r - E v_r \| + \| E v_r - E v \|$$
$$\leqq 2C \| v - v_r \|^\delta + h^p \varphi(v_r). \qquad (18)$$

Suppose further

4) Let $\mathfrak{G}_1 \subset \mathfrak{G}_2 \subset \ldots \subset \mathfrak{G}_r \subset \ldots \subset \mathfrak{G}$ be a set-series of subsets of $\mathfrak{G}$ with the following properties:
   If any $v \in \mathfrak{M}$ is approximated with respect to $\mathfrak{G}_r$ $(r = 1, 2, \ldots)$ (and with respect to the norm on $\mathfrak{B}$) a "Jackson-type" theorem

$$\rho_{\mathfrak{G}_r}(v) \leqq \frac{\omega_{\mathfrak{M}}(v)}{r^\theta} \quad (r = 1, 2, \ldots), \quad \theta > 0 \qquad (19)$$

   is valid with a functional $\omega_{\mathfrak{M}}$ (defined on $\mathfrak{M}$) which is computable.

5) There are semi-norms $\alpha_{\mathfrak{G}}^{(r)}$ (defined on $\mathfrak{G}$, which we suppose for

convenience to be a linear space) so that "Bernstein-type" inequalities are valid on the sets $\mathfrak{G}_r$:

$$\alpha_{\mathfrak{G}}^{(r)}(v_r) \leqq c_r r^{k_r} \| v_r \|, \quad \forall v_r \in \mathfrak{G}_r, \quad (r = 1, 2, \ldots) \tag{20}$$

(c_r, k_r: known constants).

The assumptions 4) and 5) are of course very incisive, but from a practical point of view these inequalities are often satisfied.

From (19) and (20) we obtain a "Zamansky-type" theorem [11]:

If v_r is the best approximation to $v \in \mathfrak{M}$ with respect to $\mathfrak{G}_r$, the inequality

$$\alpha_{\mathfrak{G}}^{(\rho)}(v_r) \leqq f_\rho(r, v) := \gamma_\rho \, \tilde{\omega}_{\mathfrak{M}}(v) \cdot \begin{cases} r^{k_\rho - \theta} & \text{for } k_\rho > \theta \\ 1 & \text{for } k_\rho < \theta \end{cases} \tag{21}$$

holds with (computable) constants γ_ρ and a (computable) functional $\tilde{\omega}_{\mathfrak{M}}$ (depending on $\omega_{\mathfrak{M}}$) (the case $k_\rho = \theta$ is of no interest in this context).

Using numerical discretization methods, the functional φ entering into the error bound is usually given by

$$\varphi(v_r) = \varphi^*(\alpha_{\mathfrak{G}}^{(1)}(v_r), \alpha_{\mathfrak{G}}^{(2)}(v_r, \ldots, \alpha_{\mathfrak{G}}^{(m)}(v_r)) \tag{22}$$

with a function φ^* which is monotonically increasing with respect to each of its variables.

Thus we find from (18), (19), (21), (22)

$$\| P_h v - Ev \| \leqq 2C \left(\frac{\omega_{\phantom{\mathfrak{M}}}(v)}{r^\theta} \right)^\delta + h^P \varphi^*(f_1(r, v), f_2(r, v), \ldots, f_m(r, v)),$$

$$\forall v \in \mathfrak{M}. \tag{23}$$

Suppose finally that there exists $\beta > 0$ such that

$$\forall r \in \mathbb{N}: r^{-\beta} \varphi^*(f_1(r, v), \ldots, f_m(r, v)) < \infty,$$

i.e.

$$\varphi^*(f_1(r, v), \ldots, f_m(r, v)) \leqq \frac{\chi(v)}{r^\beta} \tag{24}$$

with a certain functional χ (with domain $\mathfrak{M}$).

Using (21), the assumption (24) is usually satisfied because of the structure of the practically occurring functions φ^*.

Thus, it follows from (22) that

$$\| P_h v - Ev \| \leqq 2C \left(\frac{\omega_{\mathfrak{M}}(v)}{r^\theta} \right)^\delta + h^P \frac{\chi(v)}{r^\beta} =: F_r(h, v). \tag{25}$$

The error bound on the right-hand side depends on r which is available.

Therefore, (25) can be improved by minimizing with respect to r. This leads finally to the error estimate

$$\| P_h v - Ev \| \leqq \psi(v) h^{p(\theta\delta/\theta\delta + \beta)}, \quad \forall v \in \mathfrak{M} \subset \mathfrak{B},$$

where ψ is an explicitly computable functional.

The quality of the error bound (25) for data with less restrictive structure $v \in \mathfrak{M}$ depends of course on the abundance of the sequence of sets $\mathfrak{G}_1 \subset \mathfrak{G}_2 \subset \ldots$ as well as on the quality of the estimate (15) which we supposed to be valid for smooth data.

But it can be shown by examples that for sufficiently well chosen sequences of sets $\{\mathfrak{G}_r\}$, the order $p(\theta\delta/\theta\delta + \beta)$ often cannot be improved.

References

[1] Meinardus, G. und Strauer, H. D. (1963). Über Tschebyscheffsche Approximationen der Lösungen linearer Differential- und Integralgleichungen, *Arch. Rat. Mech. Anal.* 14 184–195.

[2] Ansorge, R. (1966). Tschebyscheff-Approximation der Lösungen von Differentialgleichungen bei Benutzung von Differenzenverfahren, *ZAMM* 46, 397–399.

[3] Collatz, L. (1969). Nichtlineare Approximation bei Randwertaufgaben, *Wiss. Z. Hochsch. Archit. Bau. Weimar* 169–182.

[4] Collatz, L. und Krabs, W. (1973). *Approximationstheorie*. B. G. Teubner, Stuttgart.

[5] Walsh, J. L. and Young, D. (1953). On the accuracy of the numerical solution of the Dirichlet problem by finite differences, *J. Res. Nat. Bur. Standards.* 51, 343–363.

[6] Peetre, J. and Thomée, V. (1967). On the rate of convergence for discrete initial-value problems, *Math. Scand.* 21, 159–176.

[7] Davis, P. J. and Rabinowitz, P. (1967). *Numerical Integration.* Blaisdell, Toronto-London.

[8] Bramble, J. H., Hubbard, B. E. and Thomée, V. (1969). Convergence estimates for essentially positive-type discrete Dirichlet problems, *Math. Comp.* 23, 695–710.

[9] Ansorge, R., Geiger, C. und Hass, R. (1972). Existenz und numerische Erfaßbarkeit verallgemeinerter Lösungen halblinearer Anfangswertaufgaben, *ZAMM* 52, 597–605.

[10] Redheffer, R. M. (1962). An extension of certain maximum principles, *Mh. Math. Phys.* 66, 32–34.

[11] Zamansky, M. (1949). Classes de saturation de certains procédés d'approximation des séries de Fourier des fonctions continues et application à quelques problêmes d'approximation, *Ann. sci. École Norm. Sup.* 66, 19–93.

Interior Regularity and Local Convergence of Galerkin Finite Element Approximations for Elliptic Equations

This paper establishes an inequality defining a property of interior regularity for finite element Galerkin approximations of elliptic equations. The most interesting results concern properties of local convergence for derivatives.

1. Introduction

In this paper we consider only real linear spaces. We begin by introducing some notations. $\Omega \subset \mathbb{R}^n$ is a given open bounded domain, $(u, v) = \int_\Omega uv$. For any open $\Lambda \subset \Omega$, $H_k(\Lambda)$ is the Sobolev space of functions possessing square integrable derivatives of order $\leqslant k$ on Λ; for $v \in H_k(\Lambda)$,

$$\|v\|_{k,\Lambda}^2 = \sum_{|\alpha| \leqslant k} \int_\Lambda (D^\alpha v)^2;$$

$\overset{0}{H}_k(\Lambda)$ is the closure of $\overset{0}{C}_\infty(\Lambda)$ with respect to $\|.\|_{k,\Lambda}$; for $k < 0$ we set $\|v\|_{k,\Lambda} = \|v\|_{0,\Lambda}$. $C_I(\Lambda)$ is the set of functions v such that:

(a) there exist open disjoint sets $\Lambda_1, \ldots, \Lambda_N$ depending on v with $\overline{\Lambda} = \cup_{i=1}^N \Lambda_i$,

(b) the restriction v_i of v to Λ_i is of class $C_\infty(\overline{\Lambda}_i)$; since $v \in H_k(\Lambda_i)$ we may introduce the norm $\|v\|_{k,\Lambda}^2 = \sum_{i=1}^N \|v_i\|_{k,\Lambda_i}$. $\overset{0}{C}_I(\Lambda)$ is the set of functions of $C_I(\Lambda)$ with support contained in Λ.

Let

$$a(u, v) = \int_\Omega \sum_{|\alpha|,|\beta| \leqslant m} a_{\alpha\beta} D^\alpha u \, D^\beta v$$

be a given elliptic form on $H_m(\Omega) \times H_m(\Omega)$ and $\mathscr{S}$ be a collection of finite dimensional subspaces $S \subset H_m(\Omega) \cap C_I(\Omega)$ of "finite element" type;

27

$h : S \to \mathbb{R}_+$ is a function representing the "mesh size" of S. For $S \in \mathcal{S}$ and $\Lambda \subset \Omega$, $\overset{0}{S}(\Lambda)$ is the set of functions of S with support in Λ.

Let $f \in H_0(\Omega)$ and $u \in H_m(\Omega)$ satisfy the relation $a(u, v) = (f, v)$ $\forall v \in \overset{0}{H_m}(\Omega)$. Supposing a and f sufficiently regular, the fundamental inequality relative to the regularity of u establishes the existence of constants c_k such that for $\Lambda \subset\subset \Omega$ one has

$$\| u \|_{m+k,\Lambda} \leqslant c_k (\| u \|_{0,\Omega} + \| f \|_{k-2m,\Omega}) \quad k = 1, 2, \ldots; \tag{1}$$

in particular $u \in H_k(\Lambda)$, $k = 1, 2, \ldots$. Theorem 3 is a discrete analogue of (1); however it does not imply $w \in H_k(\Lambda)$ for $k > m$. Theorem 1 may be regarded as a consistency relation and it gives the main tool for proving the other results. Theorem 2 shows that to a given Galerkin solution there corresponds locally a very close solution of the exact problem. Theorems 4 and 5 give the most important results for applications. Suppose that Ω is a subset of $\tilde{\Omega}$, $a_{\alpha\beta}$ and f are very regular and restrictions to Ω of functions $\tilde{a}_{\alpha\beta}$ and $\tilde{f}$ defined on $\tilde{\Omega}$ which can be irregular on $\tilde{\Omega} - \Omega$; let $\tilde{a}$ be the bilinear form with coefficients $\tilde{a}_{\alpha\beta}$, $\tilde{u} \in H_m(\tilde{\Omega})$ satisfy the relation $\tilde{a}(\tilde{u}, v) = (\tilde{f}, v)$, $\forall v \in \overset{0}{H_m}(\Omega)$, u be the restriction of $\tilde{u}$ to Ω, $\tilde{w}$ be a Galerkin approximation of $\tilde{u}$ with restriction w to Ω. Theorems 4 and 5 show that even if $\tilde{u}$ is irregular on $\tilde{\Omega} - \Omega$, nevertheless w gives information concerning the derivatives of u; the speeds of convergence depend only on the speed of convergence of $\| u - w \|_{0,\Omega}$ which can be estimated easily in particular when Nitsche–Aubin's trick is applicable.

[5] which considers only least-square approximations seems to be the first publication concerned with local convergence of approximations obtained by projection methods; [10] gives another approach to the same problem. The results in this paper are closely related to those of [11] but most of them have been obtained independently.

2. Hypotheses

The first hypothesis concerns the bilinear form a introduced in Section 1; the three other hypotheses relate to the finite element subspaces. The symbol r denotes a fixed integer $> m$.

$H1$ $\quad a_{\alpha\beta} \in C_\infty(\bar{\Omega})$ $|\alpha|, |\beta| \leqslant m$; a is coercive on $\overset{0}{H_m}(\Omega)$ i.e. there exists $\gamma > 0$ such that $a(v, v) \geqslant \gamma \| v \|_{m,\Omega}$ for all $v \in \overset{0}{H_m}(\Omega)$; a is elliptic on $\bar{\Omega}$ i.e.

$$\sum_{|\alpha|,|\beta|=m} a_{\alpha\beta}(x) \xi^\alpha \xi^\beta \neq 0 \quad \text{for all} \quad \xi \neq 0, \quad x \in \bar{\Omega}.$$

The condition $a_{\alpha\beta} \in C_\infty(\bar{\Omega})$ has been introduced for the sake of simplicity.

H2 *Let $\Lambda \subset\subset \Omega$. There exist $h_0 < 0$ and c such that for any $u \in \overset{0}{C}_\infty(\Lambda)$*
and any $S \in \mathscr{S}$ with $h(S) < h_0$ there exists $v \in \overset{0}{S}(\Omega)$ with

$$\| u - v \|_{k,\Omega} \leqslant ch^{l-k} \| u \|_{l,\Lambda} \quad 0 \leqslant k \leqslant l \leqslant r;$$

here c is an absolute constant but h_0 depends on Λ.

H2 is a particular version of the fundamental property of approximation by finite element subspaces; it is valid for wide classes of finite elements; see for example [1], [2], [3], [4], [7], [12].

H3 *Let $\Lambda \subset \Omega$, $\omega \in \overset{0}{C}_\infty(\Lambda)$. There exist $h_0 > 0$ and c such that for any*
$S \in \mathscr{S}$ with $h(S) < h_0$ and any $u \in S$ there exists $v \in \overset{0}{S}(\Lambda)$ with

$$\| \omega u - v \|_{k,\Omega} \leqslant ch \| u \|_{k,\Lambda} \quad k = 0, 1, 2, \ldots, r - 1;$$

c and h_0 depend on ω.

This hypothesis is introduced in [10]. [6] analyzes its validity for the case of triangular polynomial elements. In a forthcoming paper, Nitsche and Schatz will consider other cases.

H4 *Let $\Lambda \subset\subset \Phi \subset\subset \Omega$. There exist $h_0 > 0$ and c such that for any*
$u \in S \in \mathscr{S}$ with $H(S) < h_0$ one has

$$\| u \|_{k+p,\Lambda} \leqslant ch^{-p} \| u \|_{k,\Phi}, \quad k + p \leqslant r;$$

here c and h_0 depend on Λ and Φ.

H4 is known in the literature under the name of "inverse assumption"; for its validity see for example [9], [6].

As an example we consider a collection $\mathscr{S}$ of simplicial (triangular for $n = 2$) polynomial element subspaces of one of the types described in [2], [3], [7] (in particular linear simplicial elements). For a simplex σ let $d_1(\sigma)$ be its diameter and $d_2(\sigma)$ be the diameter of the inscribed sphere. We introduce three properties. P_1: for all σ relative to all $S \in \mathscr{S}$ one has $d_1(\sigma)/d_2(\sigma) \leqslant c$ where c is a positive constant; P2 : {polynomials of degree $\leqslant p$} $\subset S \; \forall \, S \in \mathscr{S}$; P3 : $\forall \, S \in \mathscr{S}$ if σ_1 and σ_2 are two simplexes relative to the same S then $d_1(\sigma_1)/d_1(\sigma_2) \leqslant c$ where c is a positive absolute constant. Then P1, P2, P3 are related to H2, H3, H4 by the following implications: P1 and P2 $\Rightarrow$ H2, H3 with $r \leqslant p + 1$; P1 and P3 $\Rightarrow$ H4.

3. Preliminaries

In this section we suppose that hypotheses H1 to H3 are satisfied. In particular by H1 a is continuous i.e. there exists M such that $|a(u, v)| \leqslant M \| u \|_{m,\Omega} \| v \|_{m,\Omega}$ for all $u, v \in H^m(\Omega)$.

Lemma 1

Let $\Gamma \subset\subset \Lambda \subset \Omega$, $b(u, v) = (u, \Sigma_{|s| \leqslant \delta} b_s D^s v)$ where $b_s \in C_\infty(\bar{\Omega})$, $f \in H_{k+\delta-2m}($
$(f \in H_0(\Lambda)$ if $k + \delta - 2m < 0)$, $\omega \in H_m(\Omega)$ satisfying the relation $a(\omega, \varphi) = b(f, \varphi)$ $\forall \varphi \in \overset{0}{C}_\infty(\Lambda)$. Then there exist coefficients c_k depending on b, Γ and Λ such that

$$\| w \|_{k, \Gamma} \leqslant c_k (\| w \|_{0, \Lambda} + \| f \|_{k+\delta-2m, \Lambda})$$

Proof

Lemma 1 is straightforward generalization of classical results on interior regularity. See for example [8] pp. 30–36; one replaces in the top equation of page 36

$$\int_Q \varphi f e^{-ikx} \, dx \quad \text{by} \quad b(f, \varphi e^{-ikx}).$$

Definition

Let $\Delta \subset\subset \Omega$ and $S \in \mathscr{S}$. We set

$$T(S, \Delta) = \{u + v : u \in \overset{0}{S}(\Omega), v \in \overset{0}{H}_m(\Omega - \bar{\Delta})\} \subset \overset{0}{H}_m(\Omega);$$

furthermore let $P(S, \Delta)$ and $Q(S, \Delta)$ be the projectors from $\overset{0}{H}_m(\Omega)$ on to $T(S, \Delta)$ defined by

$$a(P(S, \Delta)u - u, v) = 0 \quad \forall v \in T(S, \Delta);$$
$$a(v, Q(S, \Delta)u - u) = 0 \quad \forall v \in T(S, \Delta);$$

in the following, for the sake of simplicity, we shall write T, P, Q instead of $T(T, \Delta), P(S, \Delta), Q(S, \Delta)$. We define $h(T) = h(S)$. Clearly T is closed and by the Lax-Milgram theorem P and Q are well defined projectors which will satisfy the relation

$$a(Pu, v) = a(u, Qv) \quad \forall u, v \in \overset{0}{H}_m(\Omega).$$

In the following, the symbols Δ, S, T, P and Q will be used exclusively in connection with the preceding definitions. In particular in Lemmas 2, 3, 4, 5, 6, Δ will represent any fixed open set $\subset\subset \Omega$.

Lemma 2

Let $u \in \overset{0}{H}_m(\Omega)$; then for any $v \in T$ we have

$$\| u - Pu \|_{m, \Omega} \leqslant (M/\gamma) \| u - v \|_{m, \Omega}, \quad \| u - Qu \|_{m, \Omega}$$
$$\leqslant (M/\gamma) \| u - v \|_{m, \Omega}.$$

Proof

$$\gamma \| u - Pu \|_{m,\Omega}^2 \leqslant a(u - Pu, u - Pu) = a(u - Pu, u - v)$$
$$\leqslant M \| u - Pu \|_{m,\Omega} \| u - v \|_{m,\Omega}.$$

Lemma 3
Let Λ be such that $\Delta \subset\subset \Lambda \subset \Omega$; then there exist c and $h_0 > 0$ such that for any $u \in \overset{0}{H}_m(\Omega) \cap H_p(\Lambda)$ with $m \leqslant p \leqslant r$ and for any T with $h(T) < h_0$ one has

$$\| u - Pu \|_{m,\Omega} \leqslant ch^{p-m} \| u \|_{p,\Lambda}, \quad \| u - Qu \|_{m,\Omega} \leqslant ch^{p-m} \| u \|_{p,\Lambda}.$$

Proof

We may suppose that $\Lambda \subset\subset \Omega$; let $\omega \in \overset{0}{C}_\infty(\Lambda)$ with $\omega(x) = 1$ in some open set $\supset\supset \Delta$, $w_i \in \overset{0}{C}_\infty(\Lambda)$ with $\lim\limits_{i \to \infty} \| \omega u - w_i \|_{p,\Omega} = 0$. By H2 there exists $v_i \in \overset{0}{S}(\Omega)$ with $\| w_i - v_i \|_{m,\Omega} \leqslant c_1 h^{p-m} \| w_i \|_{p,\Omega}$ where c_1 is independent of u and w_i. Since $v_i + (1 - \omega)u \in T$ we have by Lemma 2:

$$\| u - Pu \|_{m,\Omega} \leqslant (M/\gamma) \| u - (v_i + (1 - \omega)u \|_{m,\Omega}$$
$$\leqslant (M/\gamma) \{ \| \omega u - w_i \|_{m,\Omega} + \| w_i - v_i \|_{m,\Omega} \};$$

letting $i \to \infty$ we get $\| u - Pu \|_{m,\Omega} \leqslant (M/\gamma)c_1 h^{p-m} \| \omega u \|_{p,\Omega}$ from which Lemma 3 follows.

Lemma 4
Let $\Lambda \subset \Delta$, $\omega \in \overset{0}{C}_\infty(\Lambda)$; there exists c and $h_0 > 0$ such that for any T with $h(T) < h_0$ and any $u \in T$ one has

$$\| \omega u - P(\omega u) \|_{m,\Omega} \leqslant ch \| u \|_{m,\Lambda}, \quad \| \omega u - Q(\omega u) \|_{m,\Omega} \leqslant ch \| u \|_{m,\Lambda}.$$

Proof

Let $u = u_1 + u_2$, $u_1 \in \overset{0}{S}(\Omega)$, $u_2 \in \overset{0}{H}_m(\Omega - \bar{\Delta})$; by H3 there exists $v \in \overset{0}{S}(\Delta)$ with $\| \omega u_1 - v \|_{m,\Omega} \leqslant c_1 h \| u_1 \|_{m,\Lambda}$ where c_1 is independent of u_1; from the relations $\omega u = \omega u_1$, $\| u_1 \|_{m,\Lambda} = \| u \|_{m,\Lambda}$ and $v \in T$ we get by Lemma 2:

$$\| \omega u - P(\omega u) \|_{m,\Omega} \leqslant (M/\gamma) \| \omega u - v \|_{m,\Omega} \leqslant c_1 (M/\gamma)h \| u \|_{m,\Lambda}.$$

Lemma 5
Let Λ and Γ be such that $\Gamma \subset\subset \Delta$, $\Lambda \subset\subset \Delta - \bar{\Gamma}$. Then there exist c and $h_0 > 0$ such that for any T with $h(T) < h_0$ and any $\varphi \in \overset{0}{H}_m(\Gamma)$ one has

$$\| P\varphi \|_{m,\Lambda} \leqslant ch \| \varphi \|_{m,\Omega}, \| Q\varphi \|_{m,\Lambda} \leqslant ch \| \varphi \|_{m,\Omega}.$$

Proof

c will denote a generic constant independent of φ and h. Let $\omega \in \overset{0}{C}_\infty(\Delta - \overline{\Gamma})$, $\omega(x) = 1$ for $x \in \Lambda$. One has

$$\| P\varphi \|_{m,\Lambda} \leqslant \| \omega P\varphi \|_{m,\Omega} \leqslant \| P(\omega P\varphi) \|_{m,\Omega} + \| \omega P\varphi - P(\omega P\varphi) \|_{m,\Omega}; \quad (1)$$

by Lemmas 4 and 2 we can write

$$\| \omega P\varphi - P(\omega P\varphi) \|_{m,\Omega} \leqslant ch \| P\varphi \|_{m,\Omega} \leqslant ch \| \varphi \|_{m,\Omega}. \quad (2)$$

Let b be the bilinear form defined on $\overset{0}{H}_m(\Omega) \times \overset{0}{H}_m(\Omega)$ by $b(v, u) = a(\omega v, u) - a(v, \omega u)$; if $v \in \overset{0}{C}_\infty(\Omega)$ we have $b(v, u) = (\Sigma_{|s| \leqslant 2m-1} b_s D^s v, u)$ where $b_s \in \overset{0}{C}_\infty(\Omega)$; let $\eta = P(\omega P\varphi)/\| \underset{0}{P}(\omega P\varphi) \|_{m,\Omega}$ and w be defined by the relation $a(v, w) = b(v, \eta)$ for all $v \in H_m(\Omega)$; in particular for $v = P\varphi - \varphi$, we have $a(P\varphi - \varphi, w) = a(\omega(P\varphi - \varphi), \eta) - a(P\varphi - \varphi, \omega\eta)$; by Lax-Milgram, since $\| \eta \|_{m,\Omega} = 1$, we have $\| w \|_{m,\Omega} \leqslant c$; by Lemma 1, for some $\Phi \supset\supset \Delta$, we have $\| w \|_{m+1,\Phi} \leqslant c(\| w \|_{m,\Omega} + \| \eta \|_{m,\Omega}) \leqslant c$. We can estimate the first term of the right member of (1):

$$\gamma \| P(\omega P\varphi) \|_{m,\Omega} \leqslant a(P(\omega P\varphi), \eta) = a(\omega P\varphi, \eta) = a(\omega(P\varphi - \varphi), \eta)$$
$$= a(P\varphi - \varphi, w) + a(P\varphi - \varphi, \omega\eta); \quad (3)$$

by Lemmas 2, 3 we get (4); by Lemmas 2, 4 we get (5):

$$a(P\varphi - \varphi, w) = a(P\varphi - \varphi, w - Pw) \leqslant ch \| \varphi \|_{m,\Omega} \| w \|_{m+1,\Phi}$$
$$\leqslant ch \| \varphi \|_{m,\Omega}; \quad (4)$$

$$a(P\varphi - \varphi, \omega\eta) = a(P\varphi - \varphi, \omega\eta - P(\omega\eta)) \leqslant ch \| \varphi \|_{m,\Omega} \| \eta \|_{m,\Omega}$$
$$\leqslant ch \| \varphi \|_{m,\Omega}; \quad (5)$$

then Lemma 5 follows from relations (1) to (5).

Lemma 6

Let Γ, Π *be such that* $\Gamma \subset\subset \Pi \subset\subset \Delta$. *Then there exist* c *and* h_0 *such that for any* $S \in \mathcal{S}$ *with* $h(S) \leqslant h_0$ *and any* $u \in S$ *satisfying the relation* $a(u, v) = 0$ $\forall v \in \overset{0}{S}(\Pi)$, *one has*

$$|a(u, \varphi)| \leqslant ch \| u \|_{m,\Pi} \| \varphi \|_{m,\Omega} \quad \forall \varphi \in \overset{0}{H}_m(\Gamma).$$

Proof

Let Φ, θ be such that $\Gamma \subset\subset \Phi \subset\subset \theta \subset\subset \Pi$, $\psi \in \overset{0}{C}_\infty(\theta)$ with $\psi(x) = 1$ for $x \in \Phi$, $\omega \in \overset{0}{C}_\infty(\Pi)$ with $\omega(x) = 1$ for $x \in \theta$. c will denote a generic constant independent of S and u. One has

$$a(u, \varphi) = a(\omega u, \varphi) = a(\omega u, \varphi - Q\varphi) + a(\omega u, \psi Q\varphi) +$$
$$+ a(\omega u, (\varphi - \psi)Q\varphi); \quad (1)$$

by Lemmas 2, 4 one gets

$$|a(\omega u, \varphi - Q\varphi)| = |a(\omega u - P(\omega u), \varphi - Q\varphi)| \leqslant ch\|u\|_{m,\Pi}\|\varphi\|_{m,\Omega};$$
(2)

by H3 and Lemma 2, there exists $v \in \overset{0}{S}(\theta)$ such that $\|\psi Q\varphi - v\|_{m,\Omega}$ $\leqslant ch\|Q\varphi\|_{m,\Pi} \leqslant ch\|\varphi\|_{m,\Omega}$; using the preceding relation and the hypothesis of this lemma we have

$$|a(\omega u, \psi Q\varphi)| = |a(\omega u, \psi Q\varphi - v)| \leqslant ch\|u\|_{m,\Pi}\|\varphi\|_{m,\Omega};$$
(3)

Let $\Lambda = \Pi - \Phi$; then $\Lambda \subset\subset \Delta - \bar\Gamma$ and by Lemma 5 $\|Q\varphi\|_{m,\Lambda} \leqslant ch\|\varphi\|_{m,\Omega}$ and we have

$$|a(\omega u, (1 - \psi)Q\varphi)| \leqslant c\|\omega u\|_{m,\Lambda}\|(1 - \psi)Q\varphi\|_{m,\Lambda}$$
$$\leqslant ch\|u\|_{m,\Lambda}\|\varphi\|_{m,\Omega}.$$
(4)

Then Lemma 6 follows from (1), (2), (3), (4).

Lemma 7
Let Λ and Γ be such that $\Lambda \subset\subset \Gamma \subset \Omega$. Then there exists c such that for any $u \in \overset{0}{H}_m(\Omega)$ and any $\alpha \in \mathbb{R}$ satisfying the relation

$$|a(u, \varphi)| \leqslant \alpha\|u\|_{m,\Omega}\|\varphi\|_{m,\Omega} \quad \forall \varphi \in \overset{0}{H}_m(\Gamma)$$

there exists $v \in \overset{0}{H}_m(\Omega)$ with

(a) $a(v, \varphi) = 0 \quad \forall \varphi \in \overset{0}{H}_m(\Lambda)$,   *(b)* $\|u - v\|_{m,\Omega} \leqslant \alpha c\|u\|_{m,\Omega}$.

Proof
c will denote a generic constant independent of α and u. Let $\omega \in \overset{0}{C}_\infty(\Gamma)$ with $\omega(x) = 1$ for $x \in \Lambda$. Using Lax-Milgram's theorem, we define $v \in \overset{0}{H}_m(\Omega)$ by the relation $a(v, \varphi) = a(u, (1 - \omega)\varphi) \ \forall \varphi \in \overset{0}{H}_m(\Omega)$; v satisfies relation (a); relation (b) follows from

$$\gamma\|u - v\|_{m,\Omega}^2 \leqslant a(u - v, u - v) = a(u, u - v) - a(v, u - v)$$
$$= a(u, u - v) - a(u, (1 - \omega)(u - v)) =$$
$$= a(u, \omega(u - v)) \leqslant \alpha\|u\|_{m,\Omega}\|\omega(u - v)\|_{m,\Omega}$$
$$\leqslant \alpha c\|u\|_{m,\Omega}\|u - v\|_{m,\Omega}.$$

4. Interior Regularity and Local Convergence

We first improve Lemma 6.

Theorem 1
Suppose H1, H2, H3 verified and let $\Gamma \subset\subset \Pi \subset\subset \Omega$. Then there exist c

*and h_0 such that for any $S \in \mathscr{S}$ with $h(S) < h_0$ and any $u \in S$ satisfying
the relation $a(u, v) = 0 \; \forall v \in \overset{0}{S}(\Pi)$ one has*

$$|a(u, \varphi)| \leqslant ch^{r-m} \| u \|_{m,\Pi} \| \varphi \|_{m,\Omega} \quad \forall \varphi \in \overset{0}{H}_m(\Gamma).$$

Proof
We prove that $|a(u, \varphi)| \leqslant ch^k \| u \|_{m,\Pi} \| \varphi \|_{m,\Omega} \; \forall \varphi \in \overset{0}{H}_m(\Gamma)$ successively for
$k = 1, 2, \ldots, r - m$. For $k = 0$ the result is a consequence of the continuity
of a. Suppose the result true for $k = M < r - m$ and let us prove it for
$k = M + 1$. c will denote a generic constant independent of u. Let A, B, C
be such that $\Gamma \subset\subset C \subset\subset B \subset\subset A \subset\subset \Pi$. By the induction hypothesis
applied to A instead of Γ we have

$$|a(u, \varphi)| \leqslant ch^M \| u \|_{m,\Pi} \| \varphi \|_{m,\Omega} \quad \forall \varphi \in \overset{0}{H}_m(A).$$

Let $\underset{\sim}{\omega} \in \overset{0}{C}_\infty(\Pi)$ with $\omega(x) = 1$ for $x \in A$. By Lemma 7 there exists
$v \in \overset{0}{H}_m(\Omega)$ such that $a(v, \varphi) = 0 \; \forall \varphi \in \overset{0}{H}_m(B)$ with

$$\| u - v \|_{m,A} \leqslant \| \omega u - v \|_{m,\Omega} \leqslant ch^M \| \omega u \|_{m,\Omega} \leqslant ch^M \| u \|_{m,\Pi}. \tag{1}$$

Let B_1 be such that $C \subset\subset B_1 \subset\subset B$, $\psi \in \overset{0}{C}_\infty(B_1)$ with $\psi(x) = 1$ for $x \in C$;
by Lemma 1, $\psi v \in \overset{0}{C}_\infty(B_1)$. Let also Δ be such that $\Pi \subset\subset \Delta \subset\subset \Omega$;
$P = P(S, \Delta)$ is defined in Section 3. By Lemmas 1, 3 we can write

$$\| \psi v - P(\psi v) \|_{m,\Omega} \leqslant ch^{r-m} \| \psi v \|_{r,B_1} \leqslant ch^{r-m} \| v \|_{m,B}$$

$$\leqslant ch^{r-m} \| u \|_{m,\Pi}. \tag{2}$$

For $\varphi \in \overset{0}{S}(C)$ we have $a(\psi v, \varphi) = a(v, \varphi) = 0$ and consequently $a(P(\psi v), \varphi) =$
$a(P(\psi v) - \psi v, \varphi) = 0$ and also $a(u - P(\psi v), \varphi) = 0$. Applying Lemma 6 to
$u - P(\psi v)$ we have for $\varphi \in \overset{0}{C}_\infty(\Gamma)$:

$$|a(u - P(\psi v), \varphi)| \leqslant ch \| u - P(\psi v) \|_{m,C} \| \varphi \|_{m,\Omega}. \tag{3}$$

From (1), (2), (3) we get for $\varphi \in \overset{0}{C}_\infty(\Gamma)$:

$$\begin{aligned}
|a(u, \varphi)| &= |a(u - \psi v, \varphi)| \leqslant |a(u - P(\psi v), \varphi)| + |a(P(\psi v) - \psi v, \varphi)| \\
&\leqslant ch \| u - P(\psi v) \|_{m,C} \| \varphi \|_{m,\Omega} + c \| P(\psi v) - \psi v \|_{m,C} \| \varphi \|_{m,\Omega} \\
&\leqslant c\{ h \| u - v \|_{M,C} + \| P(\psi v) - \psi v \|_{m,C} \} \| \varphi \|_{m,\Omega} \\
&\leqslant ch^{M+1} \| u \|_{m,\Pi} \| \varphi \|_{m,\Omega}.
\end{aligned}$$

Lemma 8
*Suppose H1, H2, H3, H4 satisfied and let $\Lambda \subset\subset \Omega$, $f \in C_\infty(\bar{\Omega})$. Then
there exist c and $h_0 > 0$ such that if $w \in S \in \mathscr{S}$ with $h(S) < h_0$ satisfies
the relation $a(w, \varphi) = (f, \varphi) \; \forall \varphi \in \overset{0}{S}(\Omega)$, there exists $u \in \overset{0}{H}_m(\Omega) \cap H_k(\Lambda)$
with*

$$a(u, \varphi) = (f, \varphi) \quad \forall \varphi \in \overset{0}{H}_m(\Lambda),$$

$$\| u - w \|_{k,\Lambda} \leqslant ch^{r-k} (\| w \|_{m,\Omega} + \| f \|_{k-2m,\Omega}), \quad k = m, m+1, \ldots, r.$$

Proof

c will denote a generic constant independent of u. We choose A, B, C and Δ such that $\Lambda \subset\subset C \subset\subset B \subset\subset \Delta \subset\subset A \subset\subset \Omega$. $P = P(S, \Delta)$ is the notation introduced in Section 3. Let $v \in \overset{0}{H}_m(\Omega)$ be defined by the relation: $a(v, \varphi) = (f, \varphi) \; \forall \varphi \in \overset{0}{H}_m(\Omega)$. By Lax-Milgram's theorem and by Lemma 1, we have

$$\| v \|_{m,\Omega} \leqslant c \| f \|_{0,\Omega}; \quad \| v \|_{r,A} \leqslant c \| f \|_{r-2m,\Omega}.$$

By Lemma 3:

$$\| v - Pv \|_{m,\Omega} \leqslant ch^{r-m} \| v \|_{r,A} \leqslant ch^{r-m} \| f \|_{r-2m,\Omega}.$$

One has $a(Pv, \varphi) = (f, \varphi) \; \forall \varphi \in \overset{0}{S}(\Omega)$ and consequently $a(w - Pv, \varphi) = 0$ $\forall \varphi \in \overset{0}{S}(\Omega)$. Let $\omega \in \overset{0}{C}_\infty(\Omega)$ with $\omega(x) = 1$ for $x \in A$. For all $\varphi \in \overset{0}{H}_m(B)$ we then have by Theorem 1:

$$| a(w - Pv, \varphi) | = | a(\omega w - Pv, \varphi) | \leqslant ch^{r-m} \| w - Pv \|_{m,\Delta} \| \varphi \|_{m,\Omega}$$
$$\leqslant ch^{r-m} \{ \| w \|_{m,\Omega} + \| f \|_{0,\Omega} \} \| \varphi \|_{m,\Omega}.$$

By Lemma 7, there exists $g \in \overset{0}{H}_m(\Omega)$ such that

$$a(g, \varphi) = 0 \quad \forall \varphi \in \overset{0}{H}_m(C), \quad \| g - (\omega w - Pv) \|_{m,\Omega} \leqslant ch^{r-m} \{ \| w \|_{m,\Omega} +$$
$$+ \| f \|_{0,\Omega} \}.$$

Setting $u = v + g$ we have $a(u, \varphi) = (f, \varphi) \; \forall \varphi \in \overset{0}{H}_m(C)$ and

$$\| u - w \|_{m,C} \leqslant \| v - Pv \|_{m,C} + \| g - (\omega w - Pv) \|_{m,C}$$
$$\leqslant ch^{r-m} \{ \| w \|_{m,\Omega} + \| f \|_{r-2m,\Omega} \}$$

which implies Theorem 1 for $k = m$. We now prove Lemma 8 for $m < k \leqslant r$. Let $\psi \in \overset{0}{C}_\infty(C)$ with $\psi(x) = 1$ for $x \in \Lambda$. By Lemma 1 $\| \psi u \|_{r,C} \leqslant c \{ \| w \|_{m,\Omega} + \| f \|_{r-2m,\Omega} \}$. By H2 there exists $g \in \overset{0}{S}(\Omega)$ such that for $m \leqslant j \leqslant r$ one has

$$\| u - g \|_{j,\Lambda} \leqslant \| \psi u - g \|_{j,C} \leqslant ch^{r-j} \| \psi u \|_{r,C} \leqslant ch^{r-j} \{ \| w \|_{m,\Omega} +$$
$$+ \| f \|_{r-2m,\Omega} \}.$$

By H4 $\| w - g \|_{k,\Lambda} \leqslant ch^{m-k} \| w - g \|_{m,C}$ and consequently

$$\| w - g \|_{k,\Lambda} \leqslant ch^{m-k} \{ \| w - u \|_{m,C} + \| u - g \|_{m,C} \}$$
$$\leqslant ch^{r-k} \{ \| w \|_{m,\Omega} + \| f \|_{r-2m,\Omega} \}.$$
$$\| u - w \|_{k,\Lambda} \leqslant \| u - g \|_{k,\Lambda} + \| g - w \|_{k,\Lambda} \leqslant ch^{r-k} \{ \| w \|_{m,\Omega} +$$
$$+ \| f \|_{r-2m,\Omega} \}.$$

Lemma 9

Suppose H1, H2, H3, H4 satisfied and let $\Lambda \subset\subset \Omega, f \in C_\infty(\bar\Omega)$. Then there exist c and $h_0 > 0$ such that for any $w \in S \in \mathscr{S}$ with $h(S) < h_0$ satisfying the relation $a(w, \varphi) = (f, \varphi)\ \forall \varphi \in \overset{0}{S}(\Omega)$, one has

$$\| w \|_{m,\Lambda} \leqslant c(\| w \|_{0,\Omega} + \| f \|_{0,\Omega})$$

Proof

One proves by induction that $\| w \|_{m,\Lambda} \leqslant c(\| w \|_{k,\Omega} + \| f \|_{0,\Omega}), k = m,$ $m - 1, \ldots, 0$. For $k = m$, the result is obvious. Suppose it correct for $k = M + 1$ and let us prove it for $k = M$. Let $\Lambda \subset\subset \Gamma \subset\subset \overset{0}{\theta} \subset\subset \Omega$. By Lemma 8 and the induction hypothesis there exists $u \in \overset{0}{H}_m(\Omega), a(u, \varphi) = (f, \varphi)\ \forall \varphi \in \overset{0}{H}_m(\Gamma)$ such that $\| u - w \|_{m,\Gamma} \leqslant ch^{r-m}(\| w \|_{M+1,\theta} + \| f \|_{0,\Omega})$. Then by Lemma 1:

$$\| w \|_{m,\Lambda} \leqslant \| u \|_{m,\Lambda} + \| u - w \|_{m,\Lambda} \leqslant c\{\| u \|_{M,\Gamma} + \| f \|_{0,\Omega} +$$

$$+ \| u - w \|_{m,\Lambda}\} \leqslant c\| w \|_{M,\Gamma} + \| u - w \|_{M,\Gamma} + \| f \|_{0,\Omega}\}$$

$$\leqslant c\| w \|_{M,\Gamma} + \| f \|_{0,\Omega}\};$$

From Lemma 8 (with a suitable modification of domains) and Lemma 9, one immediately gets Theorem 2.

Theorem 2

Suppose H1, H2, H3, H4 satisfied and let $\Lambda \subset\subset \Omega, f \in C_\infty(\bar\Omega)$. Then there exist c and $h_0 > 0$ such that if $w \in S \in \mathscr{S}$ with $h(S) < h_0$ satisfies the relation $a(w, \varphi) = (f, \varphi)\ \forall \varphi \in \overset{0}{S}(\Omega)$, there exists $u \in \overset{0}{H}_m(\Omega) \cap H_k(\Lambda)$ with

$$a(u, \varphi) = (f, \varphi)\quad \forall \varphi \in \overset{0}{H}_m(\Lambda)$$

$$\| u - w \|_{k,\Lambda} \leqslant ch^{r-k}(\| w \|_{0,\Omega} + \| f \|_{k-2m,\Omega})\quad k = m, m+1, \ldots, r.$$

Remark

Nitsche's trick shows that in fact the last relation is valid for $k = 0, 1, \ldots, r$, if $r \geqslant 2m$.

Theorem 3

Suppose H1, H2, H3, H4 verified and let $\Lambda \subset\subset \Omega, f \in C_\infty(\bar\Omega)$. Then there exist c and $h_0 > 0$ such that if $w \in S \in \mathscr{S}$ with $h(S) < h_0$ satisfies the relation $a(w, \varphi) = (f, \varphi)\ \forall \varphi \in S(\Omega)$ then

$$\| w \|_{k,\Lambda} \leqslant c(\| w \|_{0,\Omega} + \| f \|_{k-2m,\Omega})\quad k = 0, 1, 2, \ldots, r.$$

Proof

For $k \leqslant m$, Theorem 3 is equivalent to Lemma 9. Suppose $k > m$ and θ such that $\Lambda \subset\subset \theta \subset\subset \Omega$. By Theorem 2, there exists $u \in \overset{0}{H}_m(\Omega), a(u, \varphi) =$

$(f, \varphi) \; \forall \varphi \in \theta$ such that $\| u - w \|_{k,\theta} \leqslant ch^{r-k}(\| w \|_{0,\Omega} + \| f \|_{k-2m,\Omega})$. Then we have by Lemma 1:

$$\| w \|_{k,\Lambda} \leqslant \| u - w \|_{k,\Lambda} + \| u \|_{k,\Lambda} \leqslant c(\| u - w \|_{k,\Lambda} + \| u \|_{0,\theta} +$$
$$+ \| f \|_{k-2m,\Omega}) \leqslant c(\| u - w \|_{k,\theta} + \| w \|_{0,\Omega} + \| f \|_{k-2m,\Omega})$$
$$\leqslant c(\| w \|_{0,\Omega} + \| f \|_{k-2m,\Omega}).$$

Theorem 4

We suppose that hypotheses H1 to H4 are satisfied. Let $\Lambda \subset\subset \Omega$, $f \in C_\infty(\bar{\Omega})$, $u \in H_m(\Omega)$ such that $a(u, \varphi) = (f, \varphi) \forall \varphi \in H_m(\Omega)$, $w \in S$ such that $a(w, \varphi) = (f, \varphi) \forall \varphi \in \overset{0}{S}(\Omega)$. Then there exist c and $h_0 > 0$ such that if $h(S) < h_0$ then

$$\| u - w \|_{k,\Lambda} \leqslant c \{\| u - w \|_{0,\Omega}$$
$$+ h^{r-k}(\| u \|_{0,\Omega} + \| f \|_{k-2m,\Omega})\} \, k = m, \, m+1, \ldots, r.$$

Remark

Nitsche's trick shows that in fact the last relation is valid for $k = 0, 1, \ldots, r$, if $r \geqslant 2m$.

Proof

Let θ be such that $\Lambda \overset{0}{\subset\subset} \theta \subset\subset \Omega$. By Theorem 2, there exists $v \in \overset{0}{H}_m(\Omega)$, $a(v, \varphi) = (f, \varphi) \; \forall \varphi \in \overset{0}{H}_m(\theta)$, with $\| v - w \|_{k,\theta} \leqslant ch^{r-k}(\| w \|_{0,\Omega} + \| f \|_{k-2m,\Omega})$. One has by Lemma 1:

$$\| u - w \|_{k,\Lambda} \leqslant \| u - v \|_{k,\Lambda} + \| v - w \|_{k,\Lambda} \leqslant c \{\| u - v \|_{k,\theta} +$$
$$+ \| v - w \|_{k,\Lambda}\} \leqslant c \{\| u - w \|_{0,\Omega} + \| v - w \|_{k,\theta}\}$$
$$\leqslant c \{\| u - w \|_{0,\Omega} + h^{r-k}(\| w \|_{0,\Omega} + \| f \|_{k-2m,\Omega}\}$$
$$\leqslant c \{\| u - w \|_{0,\Omega} + h^{r-k}(\| u \|_{0,\Omega} + \| f \|_{k-2m,\Omega})\}.$$

5. Finite Differences

In this section we are interested in the particuliar case of geometrically regular elements. More specifically, we introduce the following hypothesis.

H5   Let $\Lambda \subset\subset \Gamma \subset\subset \Omega$. Then there exists $h_0 > 0$ such that for any $v \in \overset{0}{S}(\Lambda)$ with $h(S) < h_0$ the functions v_i^+ and v_i^- defined by $v_i^+(x_1, \ldots, x_n) = v(x_1, \ldots, x_{i-1}, x_i + h, x_{i+1}, \ldots, x_n)$ and $v_i^-(x_1, \ldots, x_{i-1}, x_i - h, x_{i+1}, \ldots, x_n)$ belong to $\overset{0}{S}(\Gamma)$, $i = 1, 2, \ldots, n$.

We consider difference operators of the form

$$v(x) \mapsto h^{-1}(v(x_1, \ldots, x_i + \alpha h, \ldots, x_n) - v(x_1, \ldots, x_i + \beta h, \ldots, x_n))$$

where α and β are integers. A product T of N operators of this type will be called a finite difference operator of order N. For h small enough and $\Lambda \subset\subset \Gamma \subset \Omega$, T satisfies the relation

$$\| Tu \|_{k,\Lambda} \leqslant c \| u \|_{N+k,\Gamma}, \quad u \in H_{N+k,\Omega},$$

c independent of u and h.

Theorem 5
We suppose that hypotheses H1 to H5 are satisfied. Let $\Lambda \subset\subset \Omega$, $f \in C_\infty(\Omega)$, $u \in H_m(\Omega)$ such that $a(u,\varphi) = (f,\varphi) \; \forall \varphi \in \overset{0}{H}_m(\Omega)$, $w \in S$ such that $a(w,\varphi) = (f,\varphi) \; \forall \varphi \in \overset{0}{S}(\Omega)$ and T be a finite operator of order N. Then there exist c and $h_0 > 0$ (depending on Λ and N) such that for $h(S) < h_0$ one has

$$\| Tu - Tw \|_{k,\Lambda} \leqslant c \{ \| u - w \|_{0,\Omega} + h^{r-k}(\| u \|_{0,\Omega} + \| f \|_{k+N-2m,\Omega}) \}$$

$$k = m, m+1, \dots, r.$$

Remarks
(1) *Nitsche's trick and some minor modifications of the proof show that the last relation is in fact true for $k = 0, 1, \dots, r$, if $r \geqslant 2m$.*
(2) *Theorem 5 is obviously still true if one replaces T by a linear combination of finite difference operators of order $\leqslant N$.*

Proof
We only sketch the proof. c will denote a generic constant independent of u and w. One proceeds by induction on N. For $N = 0$, Theorem 5 reduces to Theorem 4. Suppose it is true for $N \leqslant M - 1$ and let T be a finite difference operator of order M.

Let $\Lambda \subset\subset \Gamma \subset\subset \Phi \subset\subset \theta \subset\subset \Omega$. We first prove the theorem when u is replaced by $v \in \overset{0}{H}_m(\Omega)$ satisfying the relation (cf. Theorem 2)

$$a(v,\varphi) = (f,\varphi) \; \forall \varphi \in \overset{0}{H}_m(\theta), \quad \| v - w \|_{k,\theta} \leqslant ch^{r-k}(\| w \|_{0,\Omega} +$$

$$+ \| f \|_{k-2m,\Omega}),$$

$k = m, m+1, \dots, r$; more precisely we want to establish for $k = m, m+1, \dots, r$ the relation

$$\| Tv - Tw \|_{k,\Lambda} \leqslant c \{ \| v - w \|_{0,\theta} + h^{r-k}(\| v \|_{0,\theta} + \| f \|_{k+M-2m,\Omega}) \} \quad (1)$$

For h small enough and $\varphi \in \overset{0}{C}_\infty(\Phi)$, we can write

$$a(Tv,\varphi) = \sum_{|\alpha|,|\beta| \leqslant m} (D^\alpha Tv, a_{\alpha\beta} D^\beta \varphi) = \sum_{|\alpha|,|\beta| \leqslant m} (D^\alpha v, T^*(a_{\alpha\beta} D^\beta \varphi))$$

$$= \sum_p b_p(T_p v, \varphi) + a(v, T^*\varphi) = \sum_p b_p(T_p v, \varphi) + (f, T^*\varphi);$$

$$a(Tv,\varphi) = \sum_p b_p(T_p v, \varphi) + (Tf, \varphi) \quad \forall \varphi \in \overset{0}{C}_\infty(\Phi); \quad (2)$$

here T^* is the adjoint of T, the b_p's are a finite collection of bilinear forms of order m the coefficients of which depend on h but are C^∞ and uniformly bounded; the T_p's are finite difference operators of order $\leq M - 1$. By H5, we have also

$$a(Tw, \varphi) = \sum_p b_p(T_p w, \varphi) + (Tf, \varphi) \quad \forall \varphi \in \overset{0}{S}(\Phi); \tag{3}$$

we cannot compare directly v and w because the second members of (2) and (3) are different; for this reason we introduce $z \in \overset{0}{S}(\Phi)$ defined by the relation

$$a(z, \varphi) = \sum_p b_p(T_p(v - w), \varphi) \quad \forall \varphi \in \overset{0}{S}(\Phi); \tag{4}$$

a is coercive, T_p is of order $\leq M - 1$; by the induction hypothesis (u, Λ and Ω are replaced by v, Φ and θ) we have

$$\|z\|_{m,\Phi} \leq c \sum_p \|T_p(v - w)\|_{m,\Phi} \leq c\{\|v - w\|_{0,\theta} + h^{r-m}(\|v\|_{0,\theta}$$

$$+ \|f\|_{M-1-m,\Omega}\} \leq ch^{r-m}(\|w\|_{0,\Omega} + \|f\|_{M-1-m,\Omega});$$

by H4, we get furthermore

$$\|z\|_{k,\Gamma} \leq ch^{r-k}(\|w\|_{0,\Omega} + \|f\|_{M-1-m,\Omega}),$$
$$k = m, m+1, \ldots, r \tag{5}$$

Adding (3) and (4) we have for $Tw + z \in S$.

$$a(Tw + z, \varphi) = \sum_p b_p(T_p v, \varphi) + (Tf, \varphi) \quad \forall \varphi \in \overset{0}{S}(\Phi). \tag{6}$$

Using a slight generalization of Theorem 4 and comparing (2) and (6), we get

$$\|Tv - Tw - z\|_{k,\Lambda} \leq c\{\|Tv - Tw - z\|_{0,\Gamma} + h^{r-k}(\|v\|_{0,\Gamma} +$$

$$+ \sum_p \|T_p v\|_{k,\Gamma} + \|Tf\|_{k-2m,\Gamma})\}; \tag{7}$$

we have

$$\|T_p v\|_{k,\Gamma} \leq c\|v\|_{M-1+k,\Phi}, \quad \|Tf\|_{k-2m,\Gamma} \leq c\|f\|_{M+k-2m,\Omega}$$

and by Lemma 1:

$$\|v\|_{M-1+k,\Phi} \leq c(\|v\|_{0,\theta} + \|f\|_{M-1+k-2m,\Omega});$$

furthermore, by writing $T = T_1 T_2$ where T_1 and T_2 are finite difference operators of order 1 and $M - 1$ respectively, we have by the induction hypothesis:

$$\| T(v - w) \|_{0,\Gamma} \leqslant c \| T_2(v - w) \|_{1,\Phi} \leqslant c \{ \| v - w \|_{0,\theta} + h^{r-m}(\| v \|_{0,\Phi}$$
$$+ \| f \|_{M-1-m,\Omega}) \} ;$$

introducing these relations and (5) in (7) we get precisely (1). We now prove the theorem itself. We have by Lemma 1:

$$\| Tu - Tv \|_{k,\Lambda} \leqslant c \| u - v \|_{M+k,\Gamma} \leqslant c \| u - v \|_{0,\Phi} \leqslant c \| u - w \|_{0,\Phi}$$
$$+ \| w - v \|_{0,\Phi} \}.$$

Then for $k + m, m + 1, \ldots, r$ we get by (1):

$$\| Tu - Tw \|_{k,\Lambda} \leqslant \| Tu - Tv \|_{k,\Lambda} + \| Tv - Tw \|_{k,\Lambda}$$
$$\leqslant c \{ \| u - w \|_{0,\Phi} + \| w - v \|_{0,\theta} + h^{r-k}(\| w \|_{0,\theta}$$
$$+ \| f \|_{k+M-2m,\Omega} \} \leqslant c \{ \| u - w \|_{0,\Phi} + h^{r-k}(\| w \|_{0,\Omega}$$
$$+ \| f \|_{k+M-2m,\Omega}) \} \leqslant c \{ \| u - w \|_{0,\Omega} + h^{r-k}(\| u \|_{0,\Omega}$$
$$+ \| f \|_{k+M-2m,\Omega}) \}.$$

References

[1] Bramble, J. H. and Hilbert, S. R. (1970). Estimation of linear functionals on Sobolev spaces with application to Fourier transforms and spline interpolation, *SIAM J. Numer. Anal.* 7 (1), 112-124.

[2] Bramble, J. H. and Zlamal, M. (1970). Triangular elements in the finite element method, *Math. Comp.* 24, 809-820.

[3] Ciarlet, P. G. and Raviart, P. A. (1972). *The Combined Effect of Curved Boundaries and Numerical Integration in Isoparametric Finite Element Method. The Mathematical Foundations of the Finite Element Method with Applications to Partial Differential Equations.* Academic Press, New York, London.

[4] Clement, Ph. and Pini, F. (1973). *Approximation by Finite Element Functions Using Regularization.* Report EPFL, Dept. Math. Lausanne.

[5] Descloux, J. (1969). *Some Properties of Approximations by Finite Elements.* Report EPFL, Dept. Math. Lausanne.

[6] Descloux, J. (1973). *Two Basic Properties of Finite Elements.* Report EPFL, Dept. Math. Lausanne.

[7] Dupuis, G. and Goél, J. J. (1969). Elements finis raffinés en élasticité bidimensionnelle, *ZAMP* 20 (6), 858-881.

[8] Fichera, G. (1965). Linear elliptic differential systems and eigenvalue problems, *Lecture Notes in Mathematics 8.* Springer-Verlag, Berlin.

[9] Nitsche, J. (1969). Umkehrsätze für Spline-Approximationen, *Compositio Mathematica* 21 (4), 400-416.

[10] Nitsche, J. and Schatz, A. (1972). On local approximation properties of L_2-projection on spline-subspaces, *Applicable Analysis* 2, 161–168.

[11] Nitsche, J. and Schatz, A. (1974). Interior estimates for Ritz-Galerkin methods, *Symposium on Finite Elements and Partial Differential Equations*. Madison.

[12] Strang, G. (1972). Approximation in the finite element method, *Num. Math.* 19, 81–98.

[13] Thomée, V. and Westergren, B. (1968). Elliptic difference equations and interior regularity, *Num. Math.* 11, 196–210.

Numerical Estimates for the Error of Gauss–Jacobi Quadrature Formulae

Géza Freud

1. Introduction

In our present paper we continue our work the first part of which we published in the volume of the previous conference (see [1]). In [1] we gave asymptotic estimates for the remainder of the quadrature formula. These estimates are, as far as the numerical factor is involved, the best possible. Their disadvantage is that they are valid only for $n \to \infty$, and have not given any numerical error estimates. In our present paper we develop our result to furnish explicit error estimates for fixed but arbitrary values of n which are not much greater than the asymptotic expression of the error.

We first recapitulate the results of our previous Dublin lecture [1]. We denote the Gauss–Jacobi quadrature formula by $Q_n(d\alpha; f)$, i.e. let

$$\int_{-1}^{1} f(x)d\alpha(x) \sim Q_n(d\alpha; f) \overset{\text{def}}{=} \sum_{k=1}^{n} \lambda_{kn} f(x_{kn}) \tag{1}$$

be the uniquely defined quadrature formula over n nodes which is precise for every polynomial of degree not exceeding $2n - 1$.

Here $d\alpha$ is a nonnegative measure the support of which is contained in $[-1, 1]$ and this support consists of infinitely many points. Let us denote the orthonormal polynomials with respect to $d\alpha$ by $p_r(d\alpha; x)$ $(r = 0, 1, \ldots)$ where the leading coefficient of $p_r(d\alpha; x)$ is $\gamma_r(d\alpha) > 0$. Let $f(z)$ be analytic in a domain D which contains the line segment $[-1, 1]$, then we proved in [1] that

$$\int_{-1}^{1} f(x)d\alpha(x) - Q_n(d\alpha; f) = \sum_{\nu=n}^{\infty} \frac{\gamma_{\nu+1}(d\alpha)}{\gamma_\nu(d\alpha)} \frac{1}{2\pi i} \int_C \frac{f(\xi)\, d\xi}{p_\nu(d\alpha; \xi)p_{\nu+1}(d\alpha; \xi)}. \tag{2}$$

Here C is a rectifiable simple closed curve which surrounds the linear segment $[-1, 1]$ in the positive sense.

We obtained in our previous lecture our asymptotic estimates by combining (2) with two results of G. Szegö (see [4]): let $g(\theta) = \log \alpha'\,(\cos \theta) \in \mathcal{L}$, then, first

$$\lim_{\nu \to \infty} \frac{\gamma_{\nu+1}(d\alpha)}{\gamma_\nu(d\alpha)} = 2$$

and, second, the orthogonal polynomials obey an asymptotic law.

In order to obtain our estimates we must replace these two asymptotic results by numerical ones.

As to the first statement of Szegö, we obtain from the recursion formula (see [2], Section I.2)

$$\frac{\gamma_{r-1}(d\alpha)}{\gamma_r(d\alpha)} = \int_{-1}^{1} x p_{r-1}(d\alpha; x) p_r(d\alpha; x) d\alpha(x) \leq \int_{-1}^{1} |p_{r-1}(d\alpha; x) p_r(d\alpha; x)| d\alpha(x)$$

$$\leq \left\{ \int_{-1}^{1} p_{r-1}^2(d\alpha; x) d\alpha(x) \int_{-1}^{1} p_r^2(d\alpha; x) d\alpha(x) \right\}^{1/2} = 1. \qquad (3)$$

We shall see in a moment that estimate (3) (and not the estimate of the reciprocal value) is what is actually needed. We observe that compared to the asymptotic value we do not lose more than a factor two.

Secondly we are going to replace Szegö's asymptotic formula by the numerical estimate

$$\frac{1}{|p_\nu(d\alpha; z)|} \leqq \frac{\gamma_{\nu-1}(d\alpha)}{\gamma_\nu(d\alpha)} \frac{\sqrt{\alpha(1) - \alpha(-1)}}{\Delta(z)|T_{\nu-1}(z)|}. \qquad (4)$$

This estimation is implicitly contained in Section III.7 of our book [2] and is explicitly stated and proved in our paper [3], Lemma 2.

In the expression on the right of (4)

$$T_r(x) = \cos r(\arccos x) = \tfrac{1}{2}[(x + \sqrt{x^2 - 1})^r + (x - \sqrt{x^2 - 1})^r] \qquad (5)$$

is the rth degree Chebyshev polynomial of the first kind and $\Delta(z)$ is the Euclidean distance of the point z from the line segment $[-1, 1]$.

For the special case $d\alpha(x) = (1 - x^2)^{-1/2}$ and large values of $|z|$ we lose, compared to the asymptotic formula obtained by the application of (4), a factor $\sqrt{2}$.

2. The Numerical Estimation of the Error Term

Let

$$r(z) = |z + \sqrt{z^2 - 1}| \qquad (6)$$

and consequently

$$|z - \sqrt{z^2 - 1}| = \frac{1}{r(z)}. \tag{6a}$$

We denote by $e(R)$ the ellipse $\{z : r(z) = R\}$ and introduce for analytic $f(z)$ the elliptic maximum modulus function

$$M_e(f; R) = \sup_{z \in e(R)} |f(z)|. \tag{7}$$

We also introduce

$$B_n(d\alpha) = \sup_{r \geq n} \frac{\gamma_{r-1}(d\alpha)}{\gamma_r(d\alpha)}. \tag{8}$$

Theorem 1
Let $f(z)$ be analytic in a domain $\mathcal{D}$ containing $e(R)$ then we have

$$\left| \int_{-1}^{1} f(x)d\alpha(x) - Q_n(d\alpha; f) \right| \leq M_e(f; r) \left[\frac{\alpha(1) - \alpha(-1)}{2\pi} B_n(d\alpha) \right] \vartheta(R) \tag{9}$$

where

$$\vartheta(R) = \int_{e(R)} \frac{\eta_n(\xi)|d\xi|}{[\Delta(\xi)]^2} \tag{10}$$

and

$$\eta_n(\xi) = \sum_{\nu=n}^{\infty} \frac{1}{|T_{\nu-1}(\xi)T_\nu(\xi)|}. \tag{11}$$

Remarks
(a) *Let us observe that on the right side of (9) only the expression in square brackets depends on the measure $d\alpha$.*
(b) *For $B_n(d\alpha)$ we know by (3) that*

$$0 < B_n(d\alpha) \leq 1. \tag{12}$$

For special weights an explicit calculation might be more precise. E.g. for $d\alpha(x) = dx$ (Legendre case) we have for $n \geq 2$

$$B_n(dx) = \frac{n}{\sqrt{4n^2 - 1}}. \tag{13}$$

(c) *By Theorem 1, the problem of the error estimate is reduced to a numerical inequality for $\vartheta(R)$. This we can do as follows.*

By (5), (6) and (6a)

$$|T_n(z)| \geq \tfrac{1}{2}\{[r(z)]^n - [r(z)]^{-n}\} \tag{14}$$

thus by (11) we have for every $\xi \in e(R)$

$$\eta_n(\xi) \leq 4 \sum_{\nu=n}^{\infty} \frac{1}{(R^{\nu-1} - R^{-\nu+1})(R^{\nu} - R^{-\nu})} \stackrel{\text{def}}{=} \chi(R). \tag{15}$$

By elementary calculation we obtain

$$\Delta(\xi) \geq \tfrac{1}{2}(R + R^{-1}) - 1 \quad for \quad \xi \in e(R). \tag{16}$$

Finally the closed convex curve $e(R)$ is inside the circle $|z| = R/2$ so that the perimeter of $e(R)$ is, by the principle of Archimedes, smaller than $2\pi \cdot R/2 = \pi R$. Inserting all these estimates in (10) we get

$$\vartheta(R) \leq \frac{\pi R \chi(R)}{[\tfrac{1}{2}(R + R^{-1}) - 1]^2} \tag{17}$$

and we observe that for large values of R the function $\chi(R)$ is very near to $4R^{-2n+1}$.

For $R \gg 1$ it might be convenient to apply, in place of the elliptic maximum modulus $M_e(f; R)$, the more popular circular maximum modulus

$$M(f; r) = \sup_{|z|=r} |f(z)|. \tag{18}$$

Assuming that $f(z)$ is analytic in the disc $\{z: |z| \leq R/2\}$ which contains $e(R)$ we have

$$M_e(f; R) \leq M(f; R/2). \tag{19}$$

Inserting (17) and (18) in (9) we obtain after substituting $R = 2r$:

Theorem 2
Let $f(z)$ be analytic on the closed disc $|z| \leq r$ then we have

$$\left| \int_{-1}^{1} f(x)d\alpha(x) - Q_n(d\alpha; f) \right|$$

$$\leq M(f; r)\left[\frac{\alpha(1) - \alpha(-1)}{2\pi} B_n(d\alpha) \right] \cdot \frac{4\pi r \chi(2r)}{[2r + (2r)^{-1} - 1]^2}$$

$$\leq M(f; r) \frac{\alpha(1) - \alpha(-1)}{2\pi} \frac{4\pi r \chi(2r)}{[2r + (2r)^{-1} - 1]^2} \tag{20}$$

where $\chi(R) \approx 4R^{-2n+1}$ is the function defined by (15).

3. On the Error of Padé Approximation

As an application of particular interest we mention the Padé approximation problem which was treated by us earlier in [3].

Let $f(z)$ be analytic on the whole complex plane with the exception of an essential singularity at $z = 0$. Let the Laurent series of $f(z)$ be

$$f(z) = \sum_{\nu=0}^{\infty} b_\nu z^{-\nu}. \tag{21}$$

To each such $f(z)$ and to every integer n we assign a rational function $\rho_n(f; z) = r_n(z)/R_n(z)$ the Padé approximant of $f(z)$, which has the following properties:

(a) $r_n(z)$ and $R_n(z)$ are polynomials with degree not higher than n;
(b) we have for $|z| \to \infty$

$$R_n(z)f(z) - r_n(z) = 0(|z|^{-n-1}). \tag{22}$$

To calculate $\rho_n(f; z)$ we set

$$R_n(z) = z^n + \sum_{l=0}^{n-1} d_l z^l. \tag{23}$$

The assumption that the coefficients of z^{-s} ($s = 1, 2, \ldots, n$) in the expansion of $R_n(z)f(z)$ vanish furnishes a system of n linear equations to determine the n unknown coefficients d_l ($l = 0, 1, \ldots, n-1$) of $R_n(z)$. After $R_n(z)$ is calculated we get $r_n(z)$ as the integer part of the power series of $R_n(z)f(z)$.

We now investigate the special case when f is the Stieltjes transform of a nonnegative measure $d\alpha$ with support contained in $[-1, 1]$:

$$f(z) = \int_{-1}^{1} \frac{d\alpha(t)}{z - t} = \sum_{k=1}^{\infty} \frac{\mu_{k-1}(d\alpha)}{z^k}. \tag{24}$$

Here

$$\mu_r(d\alpha) = \int_{-1}^{1} x^r d\alpha(x) \quad (r = 0, 1, \ldots) \tag{25}$$

are the moments of $d\alpha$.

The equations determining the coefficients of $R_n(z)$ are then

$$\sum_{l=0}^{n-1} d_l \mu_{s+l-1}(d\alpha) = -\mu_{n+s-1}(d\alpha) \quad (s = 1, 2, \ldots, n). \tag{26}$$

By (25) the determinant of the system (26) is the same as the determinant of the quadratic form

$$\sum_{l=0}^{n-1} \sum_{j=0}^{n-1} \mu_{l+j}(d\alpha) y_l y_j = \int_{-1}^{1} \left(\sum_{j=0}^{n-1} y_j x^j \right)^2 d\alpha(x). \tag{27}$$

Now if the support of $d\alpha$ contains infinitely many points then the expression (27) is positive definite (see [2], Lemma 1.1). It follows that in our case the Padé approximant exists and is uniquely determined. We observe also that the degree of the denominator $R_n(z)$ is precisely n thus by (22)

$$f(z) - \rho_n(f; z) = 0(|z|^{-2n-1}). \tag{28}$$

Lemma 1
If $f(z)$ is representable in the form (24) as a Stieltjes transform of a non-negative measure $d\alpha$ then the Padé approximant of $f(z)$ is

$$\rho_n(f; z) = Q_n(d\alpha; \mathscr{F}_z) \tag{29}$$

where

$$\mathscr{F}_z = (z - x)^{-1}. \tag{30}$$

Proof
Clearly

$$Q_n(d\alpha; \mathscr{F}_z) = \sum_{k=1}^{n} \frac{\lambda_{kn}}{z - x_{kn}} \tag{31}$$

and this is a rational function of degree n.

Now let $|z| > 3$ and let us apply Theorem 2 by setting $r = |z|/2$:

$$f(z) - Q_n(d\alpha; \mathscr{F}_z) = 0(|z|^{-2n-1}) \tag{32}$$

i.e. comparing (32) with (28)

$$\rho_n(f; z) - Q_n(d\alpha; \mathscr{F}_z) = 0(|z|^{-2n-1}).$$

Since the left-hand side is the difference of two rational functions of degree n, i.e. a rational function of degree $2n$, this is only possible if (29) is valid, Q.E.D.

Theorem 3. (See [3])
We have under the conditions of Lemma 1

$$|f(z) - \rho_n(f; z)| \leq [\alpha(1) - d(-1)] B_n(d\alpha) \frac{B_n(z)}{[\Delta(z)]^2}$$

$$\leq [\alpha(1) - \alpha(-1)] \frac{B_n(z)}{[\Delta(z)]^2}. \tag{33}$$

Proof

In formula (2) we replace $f(\xi)$ by $\mathscr{F}_z(\xi) = (z - \xi)^{-1}$ and evaluate the integrals in the sum by the calculus of residues:

$$f(z) - Q_n(d\alpha; \mathscr{F}_z) = \sum_{\nu=n}^{\infty} \frac{\gamma_{\nu+1}(d\alpha)}{\gamma_\nu(d\alpha)} \frac{1}{p_\nu(d\alpha; z) p_{\nu+1}(d\alpha; z)} \tag{34}$$

Interesting in (34) the estimate (4) we obtain

$$|f(z) - Q_n(d\alpha; \mathscr{F}_z)| \leq [\alpha(1) - \alpha(-1)] \sum_{\nu=n}^{\infty} \frac{\gamma_{\nu-1}(d\alpha)}{\gamma_\nu(d\alpha)} \times$$

$$\times \frac{1}{[\Delta(z)]^2 |T_{\nu-1}(z) T_\nu(z)|} \tag{35}$$

Then the first part of (33) is implied by (29), (8) and (11) and the second part is implied by (12), Q.E.D.

4. On the Calculation of Elliptic Integrals

Let

$$\varphi(\tau) = \int_{-1}^{1} \frac{1}{\sqrt{(1 - x^2)(1 - \tau^2 x^2)}} \, dx \tag{36}$$

where τ is a positive number smaller than one. To calculate (36) we insert in (2) $d\alpha(x) = (1 - x^2)^{-1/2}$ and $f(x) = (1 - \tau^2 x^2)^{-1/2}$. Then the Gauss–Jacobi abscissae are the zeros of Chebychev polynomials

$$x_{kn} = \cos \frac{2k - 1}{2n} \pi \tag{37}$$

the coefficients of the quadrature formula are

$$\lambda_{kn} = \frac{\pi}{n} \tag{38}$$

and the orthonormal polynomials $p_n(d\alpha; x)$ must be replaced by $\sqrt{2/\pi}\, T_n(x)$ $(n = 1, 2, \ldots)$.

By a deformation of the contour C to the twice covered lines (τ^{-1}, ∞) and $(-\infty, \tau^{-1})$ we obtain

$$\frac{1}{2\pi i} \oint_C \frac{f(z)}{p_{n-1}(d\alpha; z) p_n(d\alpha; z)} \, dz = \frac{1}{4i} \oint_C \frac{1}{T_{n-1}(z) T_n(z) \sqrt{1 - \tau^2 z^2}} \, dz$$

$$= \int_{\tau^{-1}}^{\infty} \frac{1}{T_{n-1}(x) T_n(x) \sqrt{\tau^2 x^2 - 1}} \, dx. \tag{39}$$

In this way we have

$$\varphi(\tau) = \sum_{k=1}^{n} \frac{\pi}{n} \frac{1}{\sqrt{1 - \tau^2 \cos^2{(2k - 1/2\pi)n}}} +$$

$$+ \sum_{r=n+1}^{\infty} \int_{\tau^{-i}}^{\infty} \frac{1}{T_{r-1}(x)T_r(x)\sqrt{\tau^2 x^2 - 1}} \, dx. \tag{40}$$

The advantage of formula (40) seems to be that after having calculated precisely the first sum, which is the Gauss–Jacobi quadrature expression, we can increase essentially the precision by just rough calculations of the second term since this is actually an error term only.

It seems to us that this idea, namely to increase the precision of $Q_n(d\alpha; f)$ by adding a rough evaluation of the second sum in (2) might be very useful. Our attention was called to this possibility by a lecture of Professor J. Todd (Keszthely, 1973) where he treated the calculation of the lemniscate integral

$$\int \frac{1}{\sqrt{1 - x^4}} \, dx.$$

This can be done by inserting in (2) $dx = (1 - x^2)^{-1/2}$, $f(x) = (1 + x^2)^{-1/2}$. We can then deform the contour integrals in the second term of (2) to $(i, i\infty)$.

References

[1] Freud, G. (1973). Error estimates for Gauss–Jacobi quadrature formulae, *Topics in Numerical Analysis*, editor J. Miller. Academic Press, London, pp. 113–121.
[2] Freud, G. (1969). *Orthogonale Polynome*. Birkhäuser Verlag, Basel. (English translation by L. Földes, Pergamon Press, New York–Toronto–London, 1971).
[3] Freud, G. (1974). An estimate of the error of Padé approximants, *Acta Math. Acad. Sci. Hung.* 25, 213–221.
[4] Szegö, G. (1959). *Orthogonal Polynomials*, second edition. American Mathematical Society, Colloquium Publications, Vol. XXIII, New York.

Some Remarks on the Unified Treatment of Elementary Functions by Microprogramming

Sin Hitotumatu

1. Introduction

Recently, there have been extensive developments in microprogramming techniques and we now have a quite unified treatment of elementary functions. In the field, there are several algorithms from different original ideas. However, they all are closely related with one another. Here I would like to mention some of their relations and techniques for practical implementation.

I shall begin with Chen's algorithm [1]. Roughly speaking, if it is applied to exponential and logarithmic functions, due to the addition theorem it will be quite similar to the STL method proposed by Specker [6]. If it is extended to complex variables, it will reduce to the CORDIC method, first introduced by Volder [7] and later extensively developed by Walther [8]. A slightly modified application of CORDIC will generate arcsin x and $\sqrt{1-x^2}$, as in Perle [5]. If CORDIC is applied in hyperbolic coordinates, it gives square root, hyperbolic (including exponential) functions and their inverses. Meggitt's pseudo-division and pseudo-multiplication [4] will generate similar algorithms, if they are translated into the binary system from their original decimal form. A comparison of these methods is given in a report by Koyanagi, Watanabe and Hagiwara [3], where most of them are shown to be more efficient than the usual approximation formulas. Finally, I shall give a brief mention of an algorithm for the square root discussed previously [2].

2. Chen's Algorithms

Chen [1]† proposed the following method of evaluating $z_0 = f(x_0)$. We introduce a parameter y into the function $z = f(x)$ as $z = F(x, y)$ such that $z_0 = F(x_0, y_0)$ and $z_0 = F(x_\omega, y_\omega)$, $y_\omega = z_0$. On the $z = z_0$ plane, we transform (x_k, y_k) to (x_{k+1}, y_{k+1}) in such a manner that it keeps the value of $F(x, y)$

---

† The author is deeply grateful to Dr. M. Sibuya (IBM-Japan) who kindly drew his attention to this report.

invariant, approaching the value x_ω with $F(x_\omega, z_0) = z_0$. If we stop at $x_n = x_\omega + \epsilon$, we have

$$F(x_n, y_n) = y_n + \epsilon \cdot \partial F/\partial x|_{x_\omega} + O(\epsilon^2).$$

There are two stopping rules. First we proceed until $|\epsilon| \leqq 2^{-N}$ (Complete stopping rule; in abbreviation Cs). Second we stop at $|\epsilon| \leqq 2^{-N/2}$ and modify by adding the term $\epsilon \cdot \partial F/\partial x|_{x_\omega}$ (Half stopping rule; in abbreviation Hs). Generally speaking, if the constant $\partial F/\partial x|_{x_\omega}$ is simple (such as 1 or 0), Hs will be applicable, which makes the algorithm quite efficient. Chen emphasizes Hs since the product of ϵ and $\partial F/\partial x|_{x_\omega}$ needs only half-word multiplication, in general.

Practically, F is chosen linearly in y, say $F(x, y) = y \cdot g(x) + h(x)$, where we must have $g(x_\omega) = 1$, $h(x_\omega) = 0$. Some actual examples are in Table 1.

TABLE 1

Function	Interval	$g(x)$	$h(x)$	x_ω	Transformation x	y
e^x	$0 \leqq x < \log 2$	e^x	0	0	$x - \log \alpha$	$y\alpha$
$\log x$	$1/2 \leqq x < 1$	1	$\log x$	1	$x\alpha$	$y - \log \alpha$
y/x	$1/2 \leqq x < 1$	$1/x$	0	1	$x\alpha$	$y\alpha$
$y/x^{1/2}$	$1/4 \leqq x < 1$	$x^{-1/2}$	0	1	$x\alpha^2$	$y\alpha$

In the binary system, α is usually taken as 2^{-k} or 1 ± 2^{-k}, where multiplication requires only shifting and addition.

As an example, we shall discuss the evaluation of $\log x$ in the interval $1/2 \leqq x < 1$. The following is nothing but the STL (Sequential Table Look-up) method proposed by Specker [6]. Meggitt's pseudo-division algorithm [4] will give an almost similar method if it is translated into the binary system.

Prepare the table of the constants (pre-calculated)

$\beta_k = \log (1 + 2^{-k})$ $(k = 1, 2, \ldots)$, and start with $x = x$, $y = 0$.

For $k = 1, 2, \ldots$ repeat the process:

$w := x \times (1 + 2^{-k});$ *if* $w \leqq 1$, *then begin* $x := w; y := y - \beta_k$ *end*;

The transformation keeps the function $y + \log x$ invariant, and using the stopping rule Hs, x is close enough to 1 to apply the approximation $\log x = x - 1$. Hence we only have to subtract $1 - x$ from the final value y_n. The application of Hs reduces computational time by nearly one half that for the usual microprogramming methods.

3. General Classification of Known Algorithms

Most of the algorithms for elementary functions in microprogramming are included in Chen's algorithm with suitable transformation, though some of them require slight modifications. For example, the transformation of CORDIC in a polar coordinate system is given by

$$\begin{cases} x_{k+1} = x_k - 2^{-k}y_k \\ y_{k+1} = y_k + 2^{-k}x_k \\ v_{k+1} = v_k - \gamma_k \end{cases} \quad \text{or} \quad \begin{cases} x_{k+1} = x_k + 2^{-k}y_k \\ y_{k+1} = y_k - 2^{-k}x_k \\ v_{k+1} = v_k + \gamma_k \end{cases}$$

which keeps the function F corresponding to $(x + iy) \exp(u + iv)$ invariant up to the constant factor $K_k = (1 + 2^{-2k})^{1/2}$.

Now we may classify the repetition procedure as follows:

$1°$ Repeat successively for $k = 1, 2, \ldots$ (or $0, 1, \ldots$).

$2°$ Choose suitable m at each step looking through the data.

Also for each step, there are two main ideas in the process:

(a) (Restoring or asymmetric method). According to the data, apply the transformation or not. With binary chopping, it produces the usual binary number.

(b) (Non-restoring or symmetric method). According to the data, apply one transformation in addition or another in subtraction (as in the CORDIC case). With binary chopping it produces a modified binary number where 1 stands for +1 and 0 stands for −1.

Of course, all combinations are not necessarily useful.

According to the above classification, we may classify several known algorithms for microprogramming as in Table 2. Generally speaking, the repetition $2°$ is not suitable for CORDIC except in the case where the co-efficient K is unnecessary such as arctan x or tan $x = \sin x / \cos x$, since the

TABLE 2

Function	Algorithm and author	Stopping rule	Repetition
$\log x$	STL, Specker [6]	Hs	$1°$ (a)
$\log x$	Chen [1]	Hs	$2°$ (a)
e^x	STL, Specker [6]	Cs	$1°$ (a)
e^x	Chen [1]	Cs or Hs	$2°$ (a)
e^x, $\log x$	CORDIC, Hitotumatu [2]	Cs	$1°$ (b)
$\sin x$, $\cos x$	CORDIC, Walther [8], Perle [5]	Cs	$1°$ (b)
$\tan x$	pseudo multiplication, Meggitt [4]	Cs	$1°$ (a)
$\arctan x$	CORDIC, Walther [8], Perle [5]	Cs	$1°$ (b)
$\arctan x$	pseudo division, Meggitt [4]	Cs	$1°$ (a)
$\sqrt{x}$	Chen [1]	Cs or Hs	$2°$ (a)
$\sqrt{x}$	Hitotumatu [2]	Hs	$1°$ (b)

modification of the coefficient $K = \Pi K_k$ is much more complicated under such repetition. For similar reasons the use of the restoring process (a) is limited to trigonometric functions.

For exponential and logarithmic functions, a non-restoring process such as hyperbolic CORDIC (as in Hitotumatu [2]) is not so suitable, since it causes cancellation for the argument near $\log 1 = 0$ or $\exp 0 = 1$, though it reduces slightly the table of constants ($N/3$ versus $N/2$ in the restoring case).

4. Perle's Application of CORDIC

It is well known that starting with $x_0, y_0, v_0 = t$, the successive application of CORDIC to make v as close as possible to 0, will give

$$x_\omega = K_+(x_0 \cos t - y_0 \sin t)$$
$$y_\omega = K_+(x_0 \sin t + y_0 \cos t)$$

where K_+ is the constant

$$K_+ = \Pi(1 + 2^{-2k})^{1/2}, \quad K_+^{-2} = 0.36875612708.$$

Perle [5] proposed the following inverse procedure to evaluate arcsin x. Suppose that t is a real number with $|t| < 1$. Prepare the table of constants $\beta_k = \arctan 2^{-k}$ ($k = 0, 1, 2, \ldots$), and start with $x_0 = 0$, $y_0 = 1/K_+$, $v_0 = 0$. For $k = 0, 1, 2, \ldots$ repeat the following procedure:

$$\textit{if } x \geq t \textit{ then begin } x: = x - 2^{-k}y; \; y: = y + 2^{-k}x; \; v: = v - \beta_k \; \textit{ end}$$
$$\textit{else begin } x: = x + 2^{-k}y; \; y: = y - 2^{-k}x; \; v: = v + \beta_k \; \textit{ end};$$

Notice that the transformation of x and y must be performed not successively but simultaneously. If we finally arrive at $x_\omega = t$, we have $v = \arcsin t$, and as a byproduct, we have $y = \sqrt{1 - t^2}$. Similarly we may obtain arccos t. Also applying a similar method in the case of hyperbolic CORDIC, we have arcsinh $t = \log (t + \sqrt{t^2 + 1})$ or arccosh $t = \log (t + \sqrt{t^2 - 1})$ (for $t > 1$) with $\sqrt{t^2 \pm 1}$ as byproducts.

5. Remark on the Square Root through Hyperbolic CORDIC

Though $\sqrt{x^2 + y^2}$ or $\sqrt{1 - t^2}$ is obtained by CORDIC in the usual polar coordinates, this is not suitable for the usual subroutine of square root. In order to compute square root through CORDIC, it is better to apply the hyperbolic case which will give $K\sqrt{x^2 - y^2}$ where K is the constant given by $K = \Pi(1 - 2^{-2k})^{1/2}$. Starting with $x = t + c$, $y = t - c$, apply the following iteration for $k = 1, 2, 3, \ldots$, with repetition using the same k once more at $k = 4$ and 13 (and also at $k = 40$ if we need more than 80 bits):

$$\textit{if } y \geq 0 \textit{ then begin } x: = x - \delta_k y; \; y: = y - \delta_k x \; \textit{ end}$$
$$\textit{else begin } x: = x + \delta_k y; \; y: = y + \delta_k x \; \textit{ end};$$

where $\delta_k = 2^{-k}$. Again the transformation for x and y must be performed simultaneously. Finally we have

$$K\sqrt{x^2 - y^2} = 2K\sqrt{c}\sqrt{t}$$

which gives directly $\sqrt{t}$ itself, if we have chosen $c = 1/4K^2 = 0.36451229212\dots$. As I have reported in the previous paper [2], the optimal value of c in the TOSBAK-3400 (37 bits in mantissa) was 0.36451229226, where the slight difference may have been caused by the error in the binary-decimal transformation of the data.

I would like to emphasize that the stopping rule Hs (stopping at $k = N/2$) is enough in this case, since we have

$$\sqrt{x_n^2 - y_n^2} = x_n \left(1 - \frac{y_n^2}{2x_n^2} - \cdots\right).$$

Here the convergence region is about $0.037 \leqq t \leqq 3.2$. If we need a wider convergence range, we may use as the first transformation the one with $\delta_0 = 3/4$. This makes the convergence region as wide as $0.0127 \leqq t \leqq 54.5$ which is enough to cover $1/16 \leqq t \leqq 16$ or $1/64 \leqq x \leqq 1$ in a hexadecimal system.

Finally I propose a mixed algorithm for square root: use the hyperbolic CORDIC for the first few steps to get a nice approximation for Newton iteration. This is in fact an efficient production of an approximation by line segments $\phi(t)$ for the square root. For example, if we stop after the first 3 steps, the approximation formula $\phi(t)$ is as in Table 3. In most cases, 3 more Newton iterations will be enough to have the square root to more than 40 bits accuracy.

For such purposes, it will be helpful to choose the constant c such that either $\phi(t)$ or the value after Newton iteration will give the best approximation, but in practice, the difference in c seems to have little effect on the final result.

TABLE 3

$\phi(t) = At + Bc$, for $x_1 = t + c$, $y_1 = t - c$		
Interval	A	B
$t \geqq 5c$	21/64	135/64
$3c \leqq t < 5c$	27/64	105/64
$9c/5 \leqq t < 3c$	35/64	81/64
$c \leqq t < 9c/5$	45/64	63/64
$5c/9 \leqq t < c$	63/64	45/64
$c/3 \leqq t < 5c/9$	81/64	35/64
$c/5 \leqq t < c/3$	105/64	27/64
$t < c/5$	135/64	21/64

References

[1] Chen, T. C. (1972). *The Automatic Computation of Exponential, Logarithms, Ratios and Square Roots*, IBM Research Report RJ970, 32p.

[2] Hitotumatu, S. A new method for the computation of square root, exponential and logarithmic function, to appear in *Bull. Computing Inst. Cluj.*

[3] Koyanagi, S., Watanabe, K. and Hagiwara, H. (1973). Approximation of elementary functions by microprogramming (in Japanese), *Proc. 14th Annual Meeting of the Information Processing Society of Japan*, 171–172.

[4] Meggitt, J. E. (1962). Pseudo-division and pseudo-multiplication processes, *IBM J. Res. Dev.* 6, 210–226.

[5] Perle, M. D. (1971). Cordic technique reduces trigonometric function look-up, *Computer Design*, June, 72–78.

[6] Specker, W. H. (1965). A class of algorithms for ln x, exp x, sin x, cos x, $\tan^{-1} x$ and $\cot^{-1} x$, *IEEE Trans. E.C.* 14, 85–86.

[7] Volder, J. E. (1956). *Binary Computation Algorithms for Coordinate Rotation and Function Generation*, Convair Report IAR-1 148 Aeroelectronics Group.

[8] Walther, J. S. (1971). A unified algorithm for elementary functions, *Spring Joint Computer Conference*, 379–385.

Addenda

The author would like to propose a further new algorithm for square root, which is classified Cs 1° (b), in the notation of Table 2. Starting from $x_0 = (t + 0.25)/K_+ (1/K_+ = 0.60725293508)$, $y_0 = 0$, apply the usual CORDIC as in Section 5 replacing the condition by *"if $y < (t - 0.25)$."* Then we have

$$\sqrt{x_\omega^2 + y_\omega^2} = K_+|x_0| = t + 0.25, \quad y_\omega = t - 0.25$$

which yields $x_\omega = \sqrt{t}$. We need no modification for convergence as in the case of hyperbolic CORDIC, and at least theoretically, it converges for $t > 0$. However, we must apply the Cs (stopping rule) for the present algorithm.

Application of Finite Difference Methods to Exploration Seismology

K. R. Kelly, R. M. Alford,[†] S. Treitel[†] and R. W. Ward[‡]

Introduction

The basic goal of geophysicists interested in petroleum exploration is to infer the geometry and physical properties of the rocks in the earth's subsurface from appropriate measurements recorded at the surface (see Dobrin, 1960). This information, when combined with that of the geologist, hopefully permits the selection of drilling sites where hydrocarbons are most likely to accumulate. The tools used by the geophysicist can be grouped into the categories of seismic methods, gravity and magnetic methods, and methods based on electrical conduction and electromagnetic induction. Of the three methods, the one treated in this study and most widely used in petroleum exploration is the seismic method.

Mathematical models are used to gain a better physical understanding of seismic wave propagation and to refine the interpretation of seismic field measurements. Many different modeling procedures utilizing various degrees of physical and mathematical approximations have been used in the past. These procedures have been adequate to treat a range of simple subsurface models. Many areas of current interest in petroleum exploration require the solution of more complex subsurface models. This, along with the increased capabilities of digital computers, has led to an increase in the use of numerical methods. Finite-difference methods are particularly well suited to the problems of exploration seismology and are currently receiving widespread interest.

In this paper, we illustrate the application of classical finite-difference methods in exploration seismology and point out some of the numerical problems encountered. Satisfactory solutions to some of these problems have been found; others require further research.

The discussion begins with a brief treatment of exploration seismology leading to a mathematical formulation of the problem to be considered. The

† Research Center, Amoco Production Co., Tulsa, Oklahoma, U.S.A.
‡ University of Texas at Dallas, Dallas, Texas, U.S.A.

57

finite-difference algorithms used and the methods employed to include the source and apply the boundary conditions will follow. Results will be presented for typical models and the effects of truncation error will be illustrated.

Formulation of Seismic Exploration Model

The basic principle of the seismic exploration method is depicted in Fig. 1. Seismic waves are induced in the earth from a location at or near the surface by means of a suitable source (dynamite, vibrators, etc.). Waves propagate into the subsurface and are partially reflected upward to the surface, producing surface motion. The vertical component of motion is detected by seismometers along a line of discrete points on the earth's surface and recorded as a function of time. A typical field record is shown in the insert where each trace represents the motion detected by a single seismometer (or group of seismometers). Through the use of finite-difference models, solutions for idealized field subsurface geometries may be found, and synthetic field records may be generated. Attention will be restricted to two-dimensional models. The earth is assumed to be a heterogeneous, linear, isotropic, perfectly elastic medium where seismic wave propagation may be described by the elastodynamic wave equation (Karal and Keller, 1959)

$$\rho \frac{\partial^2 \bar{u}}{\partial t^2} = (\lambda + \mu)\nabla(\nabla \cdot \bar{u}) + \mu\nabla^2\bar{u} + \nabla\lambda(\nabla \cdot \bar{u}) +$$

$$+ \nabla\mu \times (\nabla \times \bar{u}) + 2(\nabla\mu \cdot \nabla)\bar{u} \tag{1}$$

where $\bar{u}$ is the displacement, ρ is the density, t is the time, and λ and μ are the Lamé parameters of the particular medium.

Appropriate boundary conditions must be specified on the exterior of the model. The earth's surface is taken to be a mathematical free-surface where it is required that both the normal stress and the tangential stress vanish. The problem under consideration is of infinite extent in the lateral and downward directions. At large distances from the source, the medium may be considered to be homogeneous. The appropriate mathematical boundary condition is the so-called "radiation condition", whose application results in only outward traveling waves.

Finite-Difference Models

Two different approaches are used for constructing finite-difference representations for the elastic equations. The most common approach is to treat the elastic medium as a collection of homogeneous regions, each characterized by constant values of density and elastic parameters. Motion in each region is described by a finite-difference approximation to the elastic wave equation for a homogeneous region. When using this method, boundary conditions across all interfaces separating different regions must be satisfied

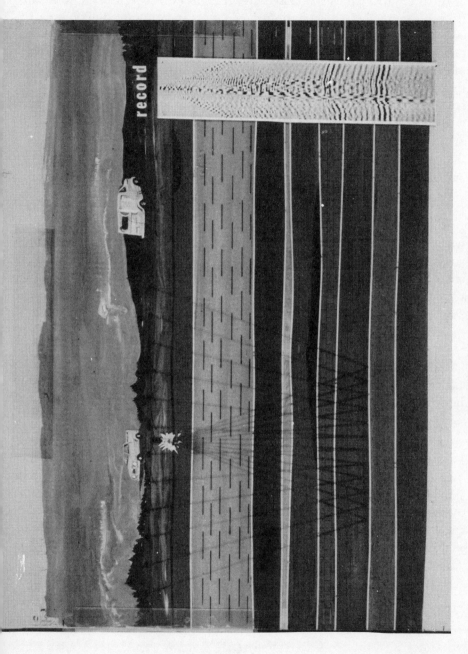

Fig. 1 Seismic exploration method.

explicitly. This approach has been used in previous studies (cf. Alterman and Karal, 1968; Ottaviani, 1971).

An alternative approach, based on finite-difference representations for the more general elastic equation given by (1) incorporates the boundary conditions implicitly. Boore (1972) has applied this idea to a scalar wave propagation problem. This approach allows one to associate different values of density and the elastic parameters with every grid point. Such a formulation permits the treatment of complex subsurface geometries. For this reason, the emphasis of our work has been placed on this method. If density is assumed constant, equation (1) in rectangular coordinates gives

$$\frac{\partial^2 u}{\partial t^2} = \frac{\partial}{\partial x}\left[\alpha^2 \frac{\partial u}{\partial x} + (\alpha^2 - 2\beta^2)\frac{\partial w}{\partial z}\right] + \frac{\partial}{\partial z}\left[\beta^2\left(\frac{\partial w}{\partial x} + \frac{\partial u}{\partial z}\right)\right],$$

$$\frac{\partial^2 w}{\partial t^2} = \frac{\partial}{\partial z}\left[\alpha^2 \frac{\partial w}{\partial z} + (\alpha^2 - 2\beta^2)\frac{\partial u}{\partial x}\right] + \frac{\partial}{\partial x}\left[\beta^2\left(\frac{\partial w}{\partial x} + \frac{\partial u}{\partial z}\right)\right], \qquad (2)$$

where u and w denote the components of displacement in the x (horizontal) and z (vertical) directions, respectively. The spatially varying parameters α and β represent the compressional and shear wave velocities. They are related to the Lamé parameters by

$$\alpha = \sqrt{\frac{\lambda + 2\mu}{\rho}} \quad \text{and} \quad \beta = \sqrt{\frac{\mu}{\rho}}.$$

The finite-difference scheme used to represent (2) is explicit, with a truncation error of second order. The time differentials on the left-hand side can be represented by the usual centered difference representation

$$\frac{\partial^2 u}{\partial t^2} \simeq \frac{u(m, n, l+1) - 2u(m, n, l) + u(m, n, l-1)}{(\Delta t)^2}, \qquad (3)$$

where $x = mh$, $z = nh$, and $t = l\Delta t$. The time step is Δt, h is the grid interval in both the x and z directions, and m, n, and l are defined to be integers.

Two types of terms occur on the right-hand side of (2): Those having partial derivatives with respect to one spatial coordinate only, and those containing partial derivatives with respect to both spatial variables (i.e., mixed derivatives).

Consider a term typical of the first type,

$$\frac{\partial}{\partial x}\left[\alpha^2(x, z)\frac{\partial u}{\partial x}\right]. \qquad (4)$$

Let $\alpha^2(x, z)$ be replaced by its discrete value $\alpha^2(m, n)$ at the grid point (m, n). We define $\alpha^2(m, n)$ to be the average value of $\alpha^2(x, z)$ over a square of dimension h centered at the grid point (mh, nh). Mitchell (1969, pp. 22–25) and Boore (1972) discuss various approximations to (4). Our experience indicates that the various formulas yield comparable results. An approximation which has been found to perform satisfactorily is

$$\frac{\alpha^2(m + 1/2, n)[u(m + 1, n, l) - u(m, n, l)] - \alpha^2(m - 1/2, n) \times}{(h)^2} \times [u(m, n, l) - u(m - 1, n, l)]$$

(5)

where the averages $\alpha^2(m + 1/2, n)$ and $\alpha^2(m - 1/2, n)$ are defined in the form,

$$\alpha^2(m + 1/2, n) = \frac{\alpha^2(m + 1, n) + \alpha^2(m, n)}{2}$$

$$\alpha^2(m - 1/2, n) = \frac{\alpha^2(m - 1, n) + \alpha^2(m, n)}{2}.$$

Next, consider a term typical of a mixed derivative,

$$\frac{\partial}{\partial z}\left[\alpha^2(x, z)\frac{\partial}{\partial x}u(x, z, t)\right] \equiv \frac{\partial}{\partial z}[c(x, z, t)],$$

(6)

where the function $c(x, z, t)$ has been introduced for convenience. The right member of (6) may be approximated by the centered first-order difference,

$$\frac{\partial}{\partial z}[c(x, z, t)] \simeq \frac{c(m, n + 1, l) - c(m, n - 1, l)}{2h}.$$

(7)

Then, let $c(x, z, t) = \alpha^2(x, z)(\partial/\partial x)u(x, z, t)$ be approximated by the centered first-order difference,

$$c(m, n, l) \simeq \alpha^2(m, n)\frac{u(m + 1, n, l) - u(m - 1, n, l)}{2h}.$$

(8)

Substitution of (8) into (7) finally yields the expression

$$\frac{\partial}{\partial z}\left[\alpha^2(x, z)\frac{\partial}{\partial x}u(x, z, t)\right] \simeq \frac{1}{4h^2}[\alpha^2(m, n + 1)\{u(m + 1, n + 1, l) -$$

$$- u(m - 1, n + 1, l)\} - \alpha^2(m, n - 1) \times$$

$$\times \{u(m + 1, n - 1, l) - u(m - 1, n - 1, l)\}].$$

(9)

   Treatment of the remaining terms of (2) in a similar manner produces an explicit difference algorithm, which permits the solution for $u(m, n, l + 1)$ and $w(m, n, l + 1)$ in terms of the values at the two previous times $l\Delta t$ and $(l - 1)\Delta t$. For homogeneous regions, i.e., regions in which the wave velocities are not functions of position, the difference algorithm under consideration reduces to the simple second-order algorithm most often used for calculations in a homogeneous region (cf. Ottaviani, 1971). Alterman and

Loewenthal (1970) have shown that this homogeneous system is stable provided that

$$P \equiv \frac{\alpha \Delta t}{h} \leqslant \left(1 + \frac{\beta^2}{\alpha^2}\right)^{-1/2}. \tag{10}$$

Our approach is to choose Δt as large as possible without violating the above condition at any point in the model.

The boundary conditions at the free surface require the stress to vanish, which give

$$(\alpha^2 - 2\beta^2) \frac{\partial u}{\partial x} + \alpha^2 \frac{\partial w}{\partial z} = 0$$

$$\frac{\partial u}{\partial z} + \frac{\partial w}{\partial x} = 0, \tag{11}$$

where the first relation refers to the normal stress and the second to the tangential stress. Following Alterman and Rotenberg (1969) one-sided differences are used to approximate normal derivatives and centered differences to approximate tangential derivatives. At the free surface ($z = 0$) one obtains the differenced boundary conditions

$$(\alpha^2 - 2\beta^2) \left[\frac{u(m+1, 0, l) - u(m-1, 0, l)}{2h} \right] +$$

$$+ \alpha^2 \left[\frac{w(m, 0, l) - w(m, -1, l)}{h} \right] = 0$$

$$\frac{u(m, 0, l) - u(m, -1, l)}{h} + \frac{w(m+1, 0, l) - w(m-1, 0, l)}{2h} = 0, \tag{12}$$

where the free surface occurs at $n = 0$. The index $n = -1$ arises because of the need to introduce a row of artificial grid points at a distance h above the free surface (Alterman and Karal, 1968). The numerical model must, of course, be limited in spatial extent. Reflections will arise at such artificial boundaries. Methods that can cope with this difficulty will be discussed later.

Inclusion of a singular source in the discrete model requires special consideration since difficulties are encountered with numerical calculations in the vicinity of the source point. The problem is avoided by subtracting the displacement due directly to the source from the total displacement in the rectangular region surrounding the source point.

Effectively, this inserts the source on the boundary of the rectangular region and avoids difficulties due to the source singularity. The direct source contribution is calculated analytically from known solutions for a source in an infinite region. The method used for matching the displacements at the boundary of the rectangular region is essentially the same as one outlined by Alterman and Karal (1968).

An analytic expression for a compressional line source varying with time as the first derivative of the error density function (Gaussian) is given by (Kelly et al., 1975)

$$\bar{u}_s = \bar{a}_\rho \frac{1}{2\pi} \int_{-\infty}^{\infty} \pi^{3/2} \xi^{-1/2} \frac{\omega^2}{\alpha^3} H_1^{(2)} \left(\frac{\omega}{\alpha} |\bar{\rho} - \bar{\rho}_0| \right) e^{-\omega^2/4\xi} e^{-i\omega t_s} \times$$

$$\times e^{i\omega t} d\omega, \tag{13}$$

where the line source is located at $\bar{\rho} = \bar{\rho}_0$ in a polar coordinate system. The unit vector $\bar{a}_\rho$ is in the direction of $\bar{\rho} - \bar{\rho}_0$, $H_1^{(2)}(z)$ is the first-order Hankel function of the second kind, t_s is selected such that the time function is approximately zero at $t = 0$, and the parameter ξ governs the pulse width.

When evaluated, the time variation of the source motion is similar to the first derivative of a Gaussian, but includes the added two-dimensional dispersion associated with a line source. It is tempting to use approximate or asymptotic source solutions for the sake of simplicity. We have found that the use of an analytic solution to the elastic wave equation, as opposed to an asymptotic approximation, results in a wavefront which propagates with the proper cylindrical symmetry and low amplitude "source noise". The source noise may be attributed to the fact that solution to the continuous analytical problem and the solution to the discrete finite-difference problem are not identical. As such, the difference between the two is propagated through the grid as noise. A further advantage in using an analytical source field is that it permits a direct comparison with an analytical solution, if available (Alford et al., 1974).

Grid Dispersion

Given the finite-difference equations, the boundary conditions, and the source, the solution to a problem of interest may be obtained. However, such solutions may contain several artifacts inherent to the finite-difference procedure. One such artifact is grid dispersion.

Waves propagating on a discrete grid become progressively dispersed with increasing traveltime. This phenomenon, called grid dispersion, has been studied analytically, as well as numerically (Boore, 1972; Alford et al., 1974). The effect gains prominence as the grid interval h becomes larger.

The effects of grid dispersion may be studied analytically by introducing a harmonic plane wave into the finite-difference algorithm. Let us consider dispersion of an acoustic wave by a second-order accuracy finite-difference algorithm for the scalar wave equation. Grid dispersion analysis for the more complicated elastic equations follows in a tedious, but straightforward, manner. The plane wave is represented by

$$u = u_0 e^{i(\omega t - kx \cos \theta - kz \sin \theta)},\tag{14}$$

where θ is the angle between the direction of propagation and the x axis, and where k is the wave number. The finite-difference algorithm is given by

$$u(m, n, l+1) = 2(1 - 2p^2)u(m, n, l) + p^2[u(m+1, n, l) + u(m-1, n, l) + $$
$$+ u(m, n+1, l) + u(m, n-1, l)] - u(m, n, l-1) + $$
$$+ 0(h^2 + \Delta t^2),\tag{15}$$

where $p = C_0 \Delta t / h$.

If (14) is substituted into (15) one obtains

$$\sin^2 \frac{\omega \Delta t}{2} = p^2 \left[\sin^2 \left(\frac{kh \cos \theta}{2} \right) + \sin^2 \left(\frac{kh \sin \theta}{2} \right) \right].\tag{16}$$

Let us define G as the number of grid points per wavelength. Then, since $kh/2 = \pi/G$, (16) can be written as

$$\frac{C_p}{C_0} \equiv \frac{\omega}{kC_0} = \frac{G}{p\pi} \sin^{-1} \left\{ p \left[\sin^2 \left(\frac{\pi \cos \theta}{G} \right) + \sin^2 \left(\frac{\pi \sin \theta}{G} \right) \right]^{1/2} \right\},\tag{17}$$

where C_p/C_0 is the normalized phase velocity of the plane wave. Group velocity, defined by $C_g = d\omega/dk$, can be obtained by an appropriate differentiation. Plots of phase and group velocity vs $1/G$ (or effectively frequency) are shown in Fig. 2 for various values of θ. The dispersion under consideration is normal; that is, the higher signal frequencies are delayed relative to the lower frequency components producing high-frequency "tails" following the signal wavelet.

A numerical example will now be given to demonstrate the effects of grid dispersion in acoustic wave propagation. Acoustic waves were generated from a localized time-varying pressure pulse and are reflected and diffracted from a 90° wedge of zero velocity material embedded in an acoustical medium as shown in Fig. 3. The spectrum and time variation of the source is shown in Fig. 4. Figure 5 shows the displacement vs time at the three receiver locations indicated. The solid curves are obtained from an analytical solution solved by boundary value problem techniques. The dots represent the second-order finite-difference results given by (15) with $G_0 = 11$, where G_0 is the number of grid points per wavelength at the upper half power frequency of the source. As may be seen, the two solutions are in reasonably good agreement. Figure 6 shows the corresponding results using a coarser grid with $G_0 = 5.5$. Effects of dispersion are prominent and characterized by pulse broadening, time delays, and oscillatory tails.

An alternate finite-difference algorithm may be obtained by using the more accurate fourth-order representation for the Laplacian given by

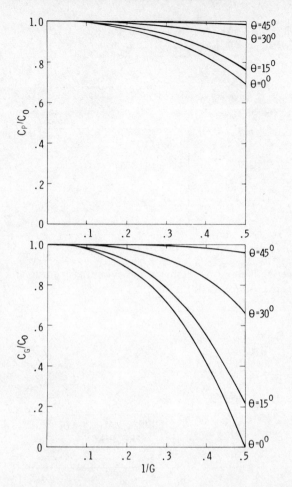

Fig. 2 Normalized phase and group velocity for different angles for the second-order scheme ($p = 0.7$). (From *Geophysics* 39).

Abramowitz and Stegun (1965). The resulting difference scheme is

$$u(m, n, l + 1) = (2 - 5p^2)u(m, n, l) + \frac{p^2}{12} \{16[u(m + 1, n, l) +$$

$$+ u(m, n + 1, l) + u(m - 1, n, l) + u(m, n - 1, l)] -$$

$$- [u(m + 2, n, l) + u(m, n + 2, l) + u(m - 2, n, l) +$$

$$+ u(m, n - 2, l)]\} - u(m, n, l - 1) + 0(h^4 + \Delta t^2). \quad (18)$$

Results computed using this algorithm for this example are shown in Fig. 7. These computations use $G_0 = 5.5$, which corresponds to the coarse grid calculations for the second-order algorithm. It is noted that dispersive effects are minimal. This observation is substantiated theoretically by considering the plots of phase and group velocity vs $1/G$ shown in Fig. 8. It is seen that the curves for the fourth-order scheme near the stability limit

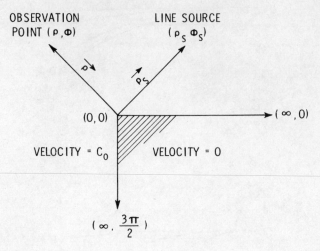

Fig. 3 Cross section illustrating geometry and cylindrical coordinates for the $90°$ wedge model. (From *Geophysics* 39).

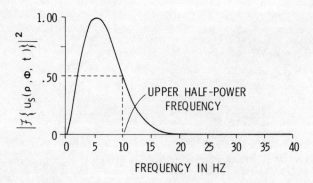

Fig. 4a Normalized source field power spectrum at $\rho = 2000'$. (From *Geophysics* 39).

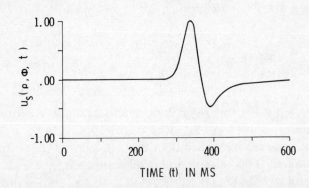

Fig. 4b Normalized source field response at $\rho = 2000'$. (From *Geophysics* 39).

($p = 0.6$) are essentially constant for much higher frequencies than the curves for the second-order scheme near the corresponding stability limit ($p = 0.7$).

This suggests that the savings in computer time and memory requirements are in some cases sufficient to justify use of the more complicated fourth-order algorithms.

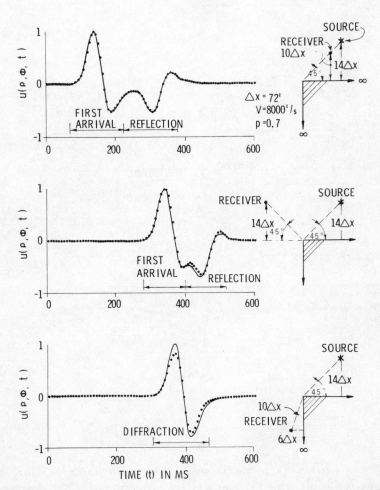

Fig. 5a, b, c Analytical solution (———) and fine grid ($G_0 \simeq 11$) finite-difference solution ($\cdots$) for the second-order scheme. (From *Geophysics* 39)

Preliminary results indicate that the degree to which grid dispersion is minimized is a function of the range of G values. In many cases the additional improvement does not justify the added complexity of still higher order schemes. The grid dispersion effect just discussed is an alternative characterization of the effects of truncation error. The minimization of grid dispersion by using the fourth-order scheme contrasted with the second-

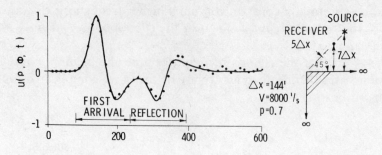

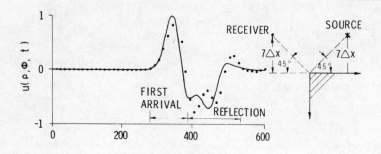

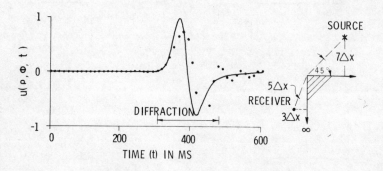

Fig. 6a, b, c Analytical solution (———) and coarse grid ($G_0 \simeq 5.5$) finite-difference solution ($\cdots$) for the second-order scheme. (From *Geophysics* 39).

order scheme can also be predicted by comparing the corresponding trunca-tion errors. We prefer the dispersion characterization as it has a physical interpretation more familiar to the physicist and engineer, and provides greater insight as to the nature of the error in finite-difference solution of wave propagation problems.

Results

Both compressional and shear waves may propagate in an elastic medium. Compressional waves produce a particle motion in the direction of propaga-

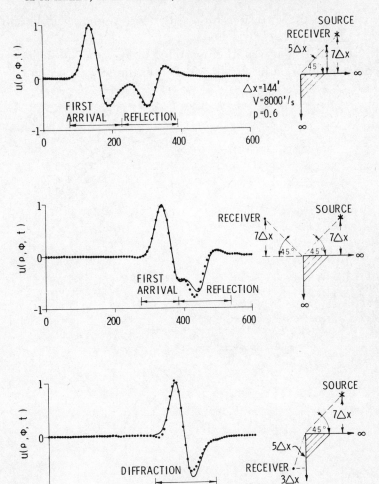

Fig. 7a, b, c Analytical solution (———) and coarse grid ($G_0 \simeq 5.5$) finite-difference solution ($\cdots$) for the fourth-order scheme. (From *Geophysics* 39).

tion, while the particle motion associated with shear waves is transverse to the direction of propagation. For a homogeneous region each type of wave motion is independent of the other. Partial conversion from one type to the other occurs at boundaries between homogeneous regions.

Examples of the numerical solutions computed using the second-order finite-difference algorithm for the elastodynamic wave equation for a heterogeneous medium will now be presented. Attention will be limited to simple examples designed to illustrate the effects mentioned earlier. The solutions will be displayed as simulations of seismic field recordings.

Figure 9a shows the geometry of the first model—a layer over a halfspace. The corresponding synthetic seismogram is shown in Fig. 9b, where each trace represents the vertical motion at the corresponding grid point as a function of time.

The early arriving waves are interpreted in Fig. 9b, and their corresponding geometrical ray paths are shown. This example was computed for a

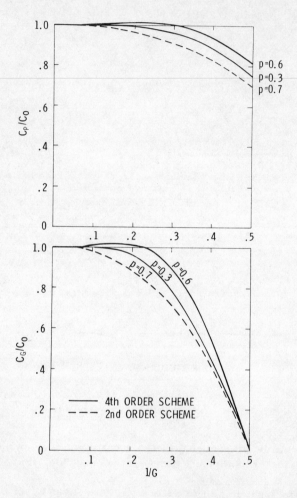

Fig. 8 Normalized phase and group velocity for different stability ratios for the fourth-order scheme ($\theta = 0$). (From *Geophysics* 39).

value of $G_0 \simeq 10$, and very little dispersive effects are present. Figure 9c shows the corresponding section for a higher frequency source obtained by doubling the frequency of each spectral component of the original source, resulting in $G_0 \simeq 5$. Although the individual waves are still identifiable, it is seen that any meaningful study of the waveforms is impossible. For a more complex model the identification of the individual events would become impossible for calculations using $G_0 = 5$.

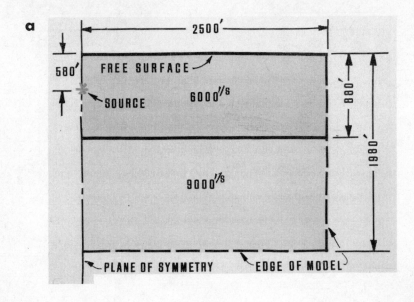

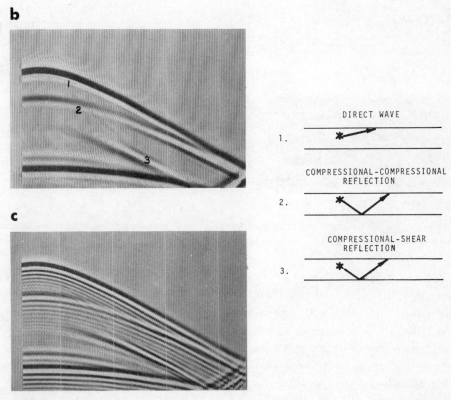

Fig. 9 Layer over a halfspace model: (a) geometry, (b) vertical motion, and (c) vertical motion using a higher frequency source.

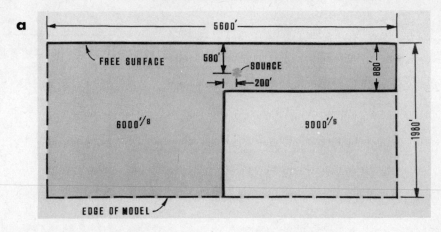

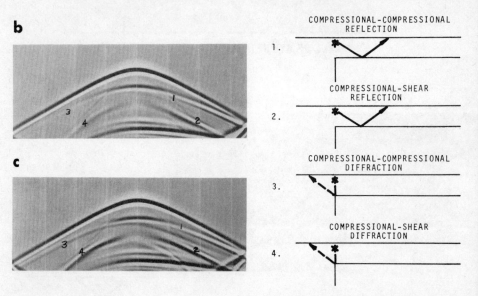

Fig. 10 90° corner model: (a) geometry, (b) vertical motion for heterogeneous formulation, and (c) vertical motion for homogeneous formulation.

The next model consists of a quarterspace embedded in a halfspace termed here the "90° corner model" (Fig. 10a). Solutions obtained by using the algorithm for a heterogeneous medium (Fig. 10b) and the more conventional algorithm for a homogeneous medium (Fig. 10c) will be compared. The embedded corner in this model will produce not only reflected waves but also diffractions. These events are identified on the seismograms and the corresponding geometrical ray paths are shown. Although the solutions obtained by using the two algorithms are in qualitative agreement, a closer examination reveals discrepancies. Amplitude differences are noticeable in the compressional-to-shear converted reflection and even more so in the compressional-to-shear diffraction. Complete agreement between the two formulations would not be expected because of the inherent difference in applying boundary conditions.

Edge Reflections

The size of the finite-difference models is limited both laterally and at depth to remain within the confines of available computer memory. These artificial boundaries produce undesirable edge reflections. To our knowledge, there is no fully satisfactory method of eliminating these reflections, although a number of possibilities have been investigated. One approach, due to Lysmer and Kuhlemeyer (1969), has proven useful in attenuating the strength of such reflections, particularly in the case of compressional waves. The method imposes a viscous behavior on the artificial boundary, and much of the incident energy is thus mathematically "dissipated". Another method has recently been proposed by Smith (1974) who has found that edge reflections can be cancelled from the results by summing appropriate combinations of solutions with either Dirichlet or Neumann boundary conditions at the artificial boundaries. Unfortunately, the method requires multiple solutions which are not always economically feasible. For many models, we have found that the edge reflections can be essentially removed by employing digital filtering techniques in the frequency–wavenumber (f, k_x) domain. The method will be demonstrated by means of a simple example. The synthetic seismogram representing the vertical component of displacement, $u(x, 0, t)$, is numerically Fourier transformed with respect to the time, t, and the horizontal coordinate, x, resulting in a two-dimensional spectrum, $\tilde{u}(k_x, 0, f)$. The spectrum is then pictured in the form of an (f, k_x) plot by displaying the amplitude on a frequency–horizontal wavenumber coordinate by means of a suitable display procedure, such as the variable-density plot. Figure 11a shows a typical seismogram for a model consisting of a relatively thin horizontal layer embedded in an otherwise homogeneous halfspace, and Fig. 11b shows the corresponding (f, k_x) spectrum. The transformation procedure effectively decomposes the waves in the time-space domain into monocromatic plane waves in the frequency–wavenumber domain. A plane wave is identified on the (f, k_x) plot by a straight line through the origin, whose slope is proportional to the wave's velocity. For the simple symmetrical model under consideration only the edge reflections are propagating in the negative x direction, and ideally lie entirely in the second quadrant of the (f, k_x) plot. The contribution of the spectrum due to the edge reflections is then removed by a suitable filter, and the seismogram is reconstructed by the inverse transformation. The resulting (f, k_x) plot is shown in Fig. 11d and the corresponding seismogram obtained by taking inverse transforms is shown in Fig. 11c.

The appropriate seismogram and (f, k_x) plot for a model of much wider spatial extent but otherwise identical geometry are shown in Figs. 11e and

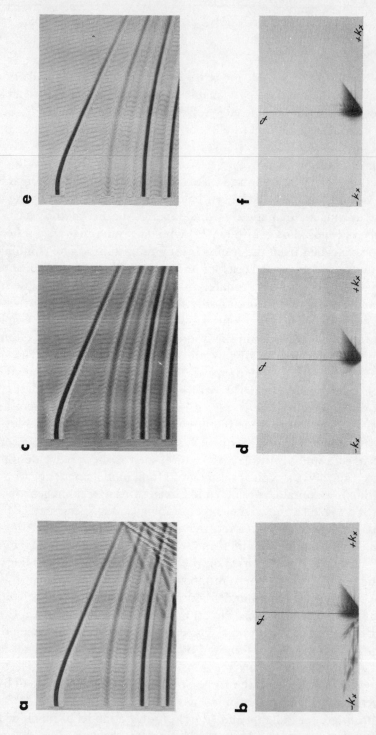

Fig. 11 Edge reflection removal: (a) vertical motion before filtering, (b) f, k_x spectrum before filtering, (c) vertical motion after filtering, (d) f, k_x spectrum after filtering, (e) vertical motion for wide model, and (f) f, k_x spectrum for wide model.

11f, respectively, for comparison purposes. It is seen that the agreement is excellent. The concept of eliminating unwanted waves may be extended to more complex situations by a more selective use of filtering.

Conclusions

The method of finite-differences enables the geophysicist to simulate the seismic responses of quite arbitrary subsurface structures. The heterogeneous finite-difference method has the versatility necessary for such simulations. From the examples presented, it is evident that the resulting seismograms produce a fineness of detail difficult to obtain by alternate procedures. Ten or more grid points per wavelength at the upper half power frequency should be adequate for calculations with the second-order finite-difference algorithms. Coarser grids may be used for higher order algorithms. Unless precautions are taken, reflections from the outer edge of the model contaminate the results. In many cases these reflections may be removed by two-dimensional digital filtering techniques. Although the discussions presented here are related to exploration geophysics, the techniques should apply to wave propagation problems in other areas of science and engineering.

The present calculations were performed on an IBM 370/165 digital computer. The computer model of Fig. 10 contained 140 horizontal and 100 vertical grid points. The model was run for about 300 time steps and the calculation required roughly 8 minutes of computer time in the Central Processing Unit. It is, of course, desirable to extend the physical dimensions of the models in order to permit the simulation of structures of more realistic complexity and size. The expense and core requirements of these calculations for the larger models on a standard large-scale computer rapidly become excessive. However, we have already established that a minicomputer with rapid-access disk storage reduce such costs to a more acceptable level.

In fact, the practical limitations of the finite-difference method are computer time and memory requirements. Most of the techniques discussed in this paper are directed toward these limitations. The discussion has been limited to formulations with a square grid spacing that is constant throughout the model. The use of variable grid sizes and rectangular grids makes the computation more efficient since the grid can be "tailored" to the model of interest. This is an area of current investigation at our laboratory.

Finally, it is mentioned that our considerations have been limited to explicit difference schemes. Other formulations should be investigated for the possibility of greater efficiency.

References

Abramowitz, M. and Stegun, I. A. (1965). *Handbook of Mathematical Functions.* Dover, New York, p. 885.

Alford, R. M., Kelly, K. R. and Boore, D. M. (1974). Accuracy of finite-difference modeling of the acoustic wave equation, *Geophysics*, **39**, 834–841.

Alterman, Z. S. and Karal, F. C., Jr. (1968). Propagation of elastic waves in layered media by finite-difference methods, *Bull. Seismological Soc. Am.* **58**, 367–398.

Alterman, Z. S. and Loewenthal, D. (1970). Seismic waves in a quarter and three-quarter plane, *Geophys. J. Roy. Astron. Soc.* **20**, 101–126.

Alterman, Z. S. and Rotenberg, A. (1969). Seismic waves in a quarter plane, *Bull. Seismological Soc. Am.* **59**, 347–368.

Boore, D. M. (1972). Finite-difference methods for seismic wave propagation in heterogeneous materials, *Methods in Computational Physics* (B. Alder, S. Fernbach and M. Rotenberg, eds.), vol. II. Academic Press, New York.

Dobrin, M. B. (1960). *Introduction to Geophysical Prospecting.* McGraw-Hill, New York.

Karal, F. C. and Keller, J. B. (1959). Elastic wave propagation in homogeneous and inhomogeneous media, *Jl. of the Acoust. Soc. Am.* **31**, 694–705.

Kelly, K. R., Ward, R. W., Treitel, S. and Alford, R. M. (1975). *Synthetic seismograms: a finite-difference approach*, to be published.

Lysmer, J. and Kuhlemeyer, R. L. (1969). Finite dynamic model for infinite media, *J. Eng. Mech. Div., Proc. Am. Soc. Civil Eng.* **95**, 859–877.

Mitchell, A. R. (1969). *Computational methods in partial differential equations.* John Wiley & Sons, New York.

Ottaviani, M. (1971). Elastic wave propagation in two evenly-welded quarter-spaces, *Bull. Seismological Soc. Am.* **61**, 1119–1152.

Smith, W. D. (1974). *A non-reflecting boundary for wave propagation problems*, *J. Comp. Physics*, **15**, 492–503.

Variable Coefficient Multistep Methods for Ordinary Differential Equations applied to Parabolic Partial Differential Equations

J. D. Lambert

1. Semi-discretization

Consider the simple parabolic test problem

$$\partial u(x, t)/\partial t = \partial^2 u(x, t)/\partial x^2, \qquad \begin{array}{l} u(x, 0) = \phi(x), \quad 0 \leqslant x \leqslant 1 \\ u(0, t) = u(1, t) = 0, \quad t \geqslant 0 \end{array} \qquad (1.1)$$

and a rectangular mesh parallel to the x- and t-axes with mesh spacings h and k, where $r = k/h^2$ is the *mesh ratio*. Let

$$x_i = ih, \quad i = 0, 1, \ldots, M + 1; \quad (M + 1)h = 1: \quad t_n = nk, \quad n = 0, 1, \ldots$$
$$\mathbf{u}(t): = [u(x_1, t), u(x_2, t), \ldots, u(x_M, t)]^T.$$

The corresponding semi-discrete problem is obtained by substituting the difference replacement $\partial^2 u/\partial x^2 \approx (\delta_x^2 u)/h^2$ in (1.1). We denote by $\mathbf{U}(t)$ the solution of the semi-discrete problem

$$d\mathbf{U}(t)/dt = B\mathbf{U}(t), \quad \mathbf{U}(0) = \boldsymbol{\phi}, \qquad (1.2)$$

where

$$\mathbf{U}(t) = [{}^1U(t), {}^2U(t), \ldots, {}^MU(t)]^T,$$
$$\boldsymbol{\phi} = [\phi(x_1), \phi(x_2), \ldots, \phi(x_M)]^T,$$

$$B = \frac{1}{h^2} \begin{bmatrix} -2 & 1 & 0 & 0 & \cdots & 0 & 0 & 0 \\ 1 & -2 & 1 & 0 & \cdots & 0 & 0 & 0 \\ \vdots & & & & & & & \\ \vdots & & & & & & & \\ 0 & 0 & 0 & 0 & \cdots & 1 & -2 & 1 \\ 0 & 0 & 0 & 0 & \cdots & 0 & 1 & -2 \end{bmatrix}$$

Problem (1.2) can in turn be solved by applying, with steplength k, any appropriate numerical method for the initial value problem for a system of ordinary differential equations. Let the approximation to $\mathbf{U}(t)$ so obtained at $t = t_n$ be denoted by $\mathbf{U}_n$, where

$$\mathbf{U}_n = [^1U_n, {}^2U_n, \ldots, {}^MU_n]^T.$$

We shall refer to the process of obtaining the approximation $\mathbf{U}_n$ to $\mathbf{u}(t_n)$ in the manner described above as an *overall process*. Alternatively, by using any of the standard finite difference methods for (1.1) (Mitchell [6], Richtmyer and Morton [7]), we can find an approximation to $\mathbf{u}(t_n)$ directly, without going through the stage of semi-discretization; we shall call such a process a *direct process*. The purpose of this paper is to investigate the equivalence of certain overall and direct processes, with particular reference to the rôle played by the numerical method used to solve (1.2). The situation can be summed up in the following diagram.

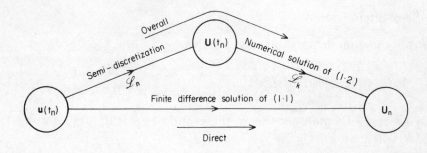

The truncation error (TE) of the overall process, which naturally depends on the TE both of the semi-discretization process and of the numerical method used to solve (1.2), will be discussed in a later section. The stability of the overall process, on the other hand, is determined, as follows, by the stability properties of the method used to solve (1.2). Consider a discretization method with steplength k for a numerical solution of the problem

$$\mathbf{y}'(t) = \mathbf{f}(t, \mathbf{y}(t)), \quad \mathbf{y}(0) = \mathbf{y}_0,$$
$$\mathbf{y} : \mathbb{R} \to \mathbb{R}^M, \quad f : \mathbb{R} \times \mathbb{R}^M \to \mathbb{R}^M. \tag{1.3}$$

Such a method is said to have *region of absolute stability* $\mathscr{R}$ in the complex plane if $\mathscr{R}$ is the largest region such that, $\forall\, k\lambda \in \mathscr{R}$, all solutions of the difference equation which results from applying the method with step length k to the test problem $\mathbf{y}' = \lambda\mathbf{y}$, $\lambda \in \mathbb{C}$, tend to zero as $n \to \infty$. It follows that all solutions of the difference equation resulting from applying the method to $\mathbf{y}' = A\mathbf{y}$, where A is an $M \times M$ matrix with distinct eigenvalues λ_ν, $\nu = 1$, $2, \ldots, M$ will tend to zero as $n \to \infty$ iff k is chosen s.t.

$$k\lambda_\nu \in \mathscr{R}, \quad \nu = 1, 2, \ldots, M.$$

If $\mathscr{R}$ contains the half-plane Re $k\lambda < 0$, the method is said to be A-stable.

Turning to the system (1.2), it is well known that the eigenvalues of B are real and are contained in the interval $(-4/h^2, 0)$. It follows that the solution U_n afforded by the overall process will tend to zero as $n \to \infty$ iff k is chosen s.t.

$$(-4r, 0) \subset \mathcal{R}. \tag{1.4}$$

Note that if (1.4) is to hold for all $r > 0$ (*unconditional stability* in the parlance of PDEs) it is not necessary for the method applied to (1.2) to be A-stable. (Indeed, it turns out that it is not even necessary for the method to be $A(0)$-stable (Widlund [9]) or A_0-stable (Cryer [2])).

The equivalence of overall processes which employ certain linear multistep methods for the solution of (1.2) to certain direct processes is well known. For example, if (1.2) is solved by Euler's Rule, the overall process is equivalent to the familar elementary explicit finite difference method (Mitchell [6], p. 20). For Euler's Rule, $\mathcal{R}$ is the circle on $[-2, 0]$ as diameter (Lambert [4], p. 227), and (1.4) is satisfied iff $0 < r \leqslant \frac{1}{2}$, which will be recognized as the correct stability condition for the direct process. If (1.2) is solved by the Trapezoidal Rule, the equivalent direct process is readily found to be the Crank–Nicolson method; the A-stability of the former corresponds to the unconditional stability of the latter.

Of more interest is the θ-method,

$$y_{n+1} - y_n = k[(1 - \theta)\mathbf{f}_{n+1} + \theta\mathbf{f}_n].$$

If this is applied to (1.2) with the particular choice

$$\theta = \frac{1}{2} + \frac{1}{12r} \tag{1.5}$$

one easily finds that the overall process is equivalent to the Douglas high-accuracy formula (Mitchell [6], p. 26). (The choice (1.5) causes a cancellation between the TE of the semi-discretization and that of the θ-method; the details can be worked out from the analysis of the next section.) The θ-method is known to be A-stable when $\theta \leqslant \frac{1}{2}$ (Lambert [4], p. 240), a condition patently not satisfied by (1.5). However, for $\theta > \frac{1}{2}$, the region $\mathcal{R}$ can be shown to be the circle on $[2/(1 - 2\theta), 0]$ as diameter. For θ given by (1.5), $\mathcal{R}$ is thus the circle on $[-12r, 0]$ as diameter, and (1.4) is satisfied for all $r > 0$, corresponding to the unconditional stability of the Douglas method. Here the region $\mathcal{R}$, though always finite, is itself a function of r, and "stretches" as r increases so as always to satisfy (1.4). We shall meet this phenomenon again later.

2. Variable Coefficient Multistep Methods (VCMM)

Some interesting equivalences between overall and direct processes emerge if we choose to solve (1.2) by one of the class of VCMM developed by

Lambert and Sigurdsson [5], and Sigurdsson [8]. The general VCMM for
the numerical solution of (1.3) is

$$\sum_{j=0}^{l} \left[a_j^{(0)} I + \sum_{s=1}^{S} a_j^{(s)} (kQ_n)^s \right] y_{n+j} = k \sum_{j=0}^{l} \left[b_j^{(0)} I + \sum_{s=1}^{S-1} b_j^{(s)} (kQ_n)^s \right] f_{n+j},$$

(2.1)

where Q_n is an arbitrary $M \times M$ matrix subject only to the conditions

$$\| Q_n \| \leqslant \kappa < \infty, \quad \left[a_l^{(0)} I + \sum_{s=1}^{S} a_l^{(s)} (kQ_n)^s \right] \text{ non-singular.}$$

Order and truncation error

We associate with (2.1) the operators $\mathcal{M}_k$, $\mathcal{N}_k$, $\mathcal{L}_k$, defined as follows.
Let $w : \mathbb{R} \to \mathbb{R}^M$ be a sufficiently smooth arbitrary function, and denote
$d^m w(t)/dt^m$ by $w^{(m)}(t)$, $m = 1, 2, \ldots$.

$$\mathcal{M}_k[w(t)] := \sum_{s=0}^{S} \left[(kQ_n)^s \sum_{j=0}^{l} a_j^{(s)} w(t + jk) \right]$$

$$\mathcal{N}_k[w(t)] := \sum_{s=0}^{S-1} \left[(kQ_n)^s \sum_{j=0}^{l} b_j^{(s)} w(t + jk) \right]$$

(2.2)

$$\mathcal{L}_k[w(t)] := \mathcal{M}_k[w(t)] - k\mathcal{N}_k[w^{(1)}(t)].$$

$\mathcal{L}_k$ is now the operator associated with (2.1) in precisely the sense that
Henrici [3], (p. 220) associates an operator with a linear multistep method.
We say that (2.1) has *order* p if p is the largest integer for which

$$\mathcal{L}_k[w(t)] = 0(k^{p+1}),$$

and that (2.1) is *consistent* if $p \geqslant 1$. It is important to observe that order is
independent of the choice made for the arbitrary matrix Q_n. The *local
truncation error*, $(TE)_k$ when (2.1) is applied to (1.3) is defined to be

$$(TE)_k := \mathcal{L}_k[y(t)],$$

(2.3)

where $y(t)$ is the solution of (1.3).

We can now define, in the following manner, an operator $\mathcal{L}_{hk}$ which is
associated in a similar manner with the overall process for the parabolic
problem in which (1.2) is solved by (2.1).

$$\mathcal{L}_{hk}[w(t)] := \mathcal{M}_k[w(t)] - k\mathcal{N}_k[Bw(t)].$$

(2.4)

The *local truncation error of the overall process* is defined to be

$$(TE)_{hk} := \mathcal{L}_{hk}[u(t)],$$

(2.5)

$u(t)$ being the solution of (1.1). A more useful form for $(TE)_{hk}$ can be obtained if we introduce a further operator $\mathcal{L}_h$ associated with the semi-

discretization process whereby (1.1) is replaced by (1.2). Let z: $\mathbb{R} \times \mathbb{R} \to \mathbb{R}$ be a sufficiently smooth arbitrary function, and define

$$z(t):= [z(x_1, t), z(x_2, t), \ldots, z(x_M, t)]^T$$

$$D_x^m z(t):= \left[\frac{\partial^m z}{\partial x^m}(x_1, t), \frac{\partial^m z}{\partial x^m}(x_2, t), \ldots, \frac{\partial^m z}{\partial x^m}(x_M, t) \right]^T, \quad m = 1, 2, \ldots$$

Then $\mathscr{L}_h$ is defined by

$$\mathscr{L}_h[z(t)]:= D_x^2 z(t) - Bz(t). \tag{2.6}$$

Clearly,

$$\mathscr{L}_h[z(t)] = -\tfrac{1}{12} h^2 D_x^4 z(t) + 0(h^4) \tag{2.7}$$

From (2.3)–(2.6) and (1.1), we obtain that

$$\begin{aligned}(TE)_{hk} &= \mathscr{M}_k[u(t)] - k\mathscr{N}_k[D_x^2 u(t) - \mathscr{L}_h[u(t)]] \\ &= \mathscr{M}_k[u(t)] - k\mathscr{N}_k[u^{(1)}(t)] + k\mathscr{N}_k[\mathscr{L}_h[u(t)]], \end{aligned}$$

or

$$(TE)_{hk} = \mathscr{L}_k[u(t)] + k\mathscr{N}_k[\mathscr{L}_h[u(t)]]. \tag{2.8}$$

The first term in (2.8) is merely the local TE of the VCMM (2.1); the second term is a function of the local TE of the semi-discretization process and the coefficients on the right side of the VCMM (2.1). Note that this analysis also holds for the case when (1.2) is solved by one of the class of linear multistep methods, since this class is a sub-class of (2.1).

By expanding the operators defined by (2.2) and using (2.7), it is possible to get explicit expressions for the principal terms in $(TE)_{hk}$. The resulting asymptotic formula is, for the general case, distressingly complicated, and we shall quote it only for the specific examples we study. However, in the case when the VCMM reduces to the linear multistep method

$$\sum_{j=0}^{l} a_j^{(0)} y_{n+j} = k \sum_{j=0}^{l} b_j^{(0)} f_{n+j}, \quad TE = C_{p+1} y^{(p+1)} k^{p+1} + 0(k^{p+2}),$$

the asymptotic formula becomes

$$(TE)_{hk} = [C_{p+1} u^{(p+1)}(t)] k^{p+1} - \left[\frac{1}{12} \left(\sum_{j=0}^{l} b_j^{(0)} \right) D_x^4 u(t) \right] kh^2 +$$

$$+ 0(k^{p+2} + k^2 h^2 + kh^4), \tag{2.9}$$

provided we assume that the coefficients of the linear multistep method are independent of h and k (but recall the situation with the Douglas formula in Section 1). For the linear multistep methods discussed in Section 1, it will be found that the ith component of (2.9) evaluated at t_n will coincide with the local TE at (x_i, t_n) of the equivalent direct method, provided that the latter is defined as in Mitchell [6]. (Some authors, e.g.

Richtmyer and Morton [7], take TE to be $1/k$ times that defined by Mitchell.) Using Mitchell's convention, a direct finite difference method is said to be consistent if $(TE)/k \to 0$ as $h, k \to 0$. It is clear from (2.8) and (2.7) that this will be so for the general overall process based on the VCMM (2.1) iff (2.1) is consistent in the sense already defined. (Note that this statement is true only if we adhere strictly to our assumption that Q_n is bounded as $n \to \infty$; see Section 3.)

Stability

Consider the test equation

$$\mathbf{y}' = A\mathbf{y}, \quad \lambda[A] < 0 \tag{2.10}$$

where A has eigenvalues (not necessarily distinct) λ_ν, $\nu = 1, 2, \ldots, M$, and

$$\lambda[A] = \max_\nu (\mathrm{Re}\, \lambda_\nu) \tag{2.11}$$

The VCMM (2.1) is said to be $\bar{A}$-*stable* if all solutions of the difference equation which results from applying (2.1) to (2.10) with $Q_n = -A$ tend to zero as $n \to \infty$, $\forall k > 0$. It is shown in Lambert and Sigurdsson [5] that there exist $\bar{A}$-stable VCMM of order $p \leq 2S$, even in the case when $b_l^{(s)} = 0$, $s = 0$, $1, \ldots, S - 1$. We hesitate to call such methods "explicit" since, although the coefficient of $\mathbf{f}_{n+l}$ is zero, it is still necessary to invert, at each time step, the matrix coefficient of $\mathbf{y}_{n+l}$; accordingly, we call such methods *linearly implicit*. While linearly implicit methods have obvious computational advantages over fully implicit methods when applied to the general system (1.3) with Q_n chosen to be $-(\partial \mathbf{f}/\partial \mathbf{y})_n$, such advantages are lost when the system is linear, as is the case for the semi-discrete problem (1.2). However, it has been found in practice that stable results for the general system can be obtained even if Q_n is chosen to be quite a "bad" approximation to the negative Jacobian. In the next section, we shall investigate some examples of overall processes where the semi-discrete problem (1.2) is solved by a VCMM in which Q_n is chosen to be a "bad" approximation to $-B$.

3. Examples

Our examples of VCMM are taken from Lambert and Sigurdsson [5].

Example 1
$S = 1$, $l = 2$, linearly implicit.

$$[I + \tfrac{1}{2}kQ_n]\mathbf{y}_{n+2} - [(1 + a)I + kQ_n]\mathbf{y}_{n+1} + [aI + \tfrac{1}{2}kQ_n]\mathbf{y}_n$$
$$= k[\tfrac{1}{2}(3 - a)\mathbf{f}_{n+1} - \tfrac{1}{2}(1 + a)\mathbf{f}_n] \tag{3.1}$$

The method is $\bar{A}$-stable if $-1 < a < 1$. It has order 2, and the asymptotic form of the local TE of the overall process derived from (2.8), (2.2) and (2.7) is

$$(TE)_{hk} = \left[\frac{5+a}{12} \mathbf{u}^{(3)}(t) + \tfrac{1}{2} Q_n \mathbf{u}^{(2)}(t) \right] k^3 - \left[\frac{1-a}{12} D_x^4 \mathbf{u}(t) \right] kh^2 +$$
$$0(k^4 + k^2 h^2 + kh^4) \tag{3.2}$$

We can make (3.1) fully explicit by choosing

$Q_n = qI$, q a scalar constant, I the unit matrix.

With this choice for Q_n, (3.1) is no longer necessarily A-stable. Indeed, Sigurdsson [8] shows that in this case, the region $\mathscr{R}(kq, a)$ of absolute stability is a simple closed region which intercepts the real axis in the interval $[-(1 + a + kq), 0]$. Thus (1.4) is satisfied if we choose q such that

$1 + a + kq \geqslant 4r$

An obvious choice is $a = -1$, $q = 4/h^2$, in which case $\mathscr{R}$ turns out to be an ellipse, and it is straightforward to show that the resulting overall process for (1.1) is equivalent to the direct process consisting of the DuFort–Frankel scheme (Mitchell [6], p. 89) which is well known to be unconditionally stable, despite its explicitness. Note that the eigenvalues of $-Q_n$ are now all $-4/h^2$, while those of B (to which $-Q_n$ is a "bad" approximation) lie in the interval $(-4/h^2, 0)$.

There is however a price to be paid for choosing Q_n to be $4I/h^2$. Had we chosen $Q_n = -B$, then $Q_n = 0(1)$ as $h \to 0$, whereas $4I/h^2 = 0(h^{-2})$. Thus the first term in $(TE)_{hk}$ given by (3.2) is no longer $0(k^3)$ but $0(k^3 + k^3/h^2)$. Thus, although (3.1) is still consistent in the ODE sense (and the solution of (3.1) will certainly converge to that of (1.2) as $k \to 0$), it is not consistent in the PDE sense that $(TE)/k \to 0$ as $h, k \to 0$ unless we assume that $k/h \to 0$ as $h, k \to 0$, the well-known *consistency restraint* associated with the DuFort–Frankel scheme (Richtmyer and Morton [7] p. 177). Note that the theory of VCMM postulates that $\| Q_n \| \leqslant \kappa < \infty$; this condition is violated by our choice of Q_n, if we allow $h \to 0$.

There exist many unconditionally stable explicit overall processes based on (3.1) and other linearly implicit VCMM. The author's investigations make it appear likely that all will suffer the consistency restraint.

Example 2
$S = 2$, $l = 1$, fully implicit.

$$[I + akQ_n + b(kQ_n)^2](\mathbf{y}_{n+1} - \mathbf{y}_n)$$
$$= k[(\tfrac{1}{2}I + ckQ_n)\mathbf{f}_{n+1} + (\tfrac{1}{2}I + (a - c)kQ_n)\mathbf{f}_n] \tag{3.3}$$

The method is $\bar{A}$-stable if $a \geqslant 0$, $2b + 2c - a \geqslant 0$. Indeed, if it is applied to (2.10) with $Q_n = -A$, we obtain

$$\mathbf{y}_{n+1} = R_2^2(kA, 2a, 4b + 4c - 2a)\mathbf{y}_n,$$

where

$$R_2^2(P, \alpha, \beta): = [I - \tfrac{1}{2}(1 + \alpha)P + \tfrac{1}{4}(\beta + \alpha)P^2]^{-1} \cdot [I + \tfrac{1}{2}(1 - \alpha)P + \tfrac{1}{4}(\beta - \alpha)P$$

is the general second order rational approximation to the matrix exponential e^P; it is well known that iff $\alpha \geqslant 0$, $\beta \geqslant 0$, then

$$\lambda[P] < 0 \Rightarrow \rho[R_2^2(P, \alpha, \beta)] < 1,$$

where $\rho[P]$ is the spectral radius of P (Lambert [4], p. 237).

Method (3.3) has order 2, and the asymptotic form of the local TE of the overall process based on it is, from (2.8), (2.2) and (2.7)

$$(TE)_{hk} = [-\tfrac{1}{12}\mathbf{u}^{(3)}(t) + (\tfrac{1}{2}a - c)Q_n\mathbf{u}^{(2)}(t) + bQ_n^2\mathbf{u}^{(1)}(t)]k^3 -$$
$$- [\tfrac{1}{12}D_x^4\mathbf{u}(t)]kh^2 + 0(k^4 + k^2h^2 + kh^4) \tag{3.4}$$

Consider the application of (3.3) to the test system

$$\mathbf{y}' = A\mathbf{y}, \quad \lambda[A] < 0, \quad A \text{ symmetric} \tag{3.5}$$

with the choice

$$Q_n = -C \tag{3.6}$$

where

$$A = C + C^T. \tag{3.7}$$

Clearly, the splitting (3.7) can be achieved in an infinite number of ways. The resulting difference equation is

$$[I - akC + b(kC)^2](\mathbf{y}_{n+1} - \mathbf{y}_n) = k[(\tfrac{1}{2}I - ckC)(C + C^T)\mathbf{y}_{n+1} +$$
$$+ (\tfrac{1}{2}I - (a - c)kC)(C + C^T)\mathbf{y}_n]. \tag{3.8}$$

We now choose coefficients so that the terms in C^2 vanish; that is, we choose $a = 0$, $b = -c$. If, further, we choose $c = \tfrac{1}{4}$, we obtain

$$[I - \tfrac{1}{2}k(C + C^T) + \tfrac{1}{4}k^2CC^T]\mathbf{y}_{n+1} = [I + \tfrac{1}{2}k(C + C^T) + \tfrac{1}{4}k^2CC^T]\mathbf{y}_n \tag{3.9}$$

which can be written in the equivalent two-stage form

$$(I - \tfrac{1}{2}kC)\mathbf{y}_{n+\frac{1}{2}} = (I + \tfrac{1}{2}kC^T)\mathbf{y}_n$$
$$(I - \tfrac{1}{2}kC^T)\mathbf{y}_{n+1} = (I + \tfrac{1}{2}kC)\mathbf{y}_{n+\frac{1}{2}} \tag{3.10}$$

(Our motivation for choosing $Q_n = -C$ is that when C is triangular—in which case the splitting (3.7) is unique—method (3.10), although derived from a fully implicit method, is effectively explicit, since it calls only for the inversion of triangular matrices.)

We still have to investigate the stability of method (3.10). We can no longer hope to do this by studying the region $\mathscr{R}$ of absolute stability. Arguments based on the requirement that the eigenvalues of kA lie within $\mathscr{R}$ essentially assume that we are able to uncouple the component equations of the difference system by applying a diagonalizing transformation which, of course, involves the eigenvalues. This is possible if the matrix coefficients in the system are polynomial in A; it is not possible for a system such as (3.9).

Let

$$R_1(P): = (I - \tfrac{1}{2}P)^{-1}(I + \tfrac{1}{2}P)$$
$$\equiv (I + \tfrac{1}{2}P)(I - \tfrac{1}{2}P)^{-1}$$

be the $(1, 1)$ Padé approximation to the matrix exponential e^P. Clearly,

$$R_1(P^T) \equiv R_1^T(P).$$

(3.10) may be written in the form

$$y_{n+1} = (I - \tfrac{1}{2}kC^T)^{-1}R_1(kC)(I + \tfrac{1}{2}kC^T)y_n,$$

which has the solution

$$y_n = (I - \tfrac{1}{2}kC^T)R_1(kC)[R_1^T(kC)R_1(kC)]^{n-1}(I + \tfrac{1}{2}kC^T)y_0, \quad n = 1, 2, \ldots$$

Hence $y_n \to 0$ as $n \to \infty$ iff

$$\rho[R_1^T(kC) . R_1(kC)] < 1 \tag{3.11}$$

We recall (Coppel [1]) the *logarithmic norm* $\mu[P]$ of an arbitrary matrix P, defined by

$$\mu[P]: = \lim_{h \to 0} \; (\|I + hP\| - 1)/h \tag{3.12}$$

Note that $\mu[\cdot]$ is norm-dependent; if the norm on the right side of (3.12) is the spectral norm $\|\cdot\|_2$, then we shall denote the corresponding logarithmic norm by $\mu_2[\cdot]$. The following are known (Coppel [1]):

(i) $\mu[P] < 0 \rightleftarrows \lambda[P] < 0$

(ii) $\mu_2[P] \equiv \lambda[\tfrac{1}{2}(P + P^T)].$ (3.13)

It can be shown (Lambert and Sigurdsson [5]) that

$$\mu_2[P] < 0 \Rightarrow \|R_1(P)\|_2 < 1 \tag{3.14}$$

(Note that similar results do not hold for higher order Padé approximations. For example, in order that the spectral norm of the $(2, 2)$ Padé approximation to e^P be less than unity, it is necessary to assume that $\mu_2[P] < 0$ *and* $\mu_2[-P^2] < 0$.)

We return now to the condition for stability (3.11). By (3.14)

$$\rho[R_1^T(kC).R_1(kC)] = \|R_1(kC)\|_2 < 1$$

if

$$\mu_2[kC] < 0.$$

But, by (3.13(ii)),

$$\mu_2[kC] = \lambda[\tfrac{1}{2}k(C + C^T)]$$
$$= \tfrac{1}{2}k\lambda[A] \qquad\qquad \text{(by (3.7))}$$
$$< 0, \forall k > 0 \qquad\qquad \text{(by (3.5))}$$

Hence, somewhat surprisingly, we find that *for all splittings* (3.7) method (3.10) is stable when applied to the symmetric test system (3.5).

Finally, let us apply (3.10) to the semi-discrete problem (1.2) and choose the unique splitting (3.7) for which C is triangular; that is,

$$C = [c_{ij}]; \quad c_{ii} = -1/h^2, \quad c_{i+1,i} = 1/h^2, \quad c_{ij} = 0, \quad j \neq i, i-1,$$
$$i, j = 1, 2, \ldots, M.$$

We find that the resulting overall process for (1.1) is equivalent to the (direct) method of Saul'ev (Richtmyer and Morton [7], p. 192), well known to be effectively explicit and unconditionally stable, in agreement with the above stability analysis.

Note, however, that with $Q_n = -C$, C specified as above,

$$Q_n u^{(2)}(t) = 0(1/h), \quad Q_n^2 u^{(1)}(t) = 0(1/h^2) \quad \text{as } h \to 0,$$

so that the first term in $(TE)_{hk}$ given by (3.4) is no longer $0(k^3)$ but $0(k^3 + k^3/h + k^3/h^2)$. Thus the method suffers precisely the same consistency restraint as does the DuFort–Frankel scheme (see Richtmyer and Morton [7], p. 192).

Acknowledgment

Much of the work of this paper was done while the author was a visitor to the Department of Mathematics, University of Kentucky. The generous support and hospitality afforded by the Department are gratefully acknowledged, as are the many useful conversations with Dr Henry J. Thacher, Dr Thad Curtz and, in particular, Dr Graeme Fairweather.

References

[1] Coppel, W. (1965). *Stability and Asymptotic Behaviour of Differential Equations.* D. C. Heath, Boston.
[2] Cryer, C. W. (1973). A new class of highly stable methods, A_0-stable methods, *BIT* 13, 153–159.

[3] Henrici, P., (1962). *Discrete Variable Methods in Ordinary Differential Equations.* John Wiley & Sons, New York.

[4] Lambert, J. D. (1973). *Computational Methods in Ordinary Differential Equations.* John Wiley & Sons, London.

[5] Lambert, J. D. and Sigurdsson, S. T., (1972). Multistep methods with variable matrix coefficients, *SIAM J. Numer. Anal.* 9, 715–733.

[6] Mitchell, A. R. (1969). *Computational Methods in Partial Differential Equations.* John Wiley & Sons, London.

[7] Richtmyer, R. D. and Morton, K. W. (1967). *Difference Methods for Initial-Value Problems* (2nd edition). Interscience, New York.

[8] Sigurdsson, S. T. (1973). *Multistep Methods with Variable Matrix Coefficients for Systems of Ordinary Differential Equations.* Chalmers Institute of Technology, Göteborg, Department of Computer Sciences Report No. 1973.04.

[9] Widlund, O. B. (1967). A note on unconditionally stable linear multistep methods, *BIT* 7, 65–70.

Realistic Estimates for Generic Constants in Multivariate Pointwise Approximation

Jean Meinguet

1. Introduction

In recent years, problems of error estimation have become the focus of much mathematical activity. There is however, one inherent shortcoming of most of the results published so far (especially in connection with the rate of convergence of the finite element method, see e.g. [5, 6, 12, 22, 30, 31, 33, 35]), namely their dependence on strictly unknown constants. To devise a *method of practical value* for finding *realistic upper bounds* of such *generic constants* in a *wide variety of situations* is the primary purpose of the research work from which this paper originates.

Based on the operator-theoretical analysis that can be found in [19], Section 2, where reference is repeatedly made to such classical tools as the Peano Kernel Theorem and the Bramble–Hilbert Lemma, is devoted to a comprehensive discussion of the general problem of error estimation. This provides basic motivations for introducing instead what might be called the *practical estimation problem*, which is actually concerned with standard classes of appraisals of type (6) rather than with specific appraisals of type (2); it should be understood that the constant $d_0 = \inf d$, since it can be defined for each standard class by the associated maximin problem (9) which properly belongs to approximation theory, has only to be estimated once and for all, which is of course a decisive advantage and justifies eventually some preliminary elaborate investigation.

Next, we turn to significant applications, namely to the wide class of *pointwise approximation* problems (the alternative class of *mean-square approximation* problems will be considered elsewhere). Taylor's formula, which clearly has to play here an important role, is reviewed in Section 3, which also contains those relevant algebraic topics we have found useful when analyzing "change of scale" effects; in particular, the importance of tensor and matrix interpretations is emphasized throughout.

In Section 4, we give sharp appraisals for the infimum d_0 of all constants d for which

$$\min_{p \in \mathscr{P}_{m-1}} \|f - p\|_\infty \leqslant d \max_{|\gamma| = m} \|D^\gamma f\|_\infty \qquad (1\mathrm{a})$$

holds for all $f \in C^m(\bar{\Omega})$ and any convex bounded subset $\bar{\Omega}$ of R^n, m and n denoting arbitrary positive integers (the other notations are self-explanatory: classical Hölder norms $\| \cdot \|$, differential operators $D^\gamma f$, space $\mathscr{P}_{m-1}$ of polynomials of total degree $m - 1$). Our *main result* in this connection is the following:

$$[\delta_1(\Omega)]^m / 2^{2m-1} m! \leqslant d_0 \leqslant [\delta_1(\Omega)]^m / 2^m m!, \qquad (1\mathrm{b})$$

where $\delta_1(\Omega)$ denotes the maximum value of $\|x - y\|_1$ for $x, y \in \bar{\Omega}$. In view of the great difficulties which are notoriously associated with multivariate Chebyshev approximation (see e.g. [7, 8, 13, 17, 23, 24, 27, 29]), it may seem surprising that sharp appraisals for the solution d_0 of the maximin problem (9) can be obtained at all; in actual fact, as was already emphasized by Golomb in [11] (see p. 113) for the one-dimensional case, there has been much success in calculating "maximin" or "minimax" quantities (or at least good bounds for them), whereas it often proves difficult to get precise results for the approximation error for an individual function; this is the more interesting as the approximation error for a class of functions is in many cases more useful to the numerical analyst than that for an individual function. It is worth remarking that the upper bound in (1b) can be obtained quite easily, whereas the lower bound requires a rather elaborate analysis.

In the final section, we consider, just to show how easily realistic estimates can be obtained in concrete problems once the corresponding d_0 has been determined (or at least carefully estimated), two specific, non-trivial examples, namely multivariate approximate integration with non-negative weights and bivariate Lagrange interpolation over a triangle.

2. The General Problem of Error Estimation

In [19], we have analyzed at length what is called here the general problem of error estimation. The underlying abstract scheme can be summarized as follows. Let there be given normed linear spaces X, Y, Z, a linear surjection $U: X \to Y$ and a linear mapping $R: X \to Z$. We assume throughout that a constant c exists such that

$$\|Rf\| \leqslant c \|Uf\| \quad \text{for all} \quad f \in X, \qquad (2)$$

which clearly implies the inclusion

$$\text{kernel } U \subset \text{kernel } R \qquad (3\mathrm{a})$$

or, equivalently, the existence of a uniquely defined linear mapping $Q : Y \to Z$ such that

$$R = QU; \tag{3b}$$

in actual fact, Q has the explicit expression:

$$Q = RV, \tag{4}$$

where $V : Y \to$ denotes an arbitrary right inverse of U, i.e., a (possibly nonlinear) mapping satisfying $UV = 1_Y$ where 1_Y is the identity mapping of Y (the existence of such V's classically follows, in view of the axiom of choice, from the surjectivity of U). The infimum c_0 of all constants c satisfying (2), which is often called the *error coefficient* associated with R, has therefore the theoretical expression:

$$c_0 = \|Q\| = \|RV\| \quad \text{for any} \quad V : Y \to X \text{ such that } UV = 1_Y. \tag{5}$$

Since the actual evaluation of c_0 can only be based on some explicit representation of Q or, equivalently, of Rf in terms of Uf, we may refer to the estimates (2) as estimates of (abstract) Peano kernel type. Indeed, as is masterfully established in [25], the classical *Peano Kernel Theorem* and its many (known or potential) variants (such as Riesz's Mass Theorem) lie at the heart of representation theory to such an extent that they often provide accessible standard forms for Rf and explicit procedures for calculating $\|Q\|$. Since suitably generalized Taylor representations with integral remainders necessarily play here a prominent role, basic difficulties are to be expected in case X is a space of functions of several variables over any region of a sufficiently general shape; as a matter of fact, it may happen that no constructive method is known for calculating $\|Q\|$ (for a far-reaching illustration, see [26], p. 400). The main reason that such difficulties may occur is that the partial derivatives of a function of several variables are somewhat dependent on one another. This actually explains the otherwise surprising fact that Taylor's formula, in its classical one-dimensional version discussed in Section 3, cannot be regarded in general as a suitable candidate for obtaining best estimates of Peano type in a multivariate context. In this respect, at least two extensions of Taylor's formula to higher dimensions are known to be permissible, namely the Sard formula and the Stancu–Simonsen formula; most unfortunately (see e.g. [33], p. 18), their use is restricted in principle to rather special regions (such as rectangular polygons in R^2) and the estimates they eventually yield are expressed in terms of somewhat unconventional norms, the underlying spaces X being not what one would naturally expect in most applications (namely spaces of continuously differentiable functions and their completions with respect to energy or Sobolev norms). Among the relatively few papers to have exploited in detail these multivariate versions of Taylor's formula, we specially mention [3] and [1]; a most readable survey of this interesting approach (whose complete theory, given by Sard in Chapter 4 of [25], is quite elaborate) can be found in [32] (see pp. 137–178), which also contains a collection of suggestive graphs of two-dimensional generalized Peano kernels (for some classical rules of double integration over a square). It should also be noticed that, even in one-dimensional applications, great technical difficulties (pertaining to "hard analysis") may arise in connection with the practical calculation of c_0 from the explicit definition of Q, especially in case c_0 is an error coefficient associated with a class of approximation rules depending on a number of parameters (for an interesting problem of this kind, see [18], where an explicit solution is obtained by nonlinear programming techniques).

For a numerical analyst, the essential lesson to draw from this discussion can only be the following: since c_0 usually proves so difficult to calculate, it may not be made the basis of error estimates; what is essentially needed instead is a practical method for finding a realistic upper bound of c_0 in a wide variety of situations. In order to answer this requirement, which is in fact our main purpose here, it seems that the following elementary analysis is most appropriate. We first remark that R is usually defined as the difference between a specific linear mapping and some linear rule of approximation, the latter being selected so as to be exact on kernel U (which subspace of X is finite-dimensional in the typical case where U is a standard differentiation operator). It then follows that, as a general rule, *the data X, Y, U may be regarded as essentially fixed whereas R is essentially variable*; in order to take advantage of this natural distinction between the data, which important fact is completely overlooked in (2), we shall henceforth consider appraisals of an alternative type, namely

$$\|Rf\| \leqslant \|R\| d \|Uf\| \quad \text{for all} \quad f \in X, \tag{6}$$

where d denotes some finite constant. For this to make sense at all, it is necessary and sufficient that the following *assumptions* be satisfied:

(H1) R is of the composite form (3b) and bounded;
(H2)   the set of bounded right inverses of U is non-empty;

(H2) readily follows from the identity $Rf \equiv RV(Uf)$ which is implied by (3) and (4). The infimum d_0 of all constants d satisfying (6) has then the theoretical expression:

$$d_0 = \inf \|V\| \quad \text{with} \quad V: Y \to X \text{ such that } UV = 1_Y; \tag{7}$$

since (6), for $d = d_0$, and the classical appraisal

$$\|Rf\| \leqslant \|R\| \inf_{Up=0} \|f - p\| \quad \text{for all} \quad f \in X \tag{8}$$

are both sharp and comparable, we readily get for d_0 the alternative expression:

$$d_0 = \sup_{\|Uf\|=1} \inf_{Up=0} \|f - p\|, \tag{9}$$

the sharp evaluation of which will be our main concern in the sequel.

On the right-hand side of (6) the mapping R is separated from the function f operated upon. This essential difference from (2) is evidently consistent with the natural distinction between the data we have emphasized above; since there are relatively few standard choices of wide applicability for the space X and the surjection U, it is by far more justifiable to concentrate on the accurate determination of the associated characteristic constants d_0 than to attempt to cope with the specific difficulties inherent in the precise evaluation of $\|R\|$ (the more so as (2), for $c = c_0$, is already sharper than (6), for $d = d_0$). Then, in so far as it is justified to regard the evaluation of d_0 as not belonging to the practical estimation problem itself, it may be said (see [12], p. 156) that a remainder analysis (of practical value) can be achieved for very broad function classes without any representation theory (of mappings). Moreover, it should be realized that (6) can be interpreted strictly speaking as the abstract version of what has come to be called, specially in the approximation theory of the finite element method, the *Bramble–Hilbert Lemma* (see [4], Theorem 2). Under assumption (H1), and just as (6) trivially follows from assumption (H2), this fundamental lemma is indeed an immediate consequence of a much older result of deep significance, increasingly general variants of which can be found in [20] (p. 85), [21] (p. 112) and [22] (p. VI-3), respectively; what this preliminary result essentially asserts is the existence (when Sobolev spaces are concerned) of at least one bounded right inverse of U. For complementary details about this rather unconventional interpretation of the Bramble–Hilbert lemma, the reader is referred again to [19]; for brevity, we shall not dwell here any longer on the underlying *operator-theoretical analysis* of the error estimation problem. In view of the foregoing, the occasional interpretation of that fundamental lemma as a generalized Peano kernel theorem (see e.g. [33], p. 19, and [12], p. 156) may not be taken too literally; as a matter of fact, the actual evaluation of the implied generic constants inevitably requires a constructive and not only an existential approach.

In conclusion, this comprehensive discussion has established that the *practical problem of error estimation* is actually concerned with appraisals of type (6) rather than of type (2). For each significant class of applications, i.e., for each of the standard choices of the space X and the surjection U whose practical importance can be taken for granted, it would be most useful to know once and for all the corresponding constant d_0 which is defined by (7) or (9); needless to say, the realistic estimation of d_0 is a highly interesting problem of approximation theory which should be considered very seriously. On the other hand, it is entirely up to the numerical analyst, when he is concerned with a specific application defined by any given bounded mapping R of the form (3b), to appreciate to what reasonable extent the evaluation of the associated $\|R\|$ should be eventually refined.

3. Taylor's Formula and Relevant Algebraic Topics

With n any positive integer, let Ω be a bounded (non-empty) domain in Euclidean n-space, R^n. In the following, we shall assume that $\bar{\Omega}$, the closure of Ω, is *star-shaped* with respect to (at least) one of its points, say a (in other words, for any $x \in \bar{\Omega}$, $\bar{\Omega}$ must contain the closed line segment between a and x).

With m any positive integer, let $X \equiv C^m(\bar{\Omega})$ denote the space of m-times continuously differentiable (real-valued) functions on $\bar{\Omega}$, in the sense of Whitney [34]. Then, for every $f \in X$, the *Taylor formula of the m-th order* (with integral expression of the remainder)

$$f(x) = \sum_{k=0}^{m-1} D^k f(a) \cdot (x-a)^k / k! + \int_0^1 \frac{(1-t)^{m-1}}{(m-1)!} \frac{\partial^m}{\partial t^m} f(a + t(x-a)) \, dt \quad (10)$$

is known to hold at any $x \in \bar{\Omega}$. Since this otherwise classical formula is written here in a coordinate free form, which is not necessarily the rule in multivariate differential calculus, some complementary explanations and interpretations may prove useful before we proceed further.

By $D^k f(a)$ in (10), we denote the *k-th derivative* of f at a; this is known to be a *k-linear form* on the Cartesian product $R^n \times \ldots \times R^n$ (k times) or, equivalently, a *k-fold covariant tensor* on R^n, whose coordinates with respect to the canonical basis $(e_1, \ldots, e_n)$ of R^n are the usual partial derivatives given by

$$\frac{\partial^k f}{\partial x_{i_1} \ldots \partial x_{i_k}} (a) = D^k f(a) \cdot (e_{i_1}, \ldots, e_{i_k}) \quad \text{for} \quad i_1, \ldots, i_k \in [1, n]$$

$$(11)$$

and denoted hereafter by $f_{i_1 \ldots i_k}(a)$. As for $(x-a)^k$ in (10), it is only an abbreviated notation for the otherwise general element $(x_1, \ldots, x_k)$ of $R^n \times \ldots \times R^n$ (k times) in case $x_i = x - a$ for all $1 \leqslant i \leqslant k$. According to the so-called *main theorem on tensor products* (see e.g. [15], p. 321), every k-linear form on a k-fold Cartesian product can be regarded, equivalently, as a linear form on the corresponding k-fold tensor product space. A concrete *matrix interpretation* of this basic result readily follows from (11) together with the k-linearity of the mapping $D^k f(a)$. The ensuing identity:

$$D^k f(a) \cdot (x_1, \ldots, x_k) \equiv \sum_{i_1 \ldots i_k} f_{i_1 \ldots i_k}(a) \xi_{1,i_1} \xi_{2,i_2} \cdots \xi_{k,i_k}, \qquad (12)$$

where the indices $i_1, \ldots, i_k$ run independently from 1 to n and $\xi_{i,j}$ denotes (here as in the sequel) the j-th coordinate of $x_i \in R^n$ with respect to the canonical basis, shows indeed that $D^k f(a)$ can be regarded as a n^k-covector (or *row-matrix*) of coordinates $f_{i_1 \ldots i_k}(a)$ provided that $(x_1, \ldots, x_k)$ itself is regarded as the *k-fold tensor product* of $x_1, \ldots, x_k$ (taken in this order), i.e., as a n^k-vector (or *column-matrix*) of coordinates $\xi_{1,i_1} \xi_{2,i_2} \cdots \xi_{k,i_k}$; of course, whenever they are regarded as entries of matrices, coordinates must be arranged in some consistent way, say, for definiteness and convenience, always in double lexicographic order according to row and column selections.

This matrix interpretation proves most useful on many occasions. By way of illustration, we may briefly consider here what has come to be called the *change of scale effect* in error estimation (see e.g. [12], p. 157), Ω being then regarded as the image, under a nonsingular linear (or possibly affine) transformation

$$x = A\hat{x}, \qquad (13a)$$

of some fixed bounded domain $\hat{\Omega}$ in R^n. Setting correspondingly

$$\hat{f}(\hat{x}) \equiv f(x), \qquad (13b)$$

we can readily deduce from the scalar identity

$$D^k f(a) \cdot (x_1, \ldots, x_k) \equiv D^k \hat{f}(\hat{a}) \cdot (\hat{x}_1, \ldots, \hat{x}_k) \qquad (14)$$

the linear transformations to be effected, in consequence of (13a), upon the covectors (resp. vectors) representing k-th derivative mappings (resp. k-fold tensor products); specifically, if A is consistently represented by an $n \times n$ nonsingular matrix, then the required formulas can be written as follows:

$$(x_1, \ldots, x_k) = A^{[k]}(\hat{x}_1, \ldots, \hat{x}_k), \qquad (15a)$$

$$D^k f(a) = D^k \hat{f}(\hat{a})(A^{[k]})^{-1}. \qquad (15b)$$

Here $A^{[k]}$ denotes the $n^k \times n^k$ matrix which is known in matrix theory (see e.g. [2, 14, 16]) as the *k-fold (right) Kronecker product* of A (or the *k-th product-matrix* of A) and can be constructed by the explicit recursion

$$\begin{aligned} A^{[2]} &= A \otimes A, \\ A^{[i+1]} &= A \otimes A^{[i]} \quad \text{for} \quad i = 2, 3, \ldots, k-1, \end{aligned} \qquad (16a)$$

the binary operator $\otimes$ being defined classically as follows:

$$B \otimes C = \begin{bmatrix} \beta_{11}C \ldots \beta_{1q}C \\ \vdots \qquad \vdots \\ \beta_{p1}C \ldots \beta_{pq}C \end{bmatrix}. \tag{16b}$$

for any two rectangular matrices B (of entries β_{ij}, with $1 \leqslant i \leqslant p$, $1 \leqslant j \leqslant q$) and C; it should be noticed that the tensor product $(x_1, \ldots, x_k)$ itself can be constructed in quite the same way by exploiting backwards the ordered sequence of arguments. Among the many interesting properties of Kronecker products, the identity

$$(A \otimes B)(C \otimes D) = (AC) \otimes (BD), \tag{17}$$

which is readily verified to hold for any rectangular matrices A, B, C, D, provided only that dimensions conform (which means simply that the matrix products AC and BD must exist), is of primary importance. In connection with the analysis of the "change of scale" effect in pointwise or in mean-square approximation, we mention the following identities (which are most easily proved by resorting to (17))

$$\text{lub}_i(A^{[k]}) = [\text{lub}_i(A)]^k \quad \text{for} \quad i = 1, 2, \infty, \tag{18}$$

and the sharp appraisals (which are accordingly implied by (15b))

$$\|D^k\hat{f}(\hat{a})\|_i \leqslant [\text{lub}_j(A)]^k \|D^k f(a)\|_i, \tag{19a}$$

$$\|D^k f(a)\|_i \leqslant [\text{lub}_j(A^{-1})]^k \|D^k \hat{f}(\hat{a})\|_i, \tag{19b}$$

where $i = 1, 2, \infty$ and $j = i/(i-1)$; as usual, the symbol $\text{lub}_i(\cdot)$ denotes the matrix norm (or bound) which is subordinate to the Hölder vector norm $\| \cdot \|_i$.

For any $f \in C^k(\bar{\Omega})$ and $a \in \bar{\Omega}$, the k-linear mapping $D^k f(a)$ is clearly *symmetric*; since this important property has been completely overlooked so far, the above matrix interpretation must be highly redundant. This apparent shortcoming can be covered (at the expense of simplicity, however) by substituting for (12) the following scalar identity

$$D^k f(a) . (x_1, \ldots, x_k) \equiv \sum_{|\gamma|=k} D^\gamma f(a)(x_1 . x_2 \ldots x_k)_\gamma, \tag{20}$$

where $\gamma = (\gamma_1, \ldots, \gamma_n)$ is any n-tuple of nonnegative integers,

$$D^\gamma \equiv \frac{\partial^{\gamma_1 + \ldots + \gamma_n}}{\partial x_1^{\gamma_1} \ldots \partial x_n^{\gamma_n}} \tag{21}$$

denotes the differential operator of order $|\gamma| \equiv \sum_{i=1}^n \gamma_i$, and $(x_1 . x_2 \ldots x_k)_\gamma$

denotes the γ-th coordinate, with respect to the canonical basis of R^{n_k} for

$$n_k = \binom{n+k-1}{k}, \tag{22}$$

of the vector which is known in advanced matrix theory as the *k-fold symmetric product* of the vectors $x_1, \ldots, x_k$ (for a coordinate free definition of this concept, see [10], p. 353); for completeness, we recall that this γ-th coordinate can be explicitly defined as the quotient by $\gamma! \equiv \gamma_1! \ldots \gamma_n!$ of the permanent (or "plus determinant", see e.g. [16], p. 18) of the $k \times k$ matrix whose i-th column (for every $1 \leqslant i \leqslant k$) is obtained from x_i by repeating γ_j times the j-th coordinate $\xi_{i,j}$ (for every $1 \leqslant j \leqslant n$). In case $x_i = x - a$ for all $1 \leqslant i \leqslant k$, (20) takes the most familiar form

$$D^k f(a) \cdot (x-a)^k/k! \equiv \sum_{|\gamma|=k} D^\gamma f(a)(x-a)^\gamma/\gamma!, \tag{23a}$$

where

$$(x-a)^\gamma/\gamma! \equiv (\xi_1 - \alpha_1)^{\gamma_1} \ldots (\xi_n - \alpha_n)^{\gamma_n}/(\gamma_1! \ldots \gamma_n!); \tag{23b}$$

hence it follows that Taylor's formula (10) can be rewritten in the form

$$f(x) = \sum_{|\gamma|=0}^{m-1} D^\gamma f(a)(x-a)^\gamma/\gamma! + \int_0^1 \frac{(1-t)^{m-1}}{(m-1)!} \frac{\partial^m}{\partial t^m} f(a + t(x-a)) \, dt \tag{24a}$$

and that

$$\frac{\partial^m}{\partial t^m} f(a + t(x-a)) \equiv D^m f(a + t(x-a)) \cdot (x-a)^m$$

$$\equiv m! \sum_{|\gamma|=m} D^\gamma f(a + t(x-a))(x-a)^\gamma/\gamma!. \tag{24b}$$

As for formulas (15), they can be replaced here by

$$x_1 \cdot x_2 \ldots x_k = A_{[k]}(\hat{x}_1 \cdot \hat{x}_2 \ldots \hat{x}_k), \tag{25a}$$

$$(D^\gamma f(a))_{|\gamma|=k} = (D^\gamma \hat{f}(\hat{a}))_{|\gamma|=k}(A_{[k]})^{-1}, \tag{25b}$$

where $A_{[k]}$ denotes the $n_k \times n_k$ matrix which is known in matrix theory as the *k-th induced matrix* of A (or the *k-th power-matrix* of A). From some classical properties of induced matrices (see again [2, 14, 16]), it can be proved that (18), where $A_{[k]}$ is used in lieu of $A^{[k]}$, retains its validity for $i = 1$ (but usually not for $i = 2, \infty$), so that we finally get the sharp appraisals

$$\max_{|\gamma|=k} |D^\gamma \hat{f}(\hat{a})| \leqslant [\mathrm{lub}_1(A)]^k \max_{|\gamma|=k} |D^\gamma f(a)|, \tag{26a}$$

$$\left[\sum_{|\gamma|=k} |D^\gamma \hat{f}(\hat{a})|^2/\gamma! \right]^{1/2} \leqslant [\mathrm{lub}_2(A)]^k \left[\sum_{|\gamma|=k} |D^\gamma f(a)|^2/\gamma! \right]^{1/2}, \qquad (26b)$$

the first of which clearly coincides with (19a) for $i = \infty$.

4. Sharp Appraisals for Practical Error Coefficients of Chebyshev Type

We turn now to the wide class of estimation problems for pointwise approximation criteria. Specifically, X is here the space $C^m(\bar{\Omega})$ of m-times continuously differentiable (real-valued) functions f on $\bar{\Omega}$ with the Chebyshev norm

$$\|f\| \equiv \max_{x \in \bar{\Omega}} |f(x)| \qquad (27a)$$

and U is the m-th Fréchet derivative (denoted by D^m in Section 3); as for the space $Y \equiv U(X)$ of (m-fold) tensor-valued functions $D^m f$, it is also provided with a uniform norm, namely

$$|f|_m \equiv \max_{|\gamma|=m} \max_{x \in \bar{\Omega}} |D^\gamma f(x)|; \qquad (27b)$$

we recall that m and n are given positive integers and that $\bar{\Omega}$ is a given bounded, closed subset of R^n which is star-shaped with respect to at least one of its points, say a. As emphasized above, the corresponding *practical error coefficient* d_0 is of a great interest, which justifies the present search for *sharp upper and lower bounds*; quite naturally, the following approach will be based on the welcome fact that d_0 can be regarded at will as the solution of the *minimax problem* (7) or as the solution of the *maximin problem* (9).

In view of (7), any (bounded) right inverse V of $U \equiv D^m$ yields an upper bound for d_0. As is suggested by Taylor's formula (24), a most natural choice for V is here the following:

$$V(D^m f)(x) \equiv \int_0^1 \frac{(1-t)^{m-1}}{(m-1)!} D^m f(a + t(x-a)) \cdot (x-a)^m \, dt; \qquad (28)$$

by virtue of an iterated version of identity (17), to wit, the identity $(A_1 \otimes \cdots \otimes A_m)(B_1 \otimes \cdots \otimes B_m) = (AB)^m$ where every A_i coincides with the n-covector of coordinates 1 and every B_i is the n-vector $x - a$ of coordinates $\xi_j - \alpha_j$ (for $j = 1, \ldots, n$), we immediately get from (28) the sharp appraisal

$$|V(D^m f)(x)| \leqslant \left(\sum_{j=1}^n |\xi_j - \alpha_j| \right)^m |f|_m/m! \quad \text{for any } x \in \bar{\Omega}, \qquad (29)$$

which finally gives the required result

$$\|V\| = (\max_{x \in \bar{\Omega}} \|x - a\|_1)^m/m!. \tag{30}$$

Introducing now the two diameters δ_1 and δ_2 of Ω which are defined by

$$\delta_i(\Omega) = \max_{x, y \in \bar{\Omega}} \|x - y\|_i \quad \text{for} \quad i = 1, 2 \tag{31a}$$

and accordingly satisfy the sharp inequalities

$$\delta_1(\Omega)/n^{1/2} \leqslant \delta_2(\Omega) \leqslant \delta_1(\Omega), \tag{31b}$$

we obtain the remarkably simple appraisal

$$d_0 \leqslant [\delta_1(\Omega)]^m/m! \leqslant [\delta_2(\Omega)]^m n^{m/2}/m!, \tag{32}$$

which usually proves better suited to practical needs than the possibly sharper result $d_0 \leqslant \|V\|$. It often happens that (32) can be improved significantly, by choosing suitably the reference point a among all points of R^n with respect to which $\bar{\Omega}$ is *star-shaped*; in particular, whenever Ω is *convex*, to minimize the expression (30) regarded as a function of a is a trivial unconstrained problem whose solution yields

$$d_0 \leqslant [\delta_1(\Omega)]^m/2^m m! \leqslant [\delta_2(\Omega)]^m n^{m/2}/2^m m!, \tag{33}$$

which is manifestly the sharpest result that can be obtained by resorting only to elementary Taylor techniques. To ascertain whether the upper bound assigned to the practical error coefficient d_0 by (32) (resp. (33)) may be accepted as realistic for star-shaped (resp. convex) bounded regions, we shall proceed now to the determination of a sharp lower bound from the maximin definition (9) of d_0.

As can be expected from the general theory of linear Chebyshev approximation, a prominent role will be played here by the well known *duality relation*:

$$\min_{p \in \mathscr{P}_{m-1}} \|f - p\| = \max \left| \sum_{i=1}^{N+1} \lambda_i f(x_i) \right| \quad \text{for all} \quad f \in C(\bar{\Omega}), \tag{34}$$

where the maximum on the right-hand side is to be taken over all subsets of $N + 1$ distinct points of $\bar{\Omega}$, the reals λ_i being subject to the set of conditions:

$$\sum_{i=1}^{N+1} \lambda_i p(x_i) = 0 \quad \text{for all} \quad p \in \mathscr{P}_{m-1}, \tag{35a}$$

$$\sum_{i=1}^{N+1} |\lambda_i| = 1, \tag{35b}$$

and N denoting the dimension of the space $\mathscr{P}_{m-1} = \text{kernel } D^m$; since n_k, defined by (22), can be interpreted as the number of formally distinct k-th partial derivatives, it is easily seen that N is given by

$$N = \sum_{k=0}^{m-1} n_k \tag{36a}$$

and has therefore the general expression

$$N = \binom{n+m-1}{n}. \tag{36b}$$

It should be noticed further that we may not require $\lambda_i \neq 0$ for all $1 \leqslant i \leqslant N + 1$, unless the word "max" in (34) is replaced by the word "sup"; this is due to the interesting fact (mentioned in [17], p. 126) that the classical *Haar condition* can always be satisfied in several variables by considering a suitable subset of $N + 1$ distinct points of $\bar{\Omega}$ in an arbitrarily small neighbourhood of any given finite subset of $\bar{\Omega}$. A simple proof of the above duality theorem, which is based only on elementary properties of convex sets, can be found in [27] (p. 18). However, for possible future extensions of the present approach, it should be realized that this basic result is only a concrete application of the characterization theorem of polynomials of best approximation in general normed linear spaces which is due to Singer (see [28], p. 170); we recall that this abstract characterization is deduced essentially from a simple corollary of the Hahn–Banach Theorem (see e.g. [17], p. 4) and from the Krein–Milman Theorem, the general conclusion being then restated in terms of the so-called Carathéodory Lemma (see e.g. [8, 13, 24, 27]); the particular formulation (34) is finally obtained by using the classical fact that the extreme points of the unit ball of the dual space of $C(\bar{\Omega})$ are simply the "point functionals": $f \to \epsilon f(x_i)$ with $\epsilon = \pm 1$ and $x_i \in \bar{\Omega}$.

It follows from (34) that (9) can be rewritten in the form

$$d_0 = \max_{\{x_i\},\{\lambda_i\}} \sup_f \left| \sum_{i=1}^{N+1} \lambda_i f(x_i) \right|, \tag{37}$$

where the supremum is to be taken over the unit semiball $|f|_m \leqslant 1$ of $C^m(\bar{\Omega})$ and the maximum is to be interpreted as explained in connection with (34), the original maximin problem being thus converted into a *maximization problem* of an apparently simpler form. By Taylor's formula (24) and taking into account conditions (35), we get for the sum in (37) the integral expression:

$$\sum_{i=1}^{N+1} \lambda_i f(x_i) = \int_0^1 \frac{(1-t)^{m-1}}{(m-1)!} g^{(m)}(t) \, dt, \tag{38a}$$

where we have set:

$$g(t) \equiv \sum_{i=1}^{N+1} \lambda_i f(a + t(x_i - a)) \quad \text{for} \quad 0 \leqslant t \leqslant 1, \tag{38b}$$

so that

$$g^{(m)}(t)/m! = \sum_{i=1}^{N+1} \lambda_i \sum_{|\gamma|=m} D^\gamma f(a + t(x_i - a))(x_j - a)^\gamma/\gamma!. \tag{38c}$$

Now, for any given subset of $N + 1$ distinct points x_i of $\bar{\Omega}$ and for any particular vector of coordinates λ_i satisfying the corresponding set of conditions (35), we are naturally interested in the supremum of the expression (38a) over $|f|_m \leqslant 1$, at least in so far as it can yield by (37) a *realistic lower bound* for d_0. For *star-shaped*, bounded regions of unrestricted generality, to determine the supremum of (38a) seems to be a practically unfeasible problem (of a rather questionable interest in itself); however, it can be shown easily that this supremum can never exceed the quantity $\max_i (\|x_i - a\|_1)^m/m!$ whose further maximization according to (37) yields nothing else than the upper bound given by (30). It is surprising that this quite elementary result is sharp, in the sense that (possibly degenerate) regions exist for which it cannot be improved; this is clearly the case, in particular, when the subset $\bar{\Omega}$ of R^n is defined as the union of a finite number of closed line segments joining the point a to points x_i such that $\|x_i - a\|_1$ is independent of i. In view of these comments and in order to proceed further with the solution of the optimization problem stated in (37), we are justified in imposing suitable restrictions on $\bar{\Omega}$; in actual fact, we shall assume henceforth that Ω is *convex*, which proves indeed sufficiently general to cover many applications of practical interest. By virtue of this assumption, the reference point $a \in \bar{\Omega}$ may be chosen at will without affecting in any way the quantity (38a); this clearly implies that the maximization (suggested by (37)) of $g^{(m)}(t)$ over $|f|_m \leqslant 1$ for some prescribed t may be a pointless problem, unless t is allowed to range over the whole interval $[0, 1]$. Now, for $t = 0$, it is easily seen from (38c) that

$$\max_{|f|_m \leqslant 1} g^{(m)}(0)/m! = \sum_{|\gamma|=m} \epsilon_\gamma \sum_{i=1}^{N+1} \lambda_i x_i^\gamma/\gamma!, \tag{39a}$$

where

$$\epsilon_\gamma \equiv \text{sgn}\left(\sum_{i=1}^{N+1} \lambda_i x_i^\gamma/\gamma!\right) \tag{39b}$$

for every nonnegative multi-index $\gamma = (\gamma_1, \ldots, \gamma_n)$ of order m; since the numbers ϵ_γ so defined do not depend on the arbitrary point $a \in \bar{\Omega}$, the expression (39a) can be interpreted as giving automatically the required supremum of (38a) over $|f|_m \leqslant 1$, which is thus attained for any $f_0 \in C^m(\bar\Omega)$ satisfying (identically, over $\bar\Omega$) the set of conditions:

$$D^\gamma f_0(x) \equiv \epsilon_\gamma \quad \text{for all } \gamma \text{ such that } |\gamma| = m, \tag{40a}$$

i.e., for every polynomial (of total degree m) defined by

$$f_0(x) \equiv \sum_{|\gamma|=m} \epsilon_\gamma x^\gamma/\gamma! \;(\text{mod } \mathscr{P}_{m-1}). \tag{40b}$$

From the duality relation (34), it follows consequently that the definition (37) of d_0 can be rewritten in the greatly simplified form:

$$d_0 = \max_{\{\epsilon_\gamma\}} \; \min_{p \in \mathscr{P}_{m-1}} \left\| \sum_{|\gamma|=m} \epsilon_\gamma x^\gamma/\gamma! - p(x) \right\|, \tag{41}$$

it being understood that the coefficients ϵ_γ may take independently any of the values $+1$, -1 or 0.

Whereas the theoretical solution of the complete collection of minimization problems stated in (41) remains of course out of the question (at least for general convex regions), it proves surprisingly easy to obtain for d_0 a lower bound of practical value. It turns out indeed that there exists one fairly general subset of $\{\epsilon_\gamma\}$ for which the maximin problem implied by (41) can be solved explicitly and conveniently, namely the set defined by

$$\epsilon_\gamma \equiv (\epsilon_1)^{\gamma_1} \ldots (\epsilon_n)^{\gamma_n} \quad \text{for every } \gamma = (\gamma_1, \ldots, \gamma_n) \text{ such that } |\gamma| = m, \tag{42a}$$

where the ϵ_i take independently one of the values $+1$, -1 or 0; remembering (23), we then readily get the relation:

$$\left\| \sum_{|\gamma|=m} \epsilon_\gamma x^\gamma/\gamma! - p(x) \right\| = \left\| \left(\sum_{i=1}^{n} \epsilon_i \xi_i \right)^m \middle/ m! - p(x) \right\|. \tag{42b}$$

By virtue of a general *invariance theorem* (see e.g. [17], p. 26, and [27], p. 58), in looking for the minimum of the expression (42b) over $\mathscr{P}_{m-1}$, we are actually allowed to replace $\mathscr{P}_{m-1}$ by the subspace of polynomials of degree $m-1$ in the single variable $\sum \epsilon_i \xi_i$; from elementary approximation theory, it follows that the expression (42b) is minimized if (and only if) p is such that

$$\left(\sum_{i=1}^{n} \epsilon_i \xi_i \right)^m \middle/ m! - p\left(\sum_{i=1}^{n} \epsilon_i \xi_i \right) \equiv \delta^m T_m \left(2 \sum_{i=1}^{n} \epsilon_i \xi_i/\delta \right) \middle/ 2^{2m-1} m!, \tag{43}$$

where $T_m(\cdot)$ denotes the Chebyshev polynomial (of the first kind) of degree m and δ denotes the distance, in the Hölder norm $\|\cdot\|_1$, between the two support hyperplanes of Ω which are orthogonal to the direction $(\epsilon_1, \ldots, \epsilon_n)$. It remains only to maximize over $\{\epsilon_i\}$ the corresponding value $\delta^m/2^{2m-1} m!$

of the expression (42b) to get the expected result, that is:

$$d_0 \geqslant [\delta_1(\Omega)]^m/2^{2m-1}m! \geqslant [\delta_2(\Omega)]^m/2^{2m-1}m!. \tag{44}$$

In so far as m is small, which is most often the case in practice, it may be said that the purpose of finding realistic estimates for d_0 has been satisfactorily achieved; indeed, the upper bound (33) (resp. (32)) and the lower bound (44), when they are expressed in terms of the diameter δ_1, are in the ratio $2^{m-1}:1$ (resp. $2^{2m-1}:1$) when the bounded region $\bar{\Omega}$ is convex (resp. star-shaped).

It should be finally noted that the appraisal (44) is sharp, in the sense that regions exist for which it cannot be improved. From the above analysis, this is clearly the case when $\bar{\Omega}$ is a closed line segment; this interesting result, due to J. Descloux (personal communication), can be proved directly by resorting only to elementary arguments. A more complicated example is the following; let $\bar{\Omega} \subset R^2$ denote the parallelogram whose vertices are the points $(\delta/2, -l, l)$, $(l, \delta/2 - l)$ and their images with respect to the origin; we assume that $0 \leqslant l \leqslant \delta/4$, so that δ is simply an abbreviated notation for the diameter $\delta_1(\Omega)$. Then, for $m = 2$, it can be proved fairly easily from (41) that the lower bound given by (44), which is here $\delta^2/16$, is attained for $\delta/8 \leqslant l \leqslant \delta/4$, whereas $d_0 = \delta^2/8$ in case $\bar{\Omega}$ is a regular polytope (i.e., for $l = 0$).

5. Two Examples of Application

A. Multivariate approximate integration with nonnegative weights

We consider here the class of rules of approximate integration over a given bounded subset $\bar{\Omega}$ of R^n, which is characterized by the property that the corresponding remainder:

$$Rf \equiv \int_{\bar{\Omega}} f(x) \, dx - \sum_i w_i f(x_i) \tag{45}$$

vanishes on the space $\mathscr{P}_{m-1}$ of polynomials of degree $\leqslant m - 1$, it being understood that:

the *abscissas* x_i should lie in $\bar{\Omega}$;   (46a)

the *weights* w_i should be nonnegative.   (46b)

As for the number of distinct abscissas, it need not be specified here any further. For several reasons, analyzed in detail in [9] (see p. 234), this class of rules is of great importance in numerical analysis; moreover, as it was shown first by Descloux, it can be considered as an excellent example of the straightforward application of the foregoing results, provided only that $\bar{\Omega}$ is star-shaped (or convex) and the function f is of class $C^m(\bar{\Omega})$, the Chebyshev norms (27a, b) being used throughout.

In view of the appraisals (32) and (33), the only problem with which we are apparently still faced, whether we refer to (6) or to (8), is to find a good upper bound for the specific quantity $\|R\|$. As a matter of fact, we are ultimately concerned rather with realistic upper bounds for $|Rf|$; it turns out that quite satisfactory results follow automatically, by (32) or (33), from the elementary inequality:

$$|Rf| \leqslant \int_\Omega |(f-p)(x)| \, dx + \max_i |(f-p)(x_i)| \int_\Omega dx, \qquad (47)$$

where $p \in \mathscr{P}_{m-1} \subset$ kernel R denotes any one of the polynomials of best uniform approximation of f over $\bar{\Omega}$. Whenever Ω is *convex*, we get in this way the general, non-trivial estimate:

$$|Rf| \leqslant \frac{\mu(\Omega)[\delta_1(\Omega)]^m}{2^{m-1}m!} |f|_m, \qquad (48)$$

where $\mu(\Omega)$ denotes the Lebesgue measure of the set $\bar{\Omega}$; as discussed at the end of Section 4, it happens that the factor 2^{m-1} in (48) may be replaced by 2^{2m-2}; this is the case in particular if $\Omega \equiv (a, b) \subset R^1$, so that we have, for any *one-dimensional integration* rule of the class under consideration, the sharper estimate of type (6):

$$|Rf| \leqslant \frac{(b-a)^{m+1}}{2^{2m-2}m!} \max_{a \leqslant x \leqslant b} |f^{(m)}(x)|, \qquad (49)$$

which was originally obtained by Descloux by an elementary, direct method.

In view of the extreme simplicity of this approach to error estimation, the question arises naturally whether the above appraisals may be accepted as realistic; in actual fact, it can be proved (see [18]), after a rather elaborate calculation of the corresponding error coefficient c_0, that the upper bound (49) and the smallest upper bound of type (2) for $|Rf|$ are in the asymptotic ratio $8/\pi : 1$ (for m large), which is surprisingly good.

B. Bivariate Lagrange interpolation over a triangulated domain

For brevity, we shall restrict ourselves here to the most classical case where an arbitrary function f of class C^2 over a given triangle $\bar{\Omega} \subset R^2$ is interpolated at the vertices $x_i = (\xi_{i,1}, \xi_{i,2})$ by a linear function; needless to say, the corresponding problem of pointwise error estimation arises naturally in connection with the interpolation over a triangulated domain by elements of what is sometimes referred to (in the terminology of the finite element method, see e.g. [31], p. 76) as *Courant's space*, i.e., the space of continuous, piecewise linear functions within each triangle.

From the well known definition:

$$Rf \equiv f(x) - \sum_{i=1}^{3} L_i(x)f(x_i) \quad \text{for} \quad x \in \bar{\Omega}, \qquad (50a)$$

where the $L_i(x)$ are simply the barycentric coordinates of $x = (\xi_1, \xi_2)$ with respect to the three vertices x_i and can be defined accordingly as

$$
L_i(x) = \det \begin{bmatrix} \xi_1 & \xi_2 & 1 \\ \xi_{j,1} & \xi_{j,2} & 1 \\ \xi_{k,1} & \xi_{k,2} & 1 \end{bmatrix} \Big/ \det \begin{bmatrix} \xi_{1,1} & \xi_{1,2} & 1 \\ \xi_{2,1} & \xi_{2,2} & 1 \\ \xi_{3,1} & \xi_{3,2} & 1 \end{bmatrix} \geqslant 0, \quad (50b)
$$

with $j \equiv i + 1 \pmod 3$ and $k \equiv i + 2 \pmod 3$, it follows that $\mathscr{P}_1 \subset \text{kernel } R$, so that we have the elementary inequality:

$$
|Rf| \leqslant |(f - p)(x)| + \max_i |(f - p)(x_i)|, \tag{51}
$$

where $p \in \mathscr{P}_1$ denotes, for example, any one of the polynomials of best uniform approximation of f over $\bar{\Omega}$. Regarding then Rf as an element of the space $X \equiv C^2(\bar{\Omega})$, we immediately get from (33) and (51), and for Chebyshev norms, the estimate:

$$
\|Rf\| \leqslant (\delta_1/2)^2 |f|_2 \leqslant (\delta_2/2^{1/2})^2 |f|_2, \tag{52}
$$

which could still be somewhat improved by taking advantage of the fact that the diameter $\delta_1(\Omega)$ is not unitarily invariant whereas the definition (50a) is not altered by orthogonal transformations; however, it should be realized that, in view of (44), the coefficient of $|f|_2$ in (52) can never be made smaller than $\delta_1^2/8$.

In case we are interested rather in estimating the error in the first partial derivatives f_l of the interpolated function, we have to replace at the outset the definition (50a) by

$$
R_l f \equiv f_l(x) - \sum_{i=1}^{3} [L_i(x)]_l f(x_i) \quad \text{for} \quad l = 1, 2; \tag{53a}
$$

as readily verified from (50b), by some elementary geometrical arguments, we have here the useful auxiliary inequality:

$$
\sum_{i=1}^{3} |[L_i(x)]_l| \leqslant 2/\rho, \tag{53b}
$$

where ρ denotes the Euclidean diameter of the inscribed sphere of $\bar{\Omega}$. To proceed further, we can rely typically on the elementary inequality:

$$
|R_l f| \leqslant |(f - p)_l(x)| + (2/\rho) \max_i |(f - p)(x_i)| \quad \text{for any} \quad p \in \mathscr{P}_1; \tag{54}
$$

by defining then p as the *unique polynomial* which is *such that $f - p$ can be represented in the integral form* (28), it is easily seen, by a trivial adaptation of the proof leading to (33), that $\|f - p\| \leqslant (\delta_1^2/8)|f|_2$ and that

$\|(f - p)_t\| \leqslant (\delta_1/2)|f|_2$, so that we finally get the estimate (of the type predicted by the general theory, see e.g. [6], p. 177):

$$\|R_l f\| \leqslant (\delta_1/2)(1 + \delta_1/2\rho)|f|_2 < [(2^{1/2} + 1)\delta_2^2/2\rho]|f|_2, \tag{55}$$

which could also be improved further (at the expense of simplicity, however).

It must be emphasized that the latter choice of p always proves convenient in concrete applications and can thus be adopted systematically.

References

[1] Barnhill, R. E. and Whiteman, J. R. (1973). Error analysis of finite element methods with triangles for elliptic boundary value problems. *The Mathematics of Finite Elements and Applications* (J. R. Whiteman ed.), pp. 83–112. Academic Press, Inc., New York.

[2] Bellman, R. (1960). *Introduction to Matrix Analysis.* McGraw-Hill Book Company, New York.

[3] Birkhoff, G., Schultz, M. H. and Varga, R. S. (1968). Piecewise Hermite interpolation in one and two variables with applications to partial differential equations, *Numer. Math.* 11, 232–256.

[4] Bramble, J. H. and Hilbert, S. R. (1970). Estimation of linear functionals on Sobolev spaces with application to Fourier transforms and spline interpolation, *SIAM J. Numer. Anal.* 7, 112–124.

[5] Bramble, J. H. and Zlámal, M. (1970). Triangular elements in the finite element method, *Math. Comp.* 24, 809–820.

[6] Ciarlet, P. G. and Raviart, P. A. (1972). General Lagrange and Hermite interpolation in R^n with applications to finite element methods, *Arch. Rat. Mech. Anal.* 46, 177–199.

[7] Collatz, L. (1964). *Funktionalanalysis und numerische Mathematik.* Springer-Verlag, Berlin, Göttingen, Heidelberg.

[8] Collatz, L. and Krabs, W. (1973). *Approximationstheorie.* B. G. Teubner, Stuttgart.

[9] Davis, P. J. (1969). Approximate integration rules with non-negative weights. *Lecture Series in Differential Equations, Vol. II* (A. K. Aziz ed.), pp. 233–256. Van Nostrand Reinhold Company, New York.

[10] Dieudonné, J. (1970). *Eléments d'Analyse (Tome III).* Gauthier-Villars Editeur, Paris.

[11] Golomb, M. (1962). *Lectures on Theory of Approximation.* Argonne National Laboratory.

[12] Jerome, J. W. (1973). Topics in multivariate approximation theory. *Approximation Theory* (G. G. Lorentz ed.), pp. 151–198. Academic Press, Inc., New York.

[13] Lorentz, G. G. (1966). *Approximation of Functions.* Holt, Rinehart and Winston, Inc., New York.

[14] Mac Duffee, C. C. (1946). *The Theory of Matrices.* Chelsea Publishing Company, New York.

[15] Mac Lane, S. and Birkhoff, G. (1967). *Algebra.* The Macmillan Company, New York.

[16] Marcus, M. and Minc, H. (1964). *A Survey of Matrix Theory and Matrix Inequalities.* Allyn and Bacon, Inc., Boston.

[17] Meinardus, G. (1967). *Approximation of Functions: Theory and Numerical Methods*. Springer-Verlag, Berlin, Heidelberg.

[18] Meinguet, J. (1975). Détermination du coefficient d'erreur associé à la classe des règles d'intégration sur un intervalle, de type lagrangien et à coefficients positifs. *Séminaires de mathématique appliquée*, Université de Louvain. To appear.

[19] Meinguet, J. and Descloux, J. An operator-theoretical approach to error estimation. To appear.

[20] Morrey, C. B. (1966). *Multiple Integrals in the Calculus of Variations*. Springer-Verlag, Inc., New York.

[21] Nečas, J. (1967). *Les Méthodes Directes en Théorie des Equations Elliptiques*. Masson et Cie, Editeurs, Paris.

[22] Raviart, P. A. (1972). *Méthode des éléments finis*. Laboratoire d'Analyse Numérique, Université de Paris VI.

[23] Rice, J. R. (1969). *The Approximation of Functions (Volume II)*. Addison-Wesley Publishing Company, Inc., London.

[24] Rivlin, T. J. and Shapiro, H. S. (1961). A unified approach to certain problems of approximation and minimization, *J. Soc. Indust. Appl. Math.* 9, 670–699.

[25] Sard, A. (1963). *Linear Approximation*. Providence, R. I., American Mathematical Society.

[26] Sard, A. (1965). Function spaces, *Bull. Amer. Math. Soc.* 71, 397–418.

[27] Shapiro, H. S. (1971). *Topics in Approximation Theory*. Springer-Verlag, Berlin, Heidelberg.

[28] Singer, I. (1970). *Best Approximation in Normed Linear Spaces by Elements of Linear Subspaces*. Springer-Verlag, Berlin, Heidelberg.

[29] Stiefel, E. (1959). Ueber diskrete und lineare Tschebyscheff-Approximationen, *Numer. Math.* 1, 1–28.

[30] Strang, G. (1972). Approximation in the finite element method, *Numer. Math.* 19, 81–98.

[31] Strang, G. and Fix, G. J. (1973). *An Analysis of the Finite Element Method*. Prentice-Hall, Inc., New Jersey.

[32] Stroud, A. H. (1971). *Approximate Calculation of Multiple Integrals*. Prentice-Hall, Inc., New Jersey.

[33] Varga, R. S. (1971). *Functional Analysis and Approximation Theory in Numerical Analysis*. Regional Conference Series in Applied Math. # 3, Philadelphia, Pa., Society for Industrial and Applied Mathematics, 76 pp.

[34] Whitney, H. (1934). Functions differentiable on the boundaries of regions, *Ann. of Math.* 35, 482–485.

[35] Zlámal, M. (1968). On the finite element method, *Numer. Math.* 12, 394–409.

Corrigendum

The "conjecture" (see p. 101, last line) according to which "the expression (39a) can be interpreted as giving automatically the required supremum of (38a)" proved false (counterexamples can indeed be constructed). As a matter of fact, (39a) gives only a lower bound for d_0 and, accordingly, the symbol "=" in (41) must be replaced by "$\geqslant$". This correction is quite unimportant, however, as far as the main result of Section 4 is concerned (i.e., the inequality (44)).

Matching of Essential Boundary Conditions in the Finite Element Method

A. R. Mitchell and J. A. Marshall†

1. Introduction

The exact matching of essential (Dirichlet) boundary conditions is one of the most difficult problems in the Finite Element Method (F.E.M.). The inability of piecewise polynomials to satisfy these conditions exactly, especially on curved boundaries, has led to a variety of approximate methods. Before dealing with curved boundaries, however, we shall examine the merit or otherwise of picking up all the boundary data on polygonal regions.

2. Blending Function Interpolants

We first consider a region bounded by straight sides parallel to the x and y axes divided up into rectangular elements. There are only three types of element: (1) completely internal, (2) one side on the boundary, (3) two sides on the boundary (corner elements). For convenience, every element will be transformed into the standard square element $[0, h] \times [0, h]$, denoted by S. The interpolants used in the above three cases on S are

$$u_1(x, y) = \left(1 - \frac{x}{h}\right)\left(1 - \frac{y}{h}\right) f(0, 0) + \left(1 - \frac{x}{h}\right)\frac{x}{h} f(0, h) + \frac{x}{h}\left(1 - \frac{y}{h}\right) \times$$

$$\times f(h, 0) + \frac{x}{h}\frac{y}{h} f(h, h), \tag{2.1}$$

$$u_2(x, y) = \left(1 - \frac{y}{h}\right) f(x, 0) + \left(1 - \frac{x}{h}\right)\frac{y}{h} f(0, h) + \frac{x}{h}\frac{y}{h} f(h, h), \tag{2.2}$$

$$u_3(x, y) = \left(1 - \frac{x}{h}\right) f(0, y) + \left(1 - \frac{y}{h}\right) f(x, 0) + \frac{x}{h}\frac{y}{h} f(h, h) -$$

$$- \left(1 - \frac{x}{h}\right)\left(1 - \frac{y}{h}\right) f(0, 0) \tag{2.3}$$

† Department of Mathematics, The University, Dundee, Scotland.

respectively. These are special cases of more general blended interpolants introduced by Gordon [1] and are represented in Fig. 1, where x and − represent points and lines respectively at which the function $f(x, y)$ is matched. From (2.1), (2.2), and (2.3), $U(x, y)$, an overall interpolant for the region, is obtained which matches exactly the boundary information on the perimeter of the region, for any value of the grid spacing h, and involves U_i, the values of $U(x, y)$ at the internal grid points, as parameters.

Fig. 1

Numerical solutions are now sought for the model boundary value problem consisting of

$$\frac{\partial^2 u}{\partial x^2} + \frac{\partial^2 u}{\partial y^2} = 0 \quad (x, y) \in \Omega \equiv (0, 1) \times (0, 1)$$

$$u = g \quad (x, y) \in \partial\Omega$$

(2.4)

where first a source is located just outside the region (Problem 1), and second the boundary conditions are periodic (Problem 2). These two problems were solved by the Galerkin version of the F.E.M. using an exact boundary interpolant and then a discretized boundary interpolant. Full details of these calculations are given in Marshall and Mitchell [2]. The maximum modulus solution on the 16 element grid is quoted in each case and compared with the theoretical solution.

Problem 1

No. of rectangular elements	Discretized	Exact
16	−2.5747	−2.5918
64	−2.5875	−2.5915
Theoretical Solution	−2.5913	

Problem 2

No. of elements	Discretized	Exact
16	0.3013	0.3383
64	0.3266	0.3353
Theoretical Solution	0.3345 ($\sin 4x \, e^{-4y}$ at $x = \frac{1}{2}$, $y = \frac{1}{4}$)	

As might have been expected, a solution of improved accuracy is obtained using the exact boundary interpolant. In both problems, the solution using the exact boundary interpolant with 16 elements is more accurate than the solution using the discretized boundary interpolant with 64 elements.

An analysis of the improved accuracy using the interpolant (2.2) instead of (2.1) for elements adjacent to the boundary $y = 0$ is now given. It is easily shown from (2.1) and (2.2) that if $e_i = f(x, y) - u_i(x, y)$, $i = 1, 2$, then

$$\frac{\partial e_2}{\partial x} = \frac{\partial e_1}{\partial x} + \left(1 - \frac{y}{h}\right)\left[\frac{1}{h}\left(f(h, 0) - f(0, 0)\right) - f_x(x, 0)\right].$$

We now square both sides and integrate over the element to obtain

$$\left\|\frac{\partial e_2}{\partial x}\right\|_{L_2}^2 - \left\|\frac{\partial e_1}{\partial x}\right\|_{L_2}^2 = \iint_S \left(1 - \frac{y}{h}\right)^2 \left[\frac{1}{h}\left(f(h, 0) - f(0, 0)\right) - \right.$$

$$\left. - f_x(x, 0)\right]^2 dS + 2 \iint_S \frac{\partial e_1}{\partial x}\left(1 - \frac{y}{h}\right) \times$$

$$\times \left[\frac{1}{h}\left(f(h, 0) - f(0, 0)\right) - f_x(x, 0)\right] dS.$$

If the right-hand side is expanded by Taylor series about the origin and only the principal terms retained, we get after integration

$$\left\|\frac{\partial e_2}{\partial x}\right\|_{L_2}^2 - \left\|\frac{\partial e_1}{\partial x}\right\|_{L_2}^2 = -\,[\tfrac{1}{18}h^4 f_{xx}^2 + \tfrac{1}{36}h^5 f_{xx}(2f_{xxx} + f_{xxy}) +$$

$$+ h^6 f_{xxx}(\tfrac{2}{135}f_{xxx} + \tfrac{1}{72}f_{xxy}) + \ldots], \tag{2.5}$$

where the derivatives of f are evaluated at $x = y = 0$. In a similar manner,

$$\left\|\frac{\partial e_2}{\partial y}\right\|_{L_2}^2 - \left\|\frac{\partial e_1}{\partial y}\right\|_{L_2}^2 = +[\tfrac{1}{120}h^4 f_{xx}^2 + \tfrac{1}{120}h^5 f_{xx}(f_{xxx} + 2f_{xxy}) +$$

$$+ h^6 f_{xxx}(\tfrac{2}{945}f_{xxx} + \tfrac{1}{120}f_{xxy}) + \ldots]. \tag{2.6}$$

If we now define

$$a(u, v) = \iint_S (u_x v_x + u_y v_y)\, dS,$$

then

$$a(e_1, e_1) = \left\|\frac{\partial e_1}{\partial x}\right\|_{L_2}^2 + \left\|\frac{\partial e_1}{\partial y}\right\|_{L_2}^2$$

$$= \left\|\frac{\partial e_2}{\partial x}\right\|_{L_2}^2 + \left\|\frac{\partial e_2}{\partial y}\right\|_{L_2}^2 + [\tfrac{17}{360}h^4 f_{xx}^2 + \tfrac{1}{360}h^5 f_{xx}(17f_{xxx} + 4f_{xxy}) +$$

$$+ h^6 f_{xxx}(\tfrac{4}{315}f_{xxx} + \tfrac{1}{180}f_{xxy}) + \ldots] \tag{2.7}$$

from (2.5) and (2.6), and so we obtain the result

$$a(e_1, e_1) > a(e_2, e_2),$$

provided the quantity in the square brackets in (2.7) is positive. This is certainly the case in the two examples given.

The square region Ω is now divided up into triangles according to Fig. 2 and each triangle is mapped by a simple linear transformation onto the standard triangle with vertices at the points $(h, 0)$, $(0, h)$, $(0, 0)$. This time

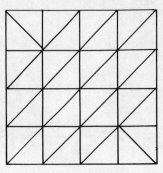

Fig. 2

there are only two types of element: (1) completely internal, (2) one side on the boundary, and these are illustrated in Fig. 3. The interpolants are

$$u_1(x, y) = \left(1 - \frac{x}{h} - \frac{y}{h}\right) f(0, 0) + \frac{x}{h} f(h, 0) + \frac{y}{h} f(0, h), \qquad (2.8)$$

$$u_2(x, y) = \left(1 - \frac{y}{h}\right) f\left(\frac{hx}{h - y}, 0\right) + \frac{y}{h} f(0, h) \qquad (2.9)$$

respectively. From (2.8) and (2.9), $U(x, y)$ an overall interpolant for the region in Fig. 2 is obtained which again matches exactly the boundary information on the perimeter of the region. Problem 2 only is solved using

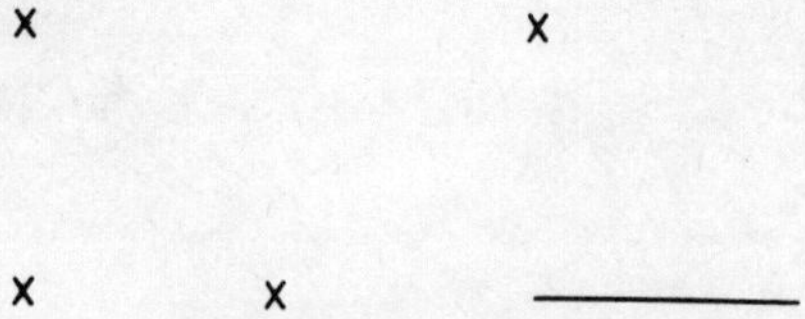

X   X

X   X   _____

Fig. 3

an exact boundary interpolant followed by a discretized boundary interpolant on triangular elements. The numerical results obtained for the solution at $x = 1/2$, $y = 1/4$ are

Problem 2

No. of triangular elements	Discretized	Exact
32	0.3634	0.3935
128	0.3421	0.3498
256	0.3379	0.3406
512	0.3364	0.3384
Theoretical Solution	0.3345	

This time we have the apparently surprising result that the discretized boundary interpolant gives more accurate results than the exact boundary interpolant.

The analysis required to explain the above numerical results for the triangular elements follows very closely that for the rectangular elements. With similar notation,

$$\frac{\partial e_2}{\partial x} = \frac{\partial e_1}{\partial x} + \left[\frac{1}{h} \left(f(h, 0) - f(0, 0) \right) - f_x \left(\frac{hx}{h - y}, 0 \right) \right]$$

which leads to

$$\left\| \frac{\partial e_2}{\partial x} \right\|_{L_2}^2 - \left\| \frac{\partial e_1}{\partial x} \right\|_{L_2}^2 = -\left[\tfrac{1}{72} h^4 f_{xx}^2 + \tfrac{1}{144} h^5 f_{xx}(f_{xxx} + 2f_{xxy}) \cdots \right] \qquad (2.10)$$

where we have shown only the principal terms of the Taylor expansion about the origin.

In a similar manner,

$$\left\| \frac{\partial e_2}{\partial y} \right\|_{L_2}^2 - \left\| \frac{\partial e_1}{\partial y} \right\|_{L_2}^2 = h^4 f_{xx}[\tfrac{1}{40} f_{xx} - \tfrac{1}{12} f_{xy} + \tfrac{1}{36} f_{yy}] +$$
$$+ h^5 [f_{xx}(\tfrac{1}{36} f_{xxx} - \tfrac{1}{40} f_{xxy} - \tfrac{1}{48} f_{xyy} + \tfrac{1}{72} f_{yyy}) +$$
$$+ f_{xxx}(-\tfrac{2}{45} f_{xy} + \tfrac{1}{72} f_{yy})] \qquad (2.11)$$

where the derivatives of f are again evaluated at the origin. From (2.10) and (2.11), we obtain the result

$$a(e_1, e_1) = a(e_2, e_2) - \tfrac{1}{6} h^4 f_{xx}[\tfrac{1}{15} f_{xx} - \tfrac{1}{2} f_{xy} + \tfrac{1}{6} f_{yy}] +$$
$$+ h^5 [f_{xx}(-\tfrac{1}{48} f_{xxx} + \tfrac{7}{180} f_{xxy} + \tfrac{1}{48} f_{xyy} - \tfrac{1}{72} f_{yyy}) +$$
$$+ f_{xxx}(\tfrac{2}{45} f_{xy} - \tfrac{1}{72} f_{yy})] \qquad (2.12)$$

and so as far as the principal term is concerned,

$$a(e_1, e_1) > a(e_2, e_2),$$

provided

$$f_{xx} < 0 \quad \text{and} \quad \tfrac{1}{15} f_{xx} - \tfrac{1}{2} f_{xy} + \tfrac{1}{6} f_{yy} > 0$$

or

$$f_{xx} > 0 \quad \text{and} \quad \tfrac{1}{15} f_{xx} - \tfrac{1}{2} f_{xy} + \tfrac{1}{6} f_{yy} < 0.$$

In Problem 2, $f_{xx}(0, 0) = f_{yy}(0, 0) = 0$, $f_{xy}(0, 0) = -16$, and $f_{xxx}(0, 0) = -64$, and so from (2.12)

$$a(e_1, e_1) = a(e_2, e_2) - \tfrac{2048}{45} h^5.$$

Hence

$$a(e_1, e_1) < a(e_2, e_2),$$

which substantiates the numerical results.

3. Transfinite Mappings

We now turn to regions with curved boundaries and examine the possibility of mapping a closed region in the (x, y) plane onto the square of side unity

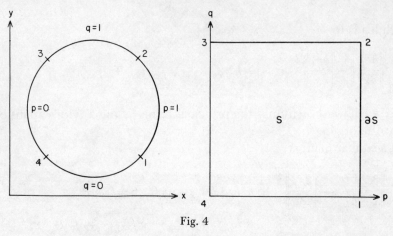

Fig. 4

in the (p, q) plane (Fig. 4). The required mapping is $T : S \to R$ where $S = [0, 1] \times [0, 1]$, and T is given by

$$T(p, q) = \frac{x(p, q)}{y(p, q)} \tag{3.1}$$

Let the four curved parts of ∂R be given by $F(1, q)$, $F(p, 1)$, $F(0, q)$ and $F(p, 0)$ respectively and so the corner points of ∂S correspond to the four points of ∂R with co-ordinates $F(1, 0)$, $F(1, 1)$, $F(0, 1)$, and $F(0, 0)$ respectively. Gordon and Hall [3] define a bilinearly blended transfinite map $T(p, q)$ given by

$$T(p, q) = (1 - p)F(0, q) + pF(1, q) + (1 - q)F(p, 0) + qF(p, 1) -$$
$$- (1 - p)(1 - q)F(0, 0) - (1 - p)qF(0, 1) -$$
$$- p(1 - q)F(1, 0) - pqF(1, 1) \tag{3.2}$$

where $T \equiv F$ for points (p, q) on ∂S. The mapping T has to be found and it is such that

$$J = \begin{vmatrix} \dfrac{\partial x}{\partial p} & \dfrac{\partial x}{\partial q} \\ \dfrac{\partial y}{\partial p} & \dfrac{\partial y}{\partial q} \end{vmatrix} \neq 0.$$

for all points in the region. Zienkiewicz and Phillips [4] use point transformations in place of (3.1) and so the original curved boundary is implicitly replaced by parabolic or cubic curves.

Transfinite mappings such as (3.2) can be used in conjunction with blend-
ing function interpolants to enable essential boundary conditions to be
matched exactly on curved boundaries.

4. Isoparametric Elements

So far the interpolants, whether of blended or discrete type, have been kept
entirely separate from the mappings. In isoparametric methods, however,
the same formula is used for the map as for the interpolant and so for a
curvilinear quadrilateral element with mapping formula (3.2), the bilinearly
blended interpolation formula is

$$f(p, q) = (1 - p)f(0, q) + pf(1, q) + (1 - q)f(p, 0) + qf(p, 1) -$$
$$-(1 - p)(1 - q)f(0, 0) - (1 - p(qf)0, 1) -$$
$$- p(1 - q)f(1, 0) - pqf(1, 1). \tag{4.1}$$

Since in most practical examples of isoparametric elements the mapping is
dictated by the interpolation formula it follows that the mapping may be
singular i.e. $J = 0$ along a curve inside the region.

In order to obtain practical interpolation formulae for use in the various
versions of the F.E.M. (Ritz, Galerkin, Least Squares, Collocation, etc.),
transfinite interpolants must be discretized in terms of a finite number of
scalar parameters. Typical examples of this are shown in Section 2 in the
form of the overall Lagrange interpolants of the rectangular region which are
labelled exact or discretized respectively. If the map follows from the interpolation formula, then it will also involve a finite number of parameters
which will be the co-ordinates of points in the region. Hence the map will
be essentially a point transformation. The points must be selected, particularly on the boundary of the region in the fully discretized case, in such a
way that the boundary curves implied by the point transformations of the
elements adjacent to the boundary constitute a close approximation to the

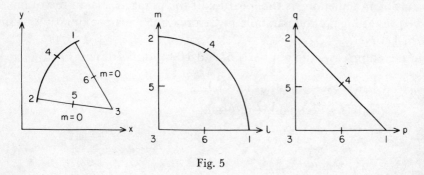

Fig. 5

given curved boundary. This will now be illustrated with respect to a tri-angular element with two straight sides and one curved side [5]. Such elements are commonplace in a triangulation of a finite region with a curved boundary. The quadratic and cubic cases are illustrated in Figs 5 and 6 respectively.

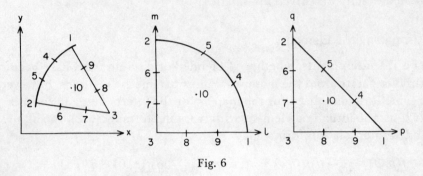

Fig. 6

In the quadratic case, if the point 4 has co-ordinates (X, Y), (L, M) and $(\frac{1}{2}, \frac{1}{2})$ respectively in the three parts of Fig. 5, then the transformation formulae are given by

$$l = p + 2(2L - 1)pq$$
$$m = q + 2(2M - 1)pq$$

(4.2)

and the implied curve passing through points 1, 4, and 2 is a parabola with equation

$$[(2M - 1)l - (2L - 1)m + (L - M + \alpha)]^2$$
$$= [2\alpha(M - 1) + 1 - L - M]l + [1 - L - M - 2\alpha(2L - 1)]m +$$
$$+ [(L - M + \alpha)^2 - (L + M - 4LM)]$$

(4.3)

where

$$\alpha = \frac{(1 - L - M)(L - M)}{(2L - 1)^2 + (2M - 1)^2}.$$

The problem remaining is the location of the point (L, M) on the original curved side to make the parabola (4.2) a reasonable approximation to the original curve.

In the cubic case, the points 4, 5, and 10 have co-ordinates

$$(X_4, Y_4), (X_5, Y_5), \text{ and } (X_{10}, Y_{10})$$
$$(L_4, M_4), (L_5, M_5), \text{ and } (L_{10}, M_{10})$$

and

$$(\tfrac{2}{3}, \tfrac{1}{3}), (\tfrac{1}{3}, \tfrac{2}{3}) \text{ and } (\tfrac{1}{3}, \tfrac{1}{3})$$

respectively. This time the transformation formulae are given by

$$l = p + \tfrac{9}{2}(6L_{10} - L_4 - L_5 - 1)pq + \tfrac{27}{2}(L_4 - 2L_{10})p^2q +$$
$$\quad + \tfrac{27}{2}(L_5 - 2L_{10} + \tfrac{1}{3})pq^2$$
$$m = q + \tfrac{9}{2}(6M_{10} - M_4 - M_5 - 1)pq + \tfrac{27}{2}(M_4 - 2M_{10} + \tfrac{1}{3})p^2q +$$
$$\quad + \tfrac{27}{2}(M_5 - 2M_{10})pq^2$$

$$(4.4)$$

and the implied curve passing through points 1, 4, 5, and 2 is a cubic curve. If we choose

$$L_4 = L_5 + \tfrac{1}{3}$$
$$M_5 = M_4 + \tfrac{1}{3}$$

the cubic curve degenerates into a *unique* parabola through the four points $(1, 0)$, (L_4, M_4), (L_5, M_5), and $(0, 1)$, and if in addition

$$L_4 = 2L_{10}$$
$$M_5 = 2M_{10},$$

the transformation formulae (4.4) reduce to

$$l = p + 9(L_{10} - \tfrac{1}{3})pq$$
$$m = q + 9(M_{10} - \tfrac{1}{3})pq$$

and the equation of the parabola is given by (4.3) where this time

$$4L = 9L_{10} - 1$$
$$4M = 9M_{10} - 1.$$

In [5], examples of arbitrary curved sides were chosen and matching parabolic arcs obtained. In all examples, a parabola was found which lay close to the original curve. Particularly in the cubic case, the isoparametric cubic curve was often a poor approximation to the original curve. Numerical results in [6] underline the fact that *isoparametric elements are extremely sensitive to distortion from the basic triangular shape*.

5. Direct Methods

In Sections 3 and 4, we have considered transformation methods for dealing with curved boundaries. Such methods suffer from two major disadvantages; (1) the Jacobian of the transformation may vanish inside the region, and (2) the boundary curves implied by the point transformation may constitute a poor approximation to the original boundary. We now look at direct methods of constructing basis functions for regions with curved boundaries, and in particular examine a triangular element with two straight sides and one curved side in the physical plane. We shall investigate basis functions which give (i) linear and (ii) higher order approximation in the triangular element.

(i) Linear approximation

A linear form has three independent parameters, and so is uniquely determined at any three non-collinear points of a curve of order higher than one. Hence four points, as illustrated in Fig. 7, where l and m are the normalized linear forms of the straight sides in the triangular element, are required at which to locate suitable basis functions for linear approximation. Also

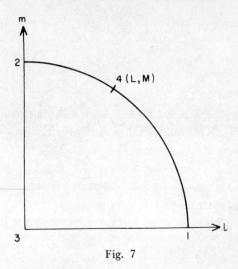

Fig. 7

quadratic arcs are usually sufficiently versatile to adequately represent material interfaces or region boundaries, and so we shall consider the curved side in Fig. 7 to be the general conic

$$f(l, m) \equiv al^2 + blm + cm^2 - (a + 1)l - (c + 1)m + 1 = 0.$$

It is shown in McLeod and Mitchell [7, 8] that basis functions W_i $(i = 1, 2, 3, 4)$ suitable for linear approximation in the triangle with 2 straight sides and one curved side are given by

$$\alpha W_3^2 + [\alpha(l + m - 1) + (al + cm - 1)]W_3 + f(l, m) = 0 \qquad (5.1)$$

together with

$$
\begin{aligned}
W_1 + W_2 + W_4 &= 1 - W_3 \\
W_1 + LW_4 \quad\;\; &= 1 \\
W_2 + MW_4 \quad\;\; &= 1,
\end{aligned}
\qquad (5.2)
$$

where α is an arbitrary parameter, and (L, M) is a node on the curved side. When $\alpha = 0$, from (5.1) and (5.2) we recover the rational basis functions of Wachspress [10]. If in addition, $a = c = 0$, then

$$W_3 = blm - l - m + 1,$$

where $b = (L + M - 1)/LM$. Hence the basis functions are polynomials, and the curve through the points 1, 4, and 2 is the hyperbola

$$blm - l - m = 0. \qquad (5.3)$$

Piecewise hyperbolic arcs, based on (5.3), can be used to approximate a curved interface or boundary, and still permit polynomial basis functions. When $\alpha \neq 0$, from (5.1),

$$W_3 = -\frac{1}{2\alpha}[\alpha(l + m - 1) + (al + cm - 1) + [\{\alpha(l + m - 1) -$$
$$- (al + cm - 1)\}^2 + 4\alpha(a + c - b)lm]^{1/2}],$$

which reduces to $1 - m$ when $l = 0$, and to $1 - l$ when $m = 0$, and is zero on the general conic.

(ii) Higher order approximation

A quadratic form has six independent parameters, and so is uniquely determined at any six points of a curve of order higher than two. Three points are required on a line to determine a quadratic form and five points on a conic. Hence eight points, as illustrated in Fig. 8a are required at which to locate suitable basis functions for quadratic approximation in a triangular element with two straight sides and one curved side, the latter being the

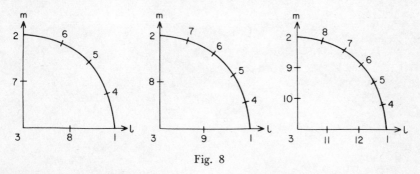

Fig. 8

general conic. If the curved side is a cubic curve, then nine points are required as illustrated in Fig. 8b. Finally for cubic approximation in a triangle with two straight sides and a general conic as a third side, twelve points are required as shown in Fig. 8c. The basis of functions for higher order approximation in curved triangular elements are given in a paper under preparation by McLeod [9].

References

[1] Gordon, W. J. (1971). Blending-function methods of bivariate and multivariate interpolation and approximation, *SIAM Numer. Anal.* **8**, 158–177.

[2] Marshall, J. A. and Mitchell, A. R. (1973). An exact boundary technique for improved accuracy in the finite element method, *JIMA* **12**, 355–362.

[3] Gordon, W. J. and Hall, C. A. (1973). Transfinite element methods: blending-function interpolation over arbitrary curved element domains, *Numer. Math.* **21**, 109–129.

[4] Zienkiewicz, O. and Phillips, D. (1971). An automatic mesh generation scheme for plane and curved surfaces by isoparmetric co-ordinates, *Int. J. Num. Meth. Eng.* **3**, 519–528.

[5] McLeod, R. and Mitchell, A. R. The use of parabolic arcs in matching curved boundaries in the finite element method. *JIMA* (to appear).

[6] Bond, T. J., Swannell, J. H., Henshell, R. D. and Warburton, G. B. (1973). A comparison of some curved two-dimensional finite elements, *Journal of Strain Analysis* **8**, 182–190.

[7] McLeod, R. and Mitchell, A. R. (1972). The construction of basis functions for curved elements in the finite element method, *JIMA* **10**, 382–393.

[8] Mitchell, A. R. and McLeod, R. (1973). Curved elements in the finite element method, *Lecture Notes in Mathematics No. 363*, Springer Verlag, 89–104.

[9] McLeod, R. Higher order approximation in curved elements (in preparation).

[10] Wachspress, E. L. (1973). A rational basis for function approximation II curved sides, *JIMA* **11**, 83–104.

Collocation, Difference Equations, and Stitched Function Representations

M. R. Osborne

Abstract

A range of difference approximations to a linear mth order differential equation is considered. These approximations are obtained by collocation methods using a class of functions which includes piecewise polynomials as a special case, and emphasis is given to choices of collocation points which permit favourable error estimates. For collocation using piecewise polynomials results of De Boor and Swartz are recovered.

1. Introduction

In this paper a class of difference schemes for approximating the solutions of ordinary differential equations is considered. In setting up these difference schemes the main tool is the construction of local approximations by collocation methods. These local approximations include piecewise polynomials as a special case.

To specify the problem more precisely let the differential equation be

$$L(y) = y^{(m)} + \sum_{i=1}^{m} a_i(t) y^{(m-i)} = f(t) \tag{1.1}$$

subject to the boundary conditions

$$B_1 \boldsymbol{\sigma}_m(y, 0) + B_2 \boldsymbol{\sigma}_m(y, 1) = \mathbf{c} \tag{1.2}$$

where we make use of the notation

$$\boldsymbol{\sigma}_\beta(y, t) = \{y(t), y^{(\alpha)}(t), \ldots, y^{((\beta-1)\alpha)}(t)\}^T, \tag{1.3}$$

and where α, β are related by

$$\alpha\beta = m. \tag{1.4}$$

121

The difference equations are defined on the partition $0 \leqslant t_1 < t_2 < \ldots$
$\ldots < t_n \leqslant 1$, and we set $\Delta = \max_i t_{i+1} - t_i$. We assume that the problem
(1.1), (1.2) has a unique solution, and that the coefficients in (1.1) are
sufficiently smooth to ensure that the possible solutions meet any smoothness requirements necessary.

The basic plan of the paper is as follows. In Section 2 we consider collocation by piecewise polynomials in $C^{m-1}(0, 1)$. The principle interest is in
the selection of collocation points to minimize the error in the difference
approximation, and we derive an expression for the error at the mesh points
first given by De Boor and Swartz [1]. In Section 3 we consider mth order
difference equations and the associated collolation schemes. In principle it
is possible to find estimates strictly comparable to those of Section 2, but
it is unlikely that these can be achieved in practice. However, there is some
scope for using approximations to the differential equation. In Section 4 a
class of intermediate problems is considered. In this case we are unable to
give such satisfactory results. However, these results must be regarded as
preliminary.

There has been quite a deal of recent interest in the numerical solution
of differential equations by collocation using piecewise polynomials, and
the best results have been obtained by De Boor and Swartz [1]. Gladwell
and Mullins [2] have considered piecewise polynomials in $C^{p-1}(0, 1)$ with
$p \geqslant m$. The use of collocation to generate higher order difference approximations has been considered by Osborne [3].

2. Collocation by Piecewise Smooth Functions

In this section we consider the construction by collocation methods of
approximations to the solution of (1.1) from the set of functions
$Z_{k,m} \cap C^{(m-1)}(0, 1)$ where

$$Z_{k,m} = \left\{ z; z = z_i = \sum_{j=1}^{m+k} \gamma_j^{(i)} \phi_j(t), \ t_i < t < t_{i+1}, \quad i = 1, 2, \ldots, n-1 \right\},$$

and where the $\phi_j(t)$ are assumed to satisfy appropriate smoothness conditions. The conditions which determine z are

(i) $\sigma_m(z_{i-1}, t_i) = \sigma_m(z_i, t_i), \quad i = 2, 3, \ldots, n,$ (2.1)

and

(ii) $L(z_i)(\xi_{ij}) = f(\xi_{ij}), \quad i = 1, 2, \ldots, n-1, j = 1, 2, \ldots, k,$ (2.2)

where the ξ_{ij} are the collocation points and satisfy

$$t_i \leqslant \xi_{i1} < \xi_{i2} < \ldots < \xi_{ik} \leqslant t_{i+1}.$$

We assume that z_i is uniquely determined by these conditions (this is true, for example, if $\phi_j(t) = t^{j-1}$ and Δ is small enough). For z to be in $C^{(m-1)}(0, 1)$ we must have

$$\sigma_m(z_i, t_{i+1}) = \sigma_m(z_{i+1}, t_{i+1}). \tag{2.3}$$

This is a relation between $\sigma_m(z_i, t_i)$ and $\sigma_m(z_{i+1}, t_{i+1})$ and is the desired difference equation. Note that this difference relation is uniquely defined by (2.3). This is a consequence of the assumption that z_i is uniquely determined by (2.1), (2.2).

Example
For the case

$$L = \frac{d}{dt} + a_1,$$

let $\xi_{i1} = t_i$, $\xi_{i2} = t_{i+1}$. We have $m = 1$, and $\sigma(z, t) = z(t)$. Choosing $\phi_j = (t - t_i)^{j-1}$, $j = 1, 2, 3$, we obtain the conditions

$$\gamma_1 = z(t_i),$$
$$\gamma_2 + a_1(t_i)z(t_i) = f(t_i),$$

and

$$\gamma_2 + 2\gamma_3(t_{i+1} - t_i) + a_i(t_{i+1})(\gamma_1 + \gamma_2(t_{i+1} - t_i) + \gamma_3(t_{i+1} - t_i)^2) = f(t_{i+1}).$$

Solving for $\gamma_1, \gamma_2, \gamma_3$ and substituting in

$$z(t_{i+1}) = \gamma_1 + \gamma_2(t_{i+1} - t_i) + \gamma_3(t_{i+1} - t_i)^2$$

gives the difference equation

$$\frac{z(t_{i+1}) - z(t_i)}{t_{i+1} - t_i} = \tfrac{1}{2}\{a_1(t_{i+1})z(t_{i+1}) + a_1(t_i)z(t_i) + f(t_i) + f(t_{i+1})\}.$$

This is the well-known trapezoidal rule formula.

Now consider the first order system equivalent to equation (1.1). We write this as

$$\frac{d}{dt}\sigma_m(y, t) = A(t)\sigma_m(y, t) + f(t)\,e_m, \tag{2.4}$$

where

$$A(t) = \begin{bmatrix} 1 & & & \\ & \ddots & & \\ & & \ddots & \\ & & & 1 \\ -a_m & \cdots & & -a_1 \end{bmatrix}.$$

In particular, on the interval (t_i, t_{i+1}) we have by the usual variation of parameters approach

$$\boldsymbol{\sigma}_m(y, t) = X_i(t)\boldsymbol{\sigma}_m(y, t_i) + X_i(t) \int_{t_i}^{t} f(u)X_i^{-1}(u)\, e_m\, du \tag{2.5}$$

where X_i, the fundamental matrix for (2.4), satisfies

$$\frac{dX_i}{dt} = A(t)X_i, \quad X_i(t_i) = I, \quad i = 1, 2, \ldots, n-1. \tag{2.6}$$

Equation (2.5) can be used to write (2.3) in an equivalent form. Let z_i satisfy

$$L(z_i) = f + r_i \tag{2.7}$$

then, by (2.2),

$$r_i(\xi_{ij}) = 0, \quad j = 1, 2, \ldots, k. \tag{2.8}$$

Substituting (2.7) into (2.5) gives the difference relation

$$\boldsymbol{\sigma}_m(z, t_{i+1}) = X_i(t_{i+1})\boldsymbol{\sigma}_m(z, t_i) + X_i(t_{i+1}) \int_{t_i}^{t_{i+1}} (f + r_i)X_i^{-1}\, e_m\, du \tag{2.9}$$

which is satisfied exactly by $\boldsymbol{\sigma}_m(z, t_j)$ for $j = 1, 2, \ldots, n$. The uniqueness of (2.3) ensures that (2.3) and (2.9) are identical. Subtracting (2.9) from (2.5) gives

$$\boldsymbol{\sigma}_m(y - z, t_{i+1}) = X_i(t_{i+1})\boldsymbol{\sigma}_m(y - z, t_i) - X_i(t_{i+1}) \int_{t_i}^{t_{i+1}} r_i X_i^{-1}\, e_m\, du$$

$$i = 1, 2, \ldots, n-1, \tag{2.10}$$

so that, using (2.10) and (1.2), the error in the approximation of y by z at the grid points $t_j, j = 1, 2, \ldots, n$, satisfies

$$\begin{bmatrix} B_1 & & & & B_2 \\ -X_1(t_2) & I & & & \\ & -X_2(t_3) & I & & \\ & \cdots\cdots\cdots\cdots\cdots\cdots\cdots\cdots & & \\ & & & -X_{n-1}(t_n) & I \end{bmatrix} \begin{bmatrix} \boldsymbol{\sigma}_m(y - z, t_1) \\ \cdot \\ \cdot \\ \cdot \\ \boldsymbol{\sigma}_m(y - z, t_n) \end{bmatrix} = \begin{bmatrix} 0 \\ s_1 \\ \cdot \\ \cdot \\ s_{n-1} \end{bmatrix} \tag{2.11}$$

where

$$s_i = -X_i(t_{i+1}) \int_{t_i}^{t_{i+1}} r_i X_i^{-1}\, e_m\, du. \tag{2.12}$$

Lemma 2.1
Provided $g(u)$ is sufficiently smooth, and the $\xi_{ij}, j = 1, 2, \ldots, k$ are chosen

as the zeros of the Legendre polynomial of degree k shifted to the interval $[t_i, t_{i+1}]$, then

$$\int_{t_i}^{t_{i+1}} r_i(u)g(u) \, du = 0(\Delta^{2k+1}). \tag{2.13}$$

Proof

As $r_i(\xi_{ij}) = 0, j = 1, 2, \ldots, k$, we have

$$r_i(u) = \prod_{j=1}^{k} (u - \xi_{ij})r_i[\xi_{i1}, \ldots, \xi_{ik}, u],$$

where $r_i[\xi_{i1}, \ldots, \xi_{ik}, u]$ is the divided difference of r_i on the points indicated. Thus

$$r_i g = \prod_{j=1}^{k} (u - \xi_{ij})r[\xi_{i1}, \ldots, \xi_{ik}, u]g$$

$$= \prod_{j=1}^{k} (u - \xi_{ij})h_i$$

$$= \prod_{j=1}^{k} (u - \xi_{ij}) \left\{ q_{k-1} + \prod_{j=1}^{k} (u - \eta_{ij})h_i[\eta_{i1}, \ldots, \eta_{ik}, u] \right\},$$

where q_{k-1} is a polynomial interpolating h_i at selected points $\eta_{ij} \, j = 1, 2, \ldots, k$, in the interval (t_i, t_{i+1}). Thus

$$\int_{t_i}^{t_{i+1}} r_i(u)g(u) \, du = \int_{t_i}^{t_{i+1}} \prod_{j=1}^{k} (u - \xi_{ij}) \prod_{j=1}^{k} (u - \eta_{ij})h_i[\eta_{i1}, \ldots, \eta_{ij}, u] \, du$$

from which the desired result follows. The argument is given in detail in [1].

We can now deduce an error estimate given in [1]. For convenience we assume the maximum norm.

Theorem 2.1

$$\max_i \| \boldsymbol{\sigma}_m(y - z, t_i) \| = 0(\Delta^{2k}). \tag{2.14}$$

Proof

It is known that the norm of the inverse of the matrix of the system of equations (2.11) is $0(\Delta^{-1})$ (Osborne [4]). Thus the result is an immediate consequence of Lemma 2.1.

Remark 1

The novelty in the above approach lies in connecting the collocation problem with the problem of setting up a difference scheme for the related first order system (2.4). This procedure works also in the case in which $z \in Z_{k,p} \cap C^{(p-1)}(0, 1)$ with $p > m$. However, here the first order system is that associated with $D^{p-m}L$ so that s_i takes the form

$$s_i = X_i(t_{i+1}) \int_{t_i}^{t_{i+1}} D^{p-m} r_i X_i^{-1} e_p \, du. \qquad (2.15)$$

By Rolles theorem $D^{p-m}r_i$ vanishes $k + m - p$ times on (t_i, t_{i+1}) so that the collocation points should be chosen so that these zeros coincide with those of the corresponding shifted Legendre polynomial. Assuming this is possible then the resulting error estimate is

$$\max_i \| \sigma_p(y - z, t_i) \| = 0(\Delta^{2k+m-p}) \qquad (2.16)$$

which is less favourable than in the case $m = p$. However, the estimate does yield uniform estimates also for higher derivatives. The main objection to this procedure is on the grounds of practicability as the choice of the ξ_{ij} appears to be problem dependent. We note that extra boundary conditions are needed to specify the problem in this case. This point is discussed in [2].

Remark 2

To compute the difference approximation on each subinterval (t_i, t_{i+1}) we can proceed directly to determine the conditions for the consistency of the linear equations which ensure

    (a) the continuity conditions at t_i,
    (b) the satisfaction of the differential equation at the collocation points,

and

    (c) the continuity conditions at t_{i+1}.

We summarize these conditions in the form

    (a) $\Phi_1^{(i)}\gamma^{(i)} = \sigma_m(z, t_i)$ where $(\Phi_1^{(i)})_{st} = \phi^{(s-1)}(t_i)$,
    (b) $\Phi_2^{(i)}\gamma^{(i)} = f_i$ where $(\Phi_2^{(i)})_{st} = L(\phi_t)(\xi_{is})$, $(f_i)_s = f(\xi_{is})$,

and

    (c) $\Phi_1^{(i+1)}\gamma^{(i)} = \sigma_m(z, t_{i+1})$.

Consider now a factorization having the form

$$\begin{bmatrix} Q_1 \\ \hline Q_2 \end{bmatrix} \begin{bmatrix} \Phi_1^{(i)} \\ \Phi_2^{(i)} \\ \hline \Phi_1^{(i+1)} \end{bmatrix} = \begin{bmatrix} U \\ \hline 0 \end{bmatrix}$$

This can be found by standard methods (for example, in the case that Q is orthogonal and U upper triangular). The conditions for the consistency of the linear systems is

$$0 = Q_2 \begin{bmatrix} \boldsymbol{\sigma}_m(z, t_i) \\ \mathbf{f}_i \\ \boldsymbol{\sigma}_m(z, t_{i+1}) \end{bmatrix}.$$

This gives the desired difference relation.

Remark 3
It is instructive to compare the cost of the collocation procedure with that of competing methods (here we consider multiple shooting and single step finite differences as in [4]). The cost, considered in terms of number of multiplications, consists of two main components: set up cost for the linear system, and solution of the linear system taking account of its structure. Order of magnitude figures are:

   (i) set up cost
multiple shooting	$K_1 m^2 n$	K_1 *large* $(10\text{-}100)$
finite differences	$K_2 mn$	K_2 *modest* $(1\text{-}10)$
collocation	$K_3(m + 2k)(m + k)^2 n$	K_3 *small* $(.1\text{-}1)$

  (ii) solution cost $K_4 m^3 n$ K_4 *small* $(.1\text{-}1)$

It is not clear that collocation is competitive with the other methods in terms of efficiency.

3. An *m*th Order Difference Approximation

In the previous section we considered collocation by the piecewise smooth function z which had $m + k$ degrees of freedom on each subinterval (t_i, t_{i+1}). The characteristic feature of our approach was the fixing of m of these degrees of freedom by using "Taylor series" data, and the resulting procedure turned out to be equivalent to approximating to the associated first order system (equation (2.4)). This suggests the question of what happens if we use other kinds of data to fix these degrees of freedom. In this section we explore the use of interpolation data for this purpose. That is, we ask that successive representations satisfy the matching conditions

$$z_i(t_j) = z_{i-1}(t_j), \quad j = i, i + 1, \ldots, i + m - 1. \tag{3.1}$$

In this case the collocation conditions are

$$L(z_i)(\xi_{ij}) = f(\xi_{ij}), \quad j = 1, 2, \ldots, k, \tag{3.2}$$

where the collocation points are now required to satisfy

$$t_i \leqslant \xi_{i1} < \xi_{i2} < \ldots < \xi_{ik} \leqslant t_{i+m}.$$

We assume this problem has a unique solution. We have $z_i(t_{i+m})$ given as a linear combination of the data on the collocation problem, and this relation gives the required difference approximation. Schemes of this kind are considered in [3].

Remark 1
In this case the successive z_i interpolate each other at the common mesh points. Thus the piecewise representations can be thought of as being stitched together.

Example
For the case

$$L = \frac{d^2}{dt^2} + a_2,$$

let $\xi_{i1} = t_i$ and $\phi_j = (t - t_i)^{j-1}$, $j = 1, 2, 3$. The conditions determining the representation are

$$z(t_{i-1}) = \gamma_1 - \gamma_2(t_i - t_{i-1}) + \gamma_3(t_i - t_{i-1})^2$$
$$z(t_i) = \gamma_1$$
$$z(t_{i+1}) = \gamma_1 + \gamma_2(t_{i+1} - t_i) + \gamma_3(t_{i+1} - t_i)^2,$$

and

$$2\gamma_3 + a_2(t_i)\gamma_1 = f(t_i).$$

Combining the first three equations gives

$$\frac{z(t_i) - z(t_{i-1})}{t_i - t_{i-1}} + \gamma_3(t_i - t_{i-1}) = \gamma_2$$

and

$$\frac{z(t_{i+1}) - z(t_i)}{t_{i+1} - t_i} - \gamma_3(t_{i+1} - t_i) = \gamma_2.$$

Eliminating γ_2 between these equations and combining with the fourth of the original equations gives

$$\frac{2}{t_{i+1} - t_{i-1}} \left\{ \frac{z(t_{i+1}) - z(t_i)}{t_{i+1} - t_i} - \frac{z(t_i) - z(t_{i-1})}{t_i - t_{i-1}} \right\} + a_2(t_i)z(t_i) = f(t_i).$$

This equation will be recognized as the familiar divided difference approximation to the differential equation.

To analyze this class of difference approximations we make use of the distribution with compact support defined by

(i) $T_i = 0, t < t_i, t > t_{i+m}$,

(ii) $T_i, T_i^{(1)}, \ldots, T_i^{(m-2)}$ continuous at $t = t_j, j = i, i+1, \ldots, i+m$,

(iii) $L(T_i) = 0, t \neq t_j, j = i, i+1, \ldots, i+m$,

and

(iv) $\displaystyle\int_{-\infty}^{\infty} T_i \, dt = 1.$

In general these conditions define T_i uniquely. Integrating by parts m times we obtain

$$\int_{-\infty}^{\infty} T_i L(y) \, dt = \int_{-\infty}^{\infty} y L(T_i) \, dt = \sum_{j=i}^{i+m} \mu_j y(t_j) \tag{3.3}$$

as $L(T_i)$ gives δ functions at the mesh points and vanishes elsewhere. Setting

$$L(z_i) = f + r_i \tag{3.4}$$

where

$$r_i(\xi_{ij}) = 0, \quad j = 1, 2, \ldots, k \tag{3.5}$$

we obtain a representation of the difference approximation in the form

$$\sum_{j=i}^{i+m} \mu_j z_i(t_j) = \int_{-\infty}^{\infty} T_i(f + r_i) \, dt. \tag{3.6}$$

Subtracting (3.6) from (3.3) gives

$$\sum_{j=i}^{i+m} \mu_j(y(t_j) - z_i(t_j)) = - \int_{-\infty}^{\infty} T_i r_i \, dt. \tag{3.7}$$

The right-hand side of (3.7) is small like $0(\Delta^k)$ at least. However, it can be made smaller if $T_i > 0$ on (t_i, t_{i+m}) by choosing the collocation points to be at the zeros of the orthogonal polynomial of degree k with respect to T_i as weight function. In this case the argument used in the previous section shows that the right-hand side is $0(\Delta^{2k})$, but this is to be considered an upper limit rather than an estimate attainable in practise because the determination of these zeros requires explicit knowledge of T_i which is as difficult to obtain as the solution to the original problem.

Remark 2
Two possibilities which could be considered are

(a) *to use a Gauss–Legendre scheme of sufficiently low order that the smoothness of T_i is not a problem,* *and*

(b) *to have a fixed number of collocation points in each interval* (t_j, t_{j+1}) *and use a Gauss–Legendre scheme for each one.*

Neither of these approaches gives a significantly improved estimate.

Now consider the splitting

$$L = L_1 + L_2 \tag{3.8}$$

where we assume

(i) that L_1 is readily invertible,

and

(ii) that L_2 can be considered as a perturbation term in some sense.

We define $T_i^{(1)}$ using L_1, and this gives the result analogous to (3.3)

$$\sum_{j=i}^{i+m} \mu_j^{(1)} y(t_j) = \int_{-\infty}^{\infty} T_i^{(1)} L_1(y) \, dt$$

$$= \int_{-\infty}^{\infty} T_i^{(1)}(f - L_2(y)) \, dt. \tag{3.9}$$

In terms of this splitting the difference scheme becomes

$$\sum_{j=i}^{i+m} \mu_j^{(1)} z_i(t_j) = \int_{-\infty}^{\infty} T_i^{(1)}(f + r_i - L_2(z_i)) \, dt \tag{3.10}$$

and the error relation is

$$\sum_{j=i}^{i+m} \mu_j^{(1)}(y(t_j) - z_i(t_j)) = - \int_{-\infty}^{\infty} T_i^{(1)}(r_i + L_2(y - z_i)) \, dt. \tag{3.11}$$

If $T_i^{(1)} > 0$ on (t_i, t_{i+m}) then choosing the Gaussian points with respect to $T_i^{(1)}$ as weight function as collocation points makes the first term on the right-hand side $0(\Delta^{2k})$. The contribution of the second term will be smaller than that due to the interpolation error alone essentially as the coefficients in L_2 are small. Examples making use of this device are given in [3]. The device of approximating to the differential equations has also been considered by Pruess [5].

4. A Class of Intermediate Schemes

In the previous two sections we have considered the cases

(i) where the difference equation relates values of y at $m + 1$ consecutive mesh points (the case of direct difference approximation to (1.1)),

and

(ii) where the difference equation relates $\boldsymbol{\sigma}_m(y, t)$ at adjacent mesh points (the case of difference approximation to the associated first order system).

Intermediate between these problems are a large number of possible approaches, and a subclass of these is considered here. We assume that (1.1) can be transformed by introducing suitable variables into the associated lower order system

$$\frac{d^\alpha}{dt^\alpha} \, \boldsymbol{\sigma}_\beta(y, t) = A_\beta \boldsymbol{\sigma}_\beta(y, t) + f \, e_\beta, \tag{4.1}$$

where $1 < \alpha, \beta < m$.

Example

For

$$L = \frac{i}{\alpha R} \, (D^2 - k^2)^2 + (u - c)(D^2 - k^2) + \frac{d^2 u}{dt^2},$$

consider the transformation

$$(D^2 - k^2)y = w,$$

$$\frac{i}{\alpha R} \, (D^2 - k^2)w + (u - c)w + \frac{d^2 u}{dt^2} \, y = 0.$$

We also assume that an analogue of (2.9) and (3.3) exists having the form

$$\sum_{j=i}^{i+\alpha} Y_j \boldsymbol{\sigma}_\beta(y, t_j) = \int_{-\infty}^{\infty} T_i L(y) \, dt, \tag{4.2}$$

where T_i is an appropriate vector of distributions, and the Y_j are $\beta \times \beta$ matrices.

Consider now the collocation problem

(i) $\boldsymbol{\sigma}_\beta(z_i, t_j) = \boldsymbol{\sigma}_\beta(z_{i-1}, t_j), \quad j = i, i+1, \ldots, i+\alpha-1,$ \hfill (4.3)

and

(ii) $L(z_i)(\xi_{ij}) = f(\xi_{ij}), \quad j = 1, 2, \ldots, k,$ \hfill (4.4)

where

$$t_i \leqslant \xi_{i1} < \xi_{i2} < \ldots < \xi_{ik} \leqslant t_{i+\alpha}.$$

Proceeding as before we construct a difference relation which is equivalent to

$$\sum_{j=i}^{i+\alpha} Y_j \boldsymbol{\sigma}_\beta(z_i, t_j) = \int_{-\infty}^{\infty} T_i(f + r_i) \, dt, \tag{4.5}$$

where r_i is given by

$$L(z_i) = f + r_i \tag{4.6}$$

and, in particular, vanishes at the points ξ_{ij}. The associated error equation is

$$\sum_{j=i}^{i=\alpha} Y_j \sigma_\beta(y - z_i, t_j) = - \int_{-\infty}^{\infty} \mathbf{T}_i r_i \, dt. \tag{4.7}$$

In this case the scope for error reduction would appear to be much reduced as it seems unlikely that the same choice of collocation points would be equally favourable to each component of the integral. For this reason the device of splitting L used in the previous section is also less useful. However, it does suggest a possible way of analysing the case in which L cannot be put into the form (4.1).

References

[1] De Boor, Carl and Swartz, Blair. (1973). Collocation at Gaussian points, *SIAM J. Numerical Analysis* 10, 582–606.

[2] Gladwell, Ian and Mullins, D. J. On the effect of boundary conditions in collocation by polynomial splines for the solution of boundary value problems in ordinary differential equations, to appear.

[3] Osborne, M. R. (1967). Minimising truncation error in finite difference approximations to ordinary differential equations, *Maths. Comp.* 21, 133–145.

[4] Osborne, M. R. (1974). On the numerical solution of boundary value problems for ordinary differential equations, *Information Processing*, 74, 3, 673–677.

[5] Pruess, Steven. (1973). Estimating the eigenvalues of Sturm–Liouville problems by approximating the differential equations, *SIAM J. Numerical Analysis* 10, 55–68.

A New Approach to Matrix Theory or Many Facets of the Matrix Decomposition Theorem

Alladi Ramakrishnan and R. Jagannathan†

Abstract

We present a new approach to matrix theory based on a theorem on the decomposition of a matrix as a linear combination of matrices generated as products of base matrices with interesting algebraic properties.

It has been shown by one of the authors [1] that any $N \times N$ matrix $M(N)$ can be expanded as

$$M(N) = \sum_{k,l=0}^{N-1} a_{kl}B(N)^k C(N)^l \tag{1}$$

where $B(N)$ is the diagonal matrix

$$\begin{bmatrix} 1 & & & & \\ & \omega(N) & & & \\ & & \omega(N)^2 & & \\ & & & \ddots & \\ & & & & \omega(N)^{N-1} \end{bmatrix} \tag{2}$$

and $C(N)$ is the $N \times N$ cyclic matrix

$$\begin{bmatrix} 0 & 1 & 0 & \ldots & 0 \\ 0 & 0 & 1 & \ldots & 0 \\ \cdot & \cdot & \cdot & \ldots & \cdot \\ 0 & 0 & 0 & \ldots & 1 \\ 1 & 0 & 0 & \ldots & 0 \end{bmatrix} \tag{3}$$

$\omega(N)$ being the primitive Nth root of unity

$$\omega(N) = \exp(2\pi i/N)$$
$$\omega(N)^N = 1. \tag{4}$$

† Matscience, Institute of Mathematical Sciences, Madras, India.

133

The matrices $C(N)$ and $B(N)$ obey the $\omega(N)$-commutation relation

$$C(N)B(N) = \omega(N)B(N)C(N) \tag{5}$$

and also

$$C(N)^N = B(N)^N = I(N) \tag{6}$$

which imply the elegant algebraic property

$$(B(N) + C(N))^N = B(N)^N + C(N)^N = 2I(N) \tag{7}$$

(a_{kl}) are the elements of a matrix $A(N)$ which is connected to the matrix M in the following manner:

To the matrix $M(N)$ there corresponds a matrix $R(N)$ obtained by rearranging the diagonal and off-diagonals as columns according to the prescription

$$M(N) = \begin{bmatrix} M_{00} & M_{01} & \cdots & M_{0,N-1} \\ M_{10} & M_{11} & \cdots & M_{1,N-1} \\ M_{20} & M_{21} & \cdots & M_{2,N-1} \\ \cdot & \cdot & \cdots & \cdot \\ M_{N-1,0} & M_{N-1,1} & \cdots & M_{N-1,N-1} \end{bmatrix}$$

$$\Rightarrow R(N) = \begin{bmatrix} M_{00} & M_{01} & \cdots & M_{0,N-1} \\ M_{11} & M_{12} & \cdots & M_{10} \\ M_{22} & M_{23} & \cdots & M_{21} \\ \cdot & \cdot & \cdots & \cdot \\ M_{N-1,N-1} & M_{N-1,0} & \cdots & M_{N-1,N-2} \end{bmatrix} \tag{8}$$

We then can write

$$R(N) = S(N)A(N) \tag{9}$$

where

$$S(N) = \begin{bmatrix} 1 & 1 & 1 & \cdots & 1 \\ 1 & \omega(N) & \omega(N)^2 & \cdots & \omega(N)^{N-1} \\ 1 & \omega(N)^2 & \omega(N)^4 & \cdots & \omega(N)^{2(N-1)} \\ \cdot & \cdot & \cdot & \cdots & \cdot \\ 1 & \omega(N)^{N-1} & \omega(N)^{2(N-1)} & \cdots & \omega(N)^{(N-1)(N-1)} \end{bmatrix} \tag{10}$$

is the Sylvester matrix associated with $\omega(N)$, the columns being the eigenvectors of $C(N)$ corresponding to the eigenvalues $(1, \omega(N), \omega(N)^2, \ldots, \omega(N)^{N-1})$ respectively. Thus the new approach to matrix theory consists in viewing a matrix not as rows and columns but as a diagonal and off-diagonals.

We shall now show that there are many facets to this decomposition theorem which strengthen our faith in the new approach. We first notice that if N is a power of another number, i.e. $N = m^n$, then any $N \times N$ matrix can be expressed as a linear combination of $N^2 = m^{2n}$ linearly independent matrices which can also be generated by taking *all possible products of all possible powers* (a.p.p. of a.p.p.) of $2n$ traceless matrices $F_1, F_2, \ldots, F_{2n}$ obeying

$$F_i^m = I \quad (i = 1, 2, \ldots, 2n) \tag{11}$$

m being the minimum integer fitting this relation. A typical matrix of the set generated by this process can be written as

$$F_1^{\beta_1} F_2^{\beta_2} \ldots F_{2n}^{\beta_{2n}} \quad (0 \leqslant \beta_i \leqslant m - 1; \forall i) \tag{12}$$

A suitable set is obtained by choosing $F_i \equiv \mathscr{L}_i, \forall i$, where the $\mathscr{L}_i$'s are matrices of dimension $m^n \times m^n$ obtained by a σ-operation, described earlier in detail by one of the authors [2], on two $m \times m$ matrices $B(m)$ and $C(m)$ which obey a $\omega(m)$-commutation relation and satisfy $C(m)^m = B(m)^m = I(m)$, $\omega(m)$ being the primitive mth root of unity.

If we define

$$\begin{aligned}
B'(N) &= B(N)^{k_1} C(N)^{l_1} \\
C'(N) &= B(N)^{k_2} C(N)^{l_2}
\end{aligned} \tag{13}$$

then they obey a relaxed ω-commutation relation

$$\begin{aligned}
C'(N)B'(N) &= \omega(N)^{b'} B'(N)C'(N) \\
b'(\mathrm{mod}.\, N) &= k_1 l_2 - k_2 l_1
\end{aligned} \tag{14}$$

This implies that all the N^2 matrices $\{B(N)^k C(N)^l \,|\, k, l = 0, 1, \ldots, N - 1\}$ used in Theorem (1) are of *"equal status"* provided that we relax the ω-commutation to include powers of $\omega(N)$. Any $N \times N$ matrix can be expressed in terms of the matrices generated from B' and C' provided that b' is coprime to N. In exactly the same way the principle of equal status can be extended to the set generated by the $2n$ matrices. This can be expressed in a striking manner by defining a *product transformation* which takes one set of $\mathscr{L}$-matrices to another set of $\mathscr{L}$-matrices through

$$\mathscr{L}'_i = \mathscr{L}_1^{u_{i1}} \mathscr{L}_2^{u_{i2}} \ldots \mathscr{L}_{2n}^{u_{i2n}} \quad (i = 1, 2, \ldots, 2n) \tag{15}$$

assuming that $\mathscr{L}$ matrices obey the "relaxed" ω-commutation relation. It is assumed that the transformation obeys the necessary and sufficient conditions which guarantee that the new set of $\mathscr{L}$-matrices generates again a set of N^2, linearly independent matrices by forming a.p.p. of a.p.p. of them. With a set of $\mathscr{L}$-matrices we define a commutation matrix $T = (t_{ij})$ by

$$\begin{aligned}
\mathscr{L}_i \mathscr{L}_j &= \omega(m)^{t_{ij}} \mathscr{L}_j \mathscr{L}_i \quad i, j = 1, 2, \ldots, 2n; \\
t_{ij} &= -t_{ji} \quad 0 \leqslant |t_{ij}| \leqslant m - 1
\end{aligned} \tag{16}$$

It should be noted that T is a $2n \times 2n$ integer matrix and is antisymmetric by definition *which implies the concept of ordered commutation relation, considered to be a characteristic feature of ω-commutation relations.* We now observe that the commutation matrices T and T' associated with the sets $\{\mathscr{L}\}$ and $\{\mathscr{L}'\}$ are related by

$$T' = UT\tilde{U} \qquad (17)$$

where $U = (u_{ij})$ is the matrix of the exponents occurring in the transformation (15). Now the condition mentioned above for the new set $\{\mathscr{L}'\}$ to generate again a set of N^2 linearly independent matrices can be stated as follows: that the dimension of the irreducible representation of the elements obeying commutation relations defined by T' should also be N (or the new set of matrices $\{\mathscr{L}'\}$ should be irreducible).

Once this principle of equality of status not only within the set but also between two different sets is realized, we find that instead of taking the $\mathscr{L}$-matrices as generators of the basis we can as well choose the set of helicity matrices earlier defined by one of the authors [2]. They are the n sets of two matrices $\{H_k^i | i = 1, 2, \ldots, n; k = 1, 2\}$. Of course in the previous work we have considered n sets of 3 matrices, but since the third one in each set is a product of powers of the other two, we need only two matrices from each set to form the set of $2n$ generators. The characteristic feature of the helicity matrices is that the members within each set ω-commute with one another while the members of different sets commute with each other. This can be expressed conveniently using the T matrix. We can adopt a simpler notation with a single index to suit the present considerations, i.e. write

$$\left. \begin{array}{l} H_{2i-1} = H_1^i \\ H_{2i} = H_2^i \end{array} \right\} \quad i = 1, 2, \ldots, n. \qquad (18)$$

Then the T matrix associated with this set has elements

$$t_{2i-1, 2i} = -t_{2i, 2i-1} = 1; \quad (i = 1, 2, \ldots, n)$$
$$\begin{cases} t_{lj} = 0 \quad \text{otherwise} \\ l, j = 1, 2, \ldots, 2n \end{cases} \qquad (19)$$

The procedure for obtaining the $\mathscr{L}$-matrices from helicity matrices by the *"tenon and mortice"* coupling method has been elaborately dealt with by one of the authors [2]. Now we recognize this as one special case of the product transformation defined above.

The advantage of taking the helicity matrices as generators is that we can obtain the coefficients in the expansion with the same ease as in the case of Theorem (1). Observing that the helicity matrices can be obtained as direct products of the m-dimensional $B(m)$ or $C(m)$, and $(n-1)$ unit matrices of the same dimension, a typical member of the set of N^2 matrices, obtained by taking a.p.p. of a.p.p. of the $2n$-helicity matrices, can be written as

$$B(m)^{k_1}C(m)^{l_1} \otimes B(m)^{k_2}C(m)^{l_2} \otimes \ldots \otimes B(m)^{k_n}C(m)^{l_n} \qquad (20)$$

As a consequence the decomposition theorem now reads

$$M(N) = \sum_{\substack{k_i, l_i = 0 \\ (l=1,2,\ldots,n)}}^{m-1} a_{k_1 l_1 k_2 l_2 \ldots k_n l_n} B(m)^{k_1} C(m)^{l_1} \otimes \ldots \otimes B(m)^{k_n} \times$$

$$\times\, C(m)^{l_n} \qquad (21)$$

Now we find that analogously to the previous case a rearranged matrix can be obtained in the following manner to enable the determination of the coefficients. We interpret the $N \times N$ matrix as an m-dimensional "supermatrix" made up of m rows and columns of "elements", each element being a matrix. The rearrangement is done by arranging the diagonal and off diagonals of the supermatrix as columns using the same prescription as before remembering that the elements themselves are matrices now. Then each "element" is treated as an m-dimensional supermatrix and rearrangement operation is performed on each of them. The rearrangement operations are continued successively and come to an end after n steps, when at the final stage "elements" are not supermatrices but are single elements. This amounts to treating the matrix M as a "chess board" of m^2 squares, each square representing a smaller chess board. Thus the chessboard is rearranged successively in n steps treating each square as an element first and then as a chessboard. Writing such a rearranged matrix as $R[N]$ we recognize that

$$R[N] = S[N]A[N] \qquad (22)$$

where

$$S[N] = S(m) \otimes S(m) \otimes \ldots \otimes S(m) \quad (n \text{ times}) \qquad (23)$$

$S(m)$ being the Sylvester matrix associated with $\omega(m)$ and $A[N]$ being the matrix of coefficients $(a_{k_1 l_1 k_2 l_2 \ldots k_n l_n})$ in which the element $a_{k_1 l_1 k_2 l_2 \ldots k_n l_n}$ occurs in the position of the element $d_{k_1 l_1} d_{k_2 l_2} \ldots d_{k_n l_n}$ in the matrix $D \otimes \ldots \otimes D$ (n times), where D is any $m \times m$ matrix $= (d_{kl})$ $k, l = 0, 1, \ldots, m - 1$.

We can extend the same considerations to

$$N = p_1^{\alpha_1} p_2^{\alpha_2} \ldots p_r^{\alpha_r} \qquad (24)$$

In this case we can take $2(\alpha_1 + \alpha_2 + \ldots + \alpha_r)$ generating base matrices. In view of the principle of equality of status we can choose them with helicity structure, i.e. as direct products of B, C, and unit matrices of dimensions $p_1, p_2, \ldots, p_r$ and obeying relaxed ω-commutation relations involving products of power of $\omega(p_1), \ldots, \omega(p_r)$. Then the set of N^2 linearly independent matrices is obtained by forming a.p.p. of a.p.p. of them, a typical member of the set being

$$B(p_1)^{k_{11}}C(p_1)^{l_{11}} \otimes \ldots \otimes B(p_1)^{k_{1\alpha_1}}C(p_1)^{l_{1\alpha_1}} \otimes B(p_2)^{k_{21}}C(p_2)^{l_{21}}$$
$$\otimes \ldots \otimes B(p_2)^{k_{2\alpha_2}}C(p_2)^{l_{2\alpha_2}} \otimes \ldots \otimes B(p_r)^{k_{r1}}C(p_r)^{l_{r1}}$$
$$\otimes \ldots \otimes B(p_r)^{k_{r\alpha_r}}C(p_r)^{l_{r\alpha_r}}$$
$$(0 \leqslant k_{i1}, k_{i2}, \ldots, k_{i\alpha_i} \leqslant p_i - 1; i = 1, 2, \ldots, r)$$

where

$$C(p_i)B(p_i) = \omega(p_i)B(p_i)C(p_i)$$
$$C(p_i)^{p_i} = B(p_i)^{p_i} = I(p_i)$$
$$(i = 1, 2, \ldots, r) \tag{25}$$

The essential distinction between choosing the generating base matrices as B and C as in (1) and the helicity matrices is that in one case we are generating the basis of N^2 matrices from matrices involving higher roots and in the other case from those involving lower roots. It is clear that while a larger number of generators are needed when lower roots are involved, a smaller number is enough when higher roots are involved, the smallest being two when the root involved is the highest.

There is one special interesting case where we can express the two matrices of dimension $N \times N$ having a $\omega(N)$-commutation relation, as a direct product of lower dimensional matrices involving lower roots. This occurs when the Nth primitive root $\omega(N)$ can be expressed as a product of lower roots corresponding to the decomposition $N = p_1^{\alpha_1}p_2^{\alpha_2} \ldots p_r^{\alpha_r}$ where $p_1, p_2, \ldots, p_r$ are primes and $\alpha_1, \alpha_2, \ldots, \alpha_r$ are positive integers. We give here just the result. When $C(N)B(N) = \omega(N)B(N)C(N)$ $C(N)$ and $B(N)$ can be represented as

$$C(N) = C(p_1^{\alpha_1})^{\beta_1} \otimes \ldots \otimes C(p_r^{\alpha_r})^{\beta_r}$$
$$B(N) = B(p_1^{\alpha_1}) \otimes \ldots \otimes B(p_r^{\alpha_r}) \tag{26}$$

where

$$C(p_i^{\alpha_i})B(p_i^{\alpha_i}) = \omega(p_i^{\alpha_i})B(p_i^{\alpha_i})C(p_i^{\alpha_i}); C(p_i^{\alpha_i})^{p_i^{\alpha_i}} = B(p_i^{\alpha_i})^{p_i^{\alpha_i}} = I(p_i^{\alpha_i})$$
$$(i = 1, 2, \ldots, r)$$

and the β_i's are positive integers subject to the condition

$$\sum_{i=1}^{r} \frac{\beta_i}{p_i^{\alpha_i}} = K + \frac{1}{N} \tag{27}$$

K being any integer. The β_i's can be obtained by the usual methods of solving linear diophantine equations. The proof follows from the observation that

$$\omega(p_1^{\alpha 1})^{\beta 1} \omega(p_2^{\alpha 2})^{\beta 2} \ldots \omega(p_r^{\alpha r})^{\beta r} = \omega(N)^{N\beta 1/p_1^{\alpha 1}} \ldots \omega(N)^{N\beta r/p_r^{\alpha r}}$$

$$= \omega(N)^{N(\Sigma_{i=1}^r \beta_i/p_i^{\alpha i})}$$

$$= \omega(N) \qquad (28)$$

when (27) is satisfied

In general we can have

$$C(N) = C(p_1^{\alpha 1})^{\mu 1} B(p_1^{\alpha 1})^{\lambda 1} \otimes \ldots \otimes C(p_r^{\alpha r})^{\mu r} B(p_r^{\alpha r})^{\lambda r}$$

$$B(N) = C(p_1^{\alpha 1})^{\nu 1} B(p_1^{\alpha 1})^{\delta 1} \otimes \ldots \otimes C(p_r^{\alpha r})^{\nu r} B(p_r^{\alpha r})^{\delta r} \qquad (29)$$

where

$$\left\| \begin{matrix} \mu_i & \lambda_i \\ \nu_i & \delta_i \end{matrix} \right\| = \beta_i \quad (i = 1, 2, \ldots, r) \qquad (30)$$

and the β_i's obey (27). It is hoped that the study of the many facets of the decomposition amounts to the introduction of a "tensor"-calculus to structures in which components are matrices.

References

[1] Ramakrishnan, Alladi. (1972). *J. Math. Analysis Applic.* 40, 36–38.
[2] Ramakrishnan, Alladi. (1972). *L-Matrix Theory or The Grammar of Dirac Matrices*. Tata McGraw-Hill Publishing Co. Ltd., Bombay, New Delhi, India.

Hybrid Finite Element Methods for Solving 2nd Order Elliptic Equations

P. A. Raviart

1. Introduction

Let Ω be a bounded open set of R^n with a sufficiently smooth boundary Γ. We consider the 2nd order elliptic model problem

$$\begin{cases} -\Delta u = f & \text{in} \quad \Omega, \\ \quad u = 0 & \text{on} \quad \Gamma, \end{cases} \tag{1.1}$$

where f is a given function which is assumed to belong to the space $L^2(\Omega)$. The standard variational formulation of problem (1.1) consists in finding the unique function $u \in H_0^1(\Omega)$ which satisfies

$$\int_\Omega \text{grad } u \cdot \text{grad } v \, dx = \int_\Omega fv \, dx \quad \text{for all} \quad v \in H_0^1(\Omega). \tag{1.2}$$

Then, the classical finite element discretization of problem (1.1) is defined in the following way: we first construct a finite-dimensional subspace V_h of the space $H_0^1(\Omega)$ made up with piecewise smooth functions which are continuous along the interelement boundaries. Next, we find the unique function $u_h \in V_h$ such that

$$\int_\Omega \text{grad } u_h \cdot \text{grad } v \, dx = \int_\Omega fv \, dx \quad \text{for all} \quad v \in V_h. \tag{1.3}$$

Such *conforming* methods have been extensively studied and convergence results are now standard: see for instance Ciarlet and Raviart [4], Strang and Fix [13].

There are other approaches based on different variational principles allowing the use of discontinuous fields at the interelement boundaries. These approaches lead to the so-called *hybrid* finite element methods which have been introduced by the Engineers (cf. Fraejs de Veubeke [6], [7], Pian [9], [10], Pian and Tong [11]).

The purpose of this paper is to give a survey of some results which have been recently obtained by Brezzi [1], Raviart and Thomas [12] and Thomas

141

[14] concerning the numerical analysis of hybrid methods for solving 2nd order elliptic problems.

An outline of the paper is as follows. Section 2 is devoted to the description of some general results of Brezzi [1] about the existence, uniqueness and approximation of the solutions of variational problems. In Section 3, we apply the results of Section 2 to the study of primal hybrid finite element methods. Various examples of primal hybrid triangular elements are given in Section 4 and the close relationship with *non conforming* triangular elements is established. Finally, we give in Section 5 a brief description of some results of Thomas [14] on dual hybrid finite element methods.

For the numerical analysis of hybrid methods for the plate bending problem, we refer to Brezzi [2], Brezzi-Marini [3].

2. Some General Results

Let X and M be two (real) Hilbert spaces with norms $\| \cdot \|_X$ and $\| \cdot \|_M$ respectively. Let us denote by X' (resp. M') the dual space of X (resp. M) and by $(\ldots)$ the duality between X' and X (resp. M' and M).

Now, we are given two bilinear forms $a(u, v)$ and $b(v, \mu)$ which are continuous over $X \times X$ and $X \times M$ respectively. We consider the following problem: Given $(f, g) \in X' \times M$, find $(u, \lambda) \in X \times M$ such that

$$\begin{cases} a(u, v) + b(v, \lambda) = (f, v) & \text{for all} \quad v \in X, \\ \quad\quad\quad b(u, \mu) = (g, \mu) & \text{for all} \quad \mu \in M. \end{cases} \tag{2.1}$$

We introduce the closed subspace V of X:

$$V = \{v | v \in X, \quad b(v, \mu) = 0 \quad \text{for all} \quad \mu \in M\}. \tag{2.2}$$

Then we have

Theorem 1
Assume that the bilinear form $a(u, v)$ is V-elliptic, i.e. there exists a constant $\alpha > 0$ such that

$$a(v, v) \geqslant \alpha \|v\|_X^2 \quad \text{for all} \quad v \in V. \tag{2.3}$$

Then, problem (2.1) has a unique solution $(u, \lambda) \in X \times M$ for any given set of data $(f, g) \in X' \times M'$ if and only if the following condition holds: there exists a constant $\beta > 0$ such that

$$\sup_{v \in X} \frac{b(v, \mu)}{\|v\|_X} \geqslant \beta \|\mu\|_M \quad \text{for all} \quad \mu \in M. \tag{2.4}$$

Assume that conditions (2.3) and (2.4) hold. We define the closed affine subspace V_g of X:

$$V_g = \{v | v \in X, \quad b(v, \mu) = (g, \mu), \quad \text{for all} \quad \mu \in M\}. \tag{2.5}$$

Then, the first element u of the solution (u, λ) of problem (2.1) may be characterized by

$$\begin{cases} u \in V_g, \\ a(u, v) = (f, v) \quad \text{for all} \quad v \in V. \end{cases} \tag{2.6}$$

Problem (2.6) is the classical abstract form of a variational problem. Moreover, problems (2.1) and (2.6) are equivalent.

When the bilinear form $a(u, v)$ is symmetric, i.e. $a(u, v) = a(v, u)$ for all $u, v \in X$, we introduce the quadratic functionals

$$J(v) = \tfrac{1}{2} a(v, v) - (f, v), \quad v \in X, \tag{2.7}$$

and

$$\mathscr{L}(v, \mu) = J(v) + b(v, \mu) - (g, \mu), \quad v \in X, \quad \mu \in M. \tag{2.8}$$

One can easily prove

Theorem 2
Assume that conditions (2.3) and (2.4) hold. Assume in addition, that the bilinear form $a(u, v)$ is symmetric. Then the solution (u, λ) of problem (2.1) is the unique saddle point of the quadratic functional $\mathscr{L}(v, \mu)$ over $X \times M$, i.e. (u, λ) satisfies

$$\mathscr{L}(u, \mu) \leqslant \mathscr{L}(u, \lambda) \leqslant \mathscr{L}(v, \lambda) \quad \text{for all} \quad v \in X \quad \text{and all} \quad \mu \in M. \tag{2.9}$$

Assuming the conditions of Theorem 2, u may be characterized as the unique element of V_g which satisfies

$$J(u) = \underset{v \in V_g}{\text{Min}} \; J(v). \tag{2.10}$$

Hence, λ appears to be the Lagrange multiplier which corresponds to the constraint $u \in V_g$.

Now, in order to approximate the solution (u, λ) of problem (2.1), we are given:

(i) a finite-dimensional subspace X_h of X;
(ii) a finite-dimensional subspace M_h of M.

We introduce the following finite-dimensional "approximation" V_h of the space V:

$$V_h = \{v | v \in X_h, \quad b(v, \mu) = 0 \quad \text{for all} \quad \mu \in M_h\}. \tag{2.11}$$

Note that, *in general, V_h is not a subspace of V.*

Next, we consider the approximate problem: Find $(u_h, \lambda_h) \in X_h \times M_h$ such that

$$\begin{cases} a(u_h, v) + b(v, \lambda_h) = (f, v) & \text{for all} \quad v \in X_h, \\ \qquad\qquad b(u_h, \mu) = (g, \mu) & \text{for all} \quad \mu \in M_h. \end{cases} \tag{2.12}$$

Concerning the existence and uniqueness of the solution of problem (2.12), we have the following finite-dimensional analogue of Theorem 1.

Theorem 3
Assume that the bilinear form $a(u, v)$ is V_h-elliptic. Then, problem (2.12) *has a unique solution $(u_h, \lambda_h) \in X_h \times M_h$ for any given set of data $(f, g) \in X' \times M'$ if and only if the following condition holds:*

$$\mu \in M_h, \quad b(v, \mu) = 0 \quad \text{for all} \quad v \in X_h \Rightarrow \mu = 0. \tag{2.13}$$

Clearly, condition (2.13) is equivalent to the following one: there exists a constant $\beta_h > 0$ such that

$$\sup_{v \in X_h} \frac{b(v, \mu)}{\|v\|_X} \geqslant \beta_h \|u\|_M \quad \text{for all} \quad \mu \in M_h.$$

Since, in general, V_h is not contained in V, the V-ellipticity of the bilinear form $a(u, v)$ does not imply its V_h-ellipticity. However, this assumption is easily checked in practical cases. On the other hand, (2.13) is a compatibility condition between the spaces X_h and M_h which can be rather tricky to check in many practical cases and which may not be satisfied for apparently reasonable choices of the spaces X_h and M_h.

As for problem (2.1), we define:

$$V_{h,g} = \{v | v \in X_h, \quad b(v, \mu) = (g, \mu) \quad \text{for all} \quad \mu \in M_h\}. \tag{2.14}$$

Then, u_h may be characterized by

$$\begin{cases} u_h \in V_{h,g}, \\ a(u_h, v) = (f, v) & \text{for all} \quad v \in V_h. \end{cases} \tag{2.15}$$

Note that the V_h-ellipticity of the bilinear form $a(u, v)$ alone ensures the existence and uniqueness of the solution u_h of problem (2.15). On the other hand, condition (2.13) ensures the existence and uniqueness of the Lagrange multiplier λ_h. Now, since $V_h \not\subset V$, problem (2.15) appears to be a *non-conforming* model for numerically solving problem (1.1).

Next, let us give some general bounds for the errors $\|u_h - u\|_X$ and $\|\lambda_h - \lambda\|_M$. Brezzi [1] has proved the following natural result.

Theorem 4

Assume that there exist two constants $\tilde{\alpha}, \tilde{\beta} > 0$ independent of h such that

$$a(v, v) \geqslant \tilde{\alpha} \|v\|_X^2 \quad \text{for all} \quad v \in V_h, \tag{2.16}$$

$$\sup_{v \in X_h} \frac{b(v, \mu)}{\|v\|_X} \geqslant \tilde{\beta} \|\mu\|_M \quad \text{for all} \quad \mu \in M_h. \tag{2.17}$$

Then, there exists a constant $c > 0$ independent of h such that

$$\|u_h - u\|_X + \|\lambda_h - \lambda\|_M \leqslant c \left\{ \inf_{v \in X_h} \|u - v\|_X + \inf_{\mu \in M_h} \|\lambda - \mu\|_M \right\}$$

$$\tag{2.18}$$

3. A Primal Hybrid Finite Element Model

Let us now give another variational form of the model problem (1.1) which leads to the first hybrid finite element method. We begin by introducing some natation.

We denote by

$$\|v\|_{0,\Omega} = \left(\int_\Omega |v(x)|^2 \, dx \right)^{1/2}$$

the L^2-norm. Given any integer $m \geqslant 1$, let

$$H^m(\Omega) = \{v | v \in L^2(\Omega), \quad \partial^\alpha v \in L^2(\Omega), \quad |\alpha| \leqslant m\}$$

be the usual Sobolev space normed by

$$\|v\|_{m,\Omega} = \left(\sum_{|\alpha| \leqslant m} \|\partial^\alpha v\|_{0,\Omega}^2 \right)^{1/2}.$$

Here $\alpha = (\alpha_1, \ldots, \alpha_n) \in \mathbb{N}^n$ is a multi-index, $|\alpha| = \alpha_1 + \ldots + \alpha_n$ and $\partial^\alpha = (\partial/\partial x_1)^{\alpha_1} \ldots (\partial/\partial x_n)^{\alpha_n}$. We shall also use the following semi-norm

$$|v|_{m,\Omega} = \left(\sum_{|\alpha| = m} \|\partial^\alpha v\|_{0,\Omega}^2 \right)^{1/2}.$$

We denote by

$$\|v\|_{0,\Gamma} = \left(\int_\Gamma |v(x)|^2 \, d\sigma(x) \right)^{1/2}$$

the L^2-norm over Γ. Let $H^{1/2}(\Gamma)$ be the space of the traces over Γ of the functions of $H^1(\Omega)$ and let $H^{-1/2}(\Gamma)$ be the dual space of $H^{1/2}(\Gamma)$. We extend the scalar product in $L^2(\Gamma)$ to represent the duality between $H^{-1/2}(\Gamma)$ and $H^{1/2}(\Gamma)$. We set

$$H_0^1(\Omega) = \{v | v \in H^1(\Omega), \quad v|_\Gamma = 0\}.$$

Finally, we introduce the space

$$Q(\Omega) = \{q \mid q \in (L^2(\Omega))^n, \quad \operatorname{div} q \in L^2(\Omega)\} \tag{3.1}$$

provided with the norm

$$\|q\|_{Q(\Omega)} = (\|q\|_{0,\Omega}^2 + \|\operatorname{div} q\|_{0,\Omega}^2)^{1/2} \tag{3.2}$$

Note that, for $q \in Q(\Omega)$, we may define $q \cdot \nu$ as an element of $H^{-1/2}(\Gamma)$ where ν is the unit outward vector normal to Γ.

Now, let $\bar{\Omega} = \cup_{i=1}^I \bar{\Omega}_i$ be a decomposition of $\bar{\Omega}$ into subdomains $\bar{\Omega}_i$ such that:

(i) Ω_i is an open subset of Ω, $1 \leq i \leq I$;

(ii) $\Omega_i \cap \Omega_j = \phi$ for $i \neq j$.

With this decomposition, we associate the space

$$X = \{v \mid v \in L^2(\Omega), \quad v|_{\Omega_i} \in H^1(\Omega_i), \quad 1 \leq i \leq I\} \approx \prod_{i=1}^I H^1(\Omega_i) \tag{3.3}$$

provided with the norm

$$\|v\|_X = \left(\sum_{i=1}^I \|v\|_{1,\Omega_i}^2 \right)^{1/2} \tag{3.4}$$

and the bilinear form

$$a(u, v) = \sum_{i=1}^I \int_{\Omega_i} \operatorname{grad} u \cdot \operatorname{grad} v \, dx, \quad u, v \in X. \tag{3.5}$$

Next, we introduce the space

$$M = \left\{ \mu \mid \mu \in \prod_{i=1}^I H^{-1/2}(\partial\Omega_i), \quad \text{there exists a function} \quad q \in Q(\Omega) \right.$$
$$\left. \text{such that} \quad q \cdot \nu_i = \mu|_{\partial\Omega_i}, \quad 1 \leq i \leq I \right\} \tag{3.6}$$

where ν_i is the unit outward vector normal to the boundary $\partial\Omega_i$ of Ω_i. We provide the space M with the norm

$$\|\mu\|_M = \inf \{\|q\|_{Q(\Omega)} \mid q \in Q(\Omega), q \cdot \nu_i = \mu|_{\partial\Omega_i}, \quad 1 \leq i \leq I\}. \tag{3.7}$$

We define the bilinear form

$$b(v, \mu) = - \sum_{i=1}^I \int_{\partial\Omega_i} \mu v \, d\sigma, \quad v \in X, \quad \mu \in M$$

where the integral $\int_{\partial\Omega_i}$ represents the duality between $H^{-1/2}(\partial\Omega_i)$ and $H^{1/2}(\partial\Omega_i)$.

One can easily check that $V = H_0^1(\Omega)$ and that

$$\sup_{v \in X} \frac{b(v, \mu)}{\|v\|_X} = \|\mu\|_M \quad \text{for all} \quad \mu \in M.$$

Therefore, by Theorem 1, *the following problem*:
"*Find* $(u, \lambda) \in X \times M$ *such that*

$$\begin{cases} a(u, v) + b(v, \lambda) = \int_\Omega fv \, dx & \text{for all } v \in X, \\[2mm] b(u, \mu) = 0 & \text{for all } \mu \in M" \end{cases} \tag{3.10}$$

has a unique solution. Moreover, u is the solution of problem (1.2) (*or problem* (1.1)) *and we have*

$$\lambda|_{\partial\Omega_i} = \frac{\partial u}{\partial \nu_i}, \quad 1 \leqslant i \leqslant I. \tag{3.11}$$

We now come to the numerical approximation of problem (1.1) by using the variational formulation (3.10). Let us assume for simplicity that $\bar\Omega$ *is a polygon of* $\mathbb{R}^2$ and that $\bar\Omega = \cup_{i=1}^I \bar\Omega_i$ is a triangulation of $\bar\Omega$ with triangles $\bar\Omega_i$ with diameters $\leqslant h$. We construct the spaces X_h and M_h as follows. Let $\hat\Omega$ be a reference triangle (usually the unit isoceles rectangular triangle). We are given a finite-dimensional subspace $\mathscr{P}(\hat\Omega)$ of $H^1(\hat\Omega)$ such that

$$P_k(\hat\Omega) \subset \mathscr{P}(\hat\Omega) \quad \text{for some integer} \quad k \geqslant 1, \tag{3.12}$$

where $P_k(\hat\Omega)$ denotes the space of the restrictions over $\hat\Omega$ of all polynomials of degree $\leqslant k$ in the two variables x_1, x_2. Now, for any $1 \leqslant i \leqslant I$, let F_i be an affine invertible mapping which maps $\hat\Omega$ onto Ω_i. We define

$$\mathscr{P}(\Omega_i) = \{v | v = \hat v \circ F_i^{-1}, \quad \hat v \in \mathscr{P}(\hat\Omega)\} \tag{3.13}$$

and

$$X_h = \{v | v \in L^2(\Omega), \quad v|_{\Omega_i} \in \mathscr{P}(\Omega_i), \quad 1 \leqslant i \leqslant I\}. \tag{3.14}$$

On the other hand, for some integer $m \geqslant 0$ and any $1 \leqslant i \leqslant I$, we introduce the space $S_m(\partial\Omega_i)$ of the functions defined over $\partial\Omega_i$ whose restrictions to any side of the triangle are polynomials of degree $\leqslant m$. Note that the functions of $S_m(\partial\Omega_i)$ are discontinuous in general at the vertices of the triangle Ω_i. We set

$$M_h = \left\{ \mu | \mu \in \prod_{i=1}^I S_m(\partial\Omega_i), \quad \mu|_{\partial\Omega_i} + \mu|_{\partial\Omega_j} = 0 \quad \text{on} \quad \partial\Omega_i \cap \partial\Omega_j \right\}$$

$$\tag{3.15}$$

It is an easy matter to check that M_h is a subspace of M.

According to this choice of the spaces X_h and M_h, the V_h-ellipticity of the bilinear form $a(u, v)$ always holds. But condition (2.13) holds if and only if

$$\mu \in S_m(\partial\Omega), \quad \int_{\partial\Omega} \mu v \, d\sigma = 0 \quad \text{for all} \quad v \in \mathscr{P}(\hat{\Omega}) \Rightarrow \mu = 0. \tag{3.16}$$

Therefore, by Theorem 3, condition (3.16) guarantees the existence and uniqueness of the solution $(u_h, \lambda_h) \in X_h \times M_h$ of

$$\begin{cases} a(u_h, v) + b(v, \lambda_h) = \displaystyle\int_{\Omega} fv \, dx & \text{for all} \quad v \in X_h, \\ \qquad b(u_h, \mu) = 0 & \text{for all} \quad \mu \in M_h. \end{cases} \tag{3.17}$$

In order to get bounds for the errors $u_h - u$ and $\lambda_h - \lambda$, we need to define more convenient norms over the spaces X and M. We denote by h_i the diameter of Ω_i, $1 \leqslant i \leqslant I$, and we set:

$$|||v|||_X = \left(\sum_{i=1}^{I} (|v|_{1,\Omega_i}^2 + h_i^{-2}\|v\|_{0,\Omega_i}^2) \right)^{1/2}, \tag{3.18}$$

$$|||\mu|||_M = \sup_{v \in X} \frac{b(\mu, v)}{|||v|||_X}. \tag{3.19}$$

By applying Theorem 4, one can prove (cf. Raviart and Thomas [12])

Theorem 5
Assume that the triangulation $\bar{\Omega} = \cup_{i=1}^{I} \bar{\Omega}_i$ is regular, i.e. there exists an angle $\theta > 0$ independent of h such that

$$\theta_i \geqslant \theta, \quad 1 \leqslant i \leqslant I, \tag{3.20}$$

where θ_i is the smallest angle of the triangle Ω_i. Assume in addition, that conditions (3.12) and (3.16) hold. Then, for all integer l such that $1 \leqslant l \leqslant \min(k, m+1)$, we get when $u \in H^{l+1}(\Omega)$

$$\|u_h - u\|_X + |||\lambda_h - \lambda|||_M \leqslant ch^l |u|_{l+1,\Omega}, \tag{3.21}$$

where the constant c is independent of h.

As a consequence of (3.21), we get a more explicit bound for the error $\lambda_h - \lambda$:

$$\left(\sum_{i=1}^{I} h_i \|\lambda_h - \lambda\|_{0,\partial\Omega_i}^2 \right)^{1/2} \leqslant ch^l |u|_{l+1,\Omega}. \tag{3.22}$$

4. Examples of Primal Hybrid Triangular Elements

We now describe some hybrid triangular elements which satisfy condition (3.16). Let $\hat{a}_i$, $1 \leqslant i \leqslant 3$, be the vertices of the reference triangle $\hat{\Omega}$. We shall denote by $\chi_i = \chi_i(x)$ the barycentric coordinates of a point $x \in \mathbb{R}^2$ with respect to the vertices $\hat{a}_i$ of $\hat{\Omega}$. In order to derive sufficient conditions for hypothesis (3.16) to hold, one can prove

Lemma 1
The conditions

$$\begin{cases} \mu \in S_m(\partial\hat{\Omega}), \\[2ex] \displaystyle\int_{\partial\Omega} \mu v \, d\sigma = 0 \quad for\ all \quad v \in P_k(\hat{\Omega}), \end{cases} \tag{4.1}$$

imply $\mu = 0$ *if and only if*

$$k \geqslant \begin{cases} m + 1 & when\ m\ is\ even, \\ m + 2 & when\ m\ is\ odd. \end{cases} \tag{4.2}$$

Example 1
Let $m \geqslant 0$ be an *even* integer and let $k = m + 1$. We consider the hybrid method associated with $(P_k(\hat{\Omega}), S_m(\hat{\Omega}))$. Then, by Lemma 1 and Theorem 5, problem (3.17) has a unique solution $(u_h, \lambda_h) \in X_h \times M_h$ and, for $u \in H^{k+1}(\Omega)$, we have the error estimate

$$\|u_h - u\|_X + \|\|\lambda_h - \lambda\|\|_M \leqslant ch^k |u|_{k+1,\Omega}. \tag{4.3}$$

In fact, this hybrid method coincides with a now classical non-conforming method. Indeed, for each edge e of the triangulation the integral $\int_e \mu v \, d\sigma$ can be exactly computed in terms of the values of μv at the k *Gauss* points of e. Thus, a function $v \in X_h$ belongs to the space V_h if and only if:

(i) v is continuous at these Gauss points along the interelement boundaries;

(ii) v vanishes at these Gauss points along the boundary Γ of Ω.

Therefore, the degrees of freedom of a function $v \in V_h$ are its values at these Gauss points which do not belong to Γ and at some other points of $\cup_{i=1}^{I} \Omega_i$. Moreover, the corresponding basis functions of V_h are easily determined. On the other hand, we can choose the degrees of freedom of a function $\mu \in M_h$ to be its values at these Gauss-points.

The first hybrid elements are described in Fig. 1 where we have used the following conventions for the degrees of freedom:

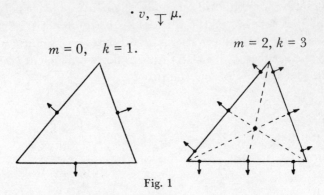

Fig. 1

These hybrid elements can be considered as non-conforming elements using Loof connections (cf. Crouzeix and Raviart [5], Irons and Razzaque [8]).

Example 2

Now, let $m \geqslant 1$ be an *odd* integer. Then, by Lemma 1, the hybrid method associated with $(P_k(\hat{\Omega}), S_m(\partial\hat{\Omega}))$ is no longer adequate for $k = m + 1$. A suitable choice is $k = m + 2$. It is possible however to construct a more interesting hybrid element $(\mathscr{P}(\hat{\Omega}), S_m(\partial\hat{\Omega}))$ such that

$$P_k(\hat{\Omega}) \subset \mathscr{P}(\hat{\Omega}) \subset P_{k+1}(\hat{\Omega}), \quad k = m + 1.$$

The main tool is

Lemma 2

Assume that $m = k - 1$ is an odd integer. Define $\mathscr{P}(\hat{\Omega})$ to be the space of polynomials spanned by $P_k(\hat{\Omega})$ and by

$$v_0 = (\chi_1 - \chi_2)(\chi_2 - \chi_3)(\chi_3 - \chi_1)[(\chi_1\chi_2)^{(k-2)/2} + (\chi_2\chi_3)^{(k-2)/2} + (\chi_3\chi_1)^{(k-2)/2}].$$

Then, the hybrid element $(\mathscr{P}(\hat{\Omega}), S_m(\partial\hat{\Omega}))$ satisfies condition (3.16).

Thus, consider the hybrid method associated with this hybrid element $(\mathscr{P}(\hat{\Omega}), S_m(\partial\hat{\Omega}))$. By Lemma 2 and Theorem 5, problem (3.17) has a unique solution and, for $u \in H^{k+1}(\Omega)$, we get the error estimate (4.3).

Note that, in this example, the integrals $\int_e \mu v \, d\sigma, \, v \in X_h, \, \mu \in M_h$ can no longer be computed exactly in terms of the values of μv at the k Gauss points of e. Hence, we are not able here to determine in a simple way the degrees of freedom of a function $v \in V_h$.

Remark

It can be of interest to use numerical quadrature for evaluating the various integrals $\int_e \mu v \, d\sigma$ which appear in the bilinear form $b(v, \mu)$. Then, we obtain

a new bilinear form $b^\star(v, \mu)$ and we replace problem (3.17) by the following
one: Find $(u_h^\star, \lambda_h^\star) \in X_h \times M_h$ such that

$$\begin{cases} a(u_h^\star, v) + b^\star(v, \lambda_h^\star) = \int\limits_\Omega fv \, dx & \text{for all} \quad v \in X_h, \\[2mm] \qquad\qquad b^\star(u_h^\star, \mu) = 0 & \text{for all} \quad \mu \in M_h. \end{cases} \tag{4.4}$$

With the bilinear form $b^\star(v, \mu)$, we associate the subspace $V_h^\star$ of X_h:

$$V_h^\star = \{v | v \in X_h, b^\star(v, \mu) = 0 \quad \text{for all} \quad \mu \in M_h\} \tag{4.5}$$

Now, the degrees of freedom of a function $v \in V_h^\star$ may be easier to determine
than those of a function $v \in V_h$. This is indeed the case in Example 2 if we
use a k-point Gauss formula for computing $\int_e \mu v \, d\sigma$. In fact, we can prove
that the corresponding problem (4.4) has a unique solution $(u_h^\star, \lambda_h^\star)$ and,
for $u \in H^k(\Omega)$, we get the error estimate

$$\|u_h^\star - u\|_X + \|\|\lambda_h^\star - \lambda\|\|_M \leqslant ch^{k-1}|u|_{k,\Omega} \tag{4.6}$$

Note that the order of accuracy of the hybrid method has been decreased
by one but, on the other hand, the corresponding space $V_h^\star$ can now be
characterized as in Example 1 (condition (i), (ii)). We describe in Fig. 2 the

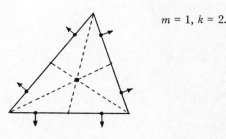

$$m = 1, k = 2.$$

Fig. 2

1st hybrid element of this type. We recognize here a non-conforming element
introduced by Irons and Razzaque [8].

For examples of hybrid *quadrilateral* elements and associated non-
conforming elements, we refer to Raviart and Thomas [12].

5. Dual Hybrid Finite Element Models

Let us introduce first the dual variational form of problem (1.1). Given
$f \in L^2(\Omega)$, we define the following closed subspace $Q_f(\Omega)$ of the space
$Q(\Omega)$:

$$Q_f(\Omega) = \{q | q \in Q(\Omega), \quad \text{div } q + f = 0 \text{ in } \Omega\}. \tag{5.1}$$

Then, the dual variational form of problem (1.1) consists in finding the unique function $w \in Q_f(\Omega)$ which satisfies

$$\int_{\Omega} w \cdot q \, dx = 0 \quad \text{for all} \quad q \in Q_0(\Omega). \tag{5.2}$$

Moreover, we have

$$w = \text{grad } u, \tag{5.3}$$

where u is the solution of problem (1.1) (or problem (1.2)).

Given a decomposition $\bar{\Omega} = \cup_{i=1}^{I} \bar{\Omega}_i$ of the set $\bar{\Omega}$, we introduce a function $p_0 \in (\dot{L}^2(\Omega))^n$ such that

$$\text{div } p_0 + f = 0 \quad \text{in each} \quad \Omega_i. \tag{5.4}$$

With this decomposition, we associate the space

$$X = \{q | q \in (L^2(\Omega))^n, \quad \text{div } q = 0 \text{ in } \Omega_i, \quad 1 \leq i \leq I\} \tag{5.5}$$

provided with the L^2-norm and the bilinear form

$$a(p, q) = \int_{\Omega} p \cdot q \, dx, \quad p, q \in X. \tag{5.6}$$

Next, we define the space

$$M = \left\{ \mu | \mu \in \prod_{i=1}^{I} H^{1/2}(\partial\Omega_i), \quad \text{there exists a function } v \in H_0^1(\Omega) \right.$$

$$\left. \text{such that} \quad v|_{\partial\Omega_i} = \mu|_{\partial\Omega_i}, \quad 1 \leq i \leq I \right\} \tag{5.7}$$

provided with the norm

$$\|\mu\|_M = \inf\{ |v|_{1,\Omega} \, | \, v \in H_0^1(\Omega), v|_{\partial\Omega_i} = \mu|_{\partial\Omega_i}, \quad 1 \leq i \leq I\}. \tag{5.8}$$

Then, we consider the bilinear form

$$b(q, \mu) = - \sum_{i=1}^{I} \int_{\partial\Omega_i} \mu q \cdot \nu_i \, d\sigma, \quad q \in X, \mu \in M. \tag{5.9}$$

It is easy to check that $V = \{q | q \in X, b(q, \mu) = 0 \text{ for all } \mu \in M\} = Q_0(\Omega)$ and that

$$\sup_{q \in X} \frac{b(q, \mu)}{\|q\|_{0,\Omega}} = \|\mu\|_M. \tag{5.10}$$

Therefore, by Theorem 1, the following problem "*Find* $(\mathbf{p}, \lambda) \in X \times M$
such that

$$a(\mathbf{p}, \mathbf{q}) + b(\mathbf{q}, \lambda) = -a(\mathbf{p}_0, \mathbf{q}) \quad \textit{for all} \quad q \in X$$
$$b(\mathbf{p}, \mu) = -b(\mathbf{p}_0, \mu) \quad \textit{for all} \quad \mu \in M'' \tag{5.11}$$

has a unique solution. Moreover, this solution $(\mathbf{p}, \lambda)$ *is related to the solution u of problem* (1.1) *by*

$$\begin{cases} \mathbf{p} + \mathbf{p}_0 = \operatorname{grad} u, \\ \lambda|_{\partial\Omega_i} = u|_{\partial\Omega_i}, \quad 1 \leqslant i \leqslant I. \end{cases} \tag{5.12}$$

Let us give now a short description of the numerical approximation of problem (1.1) by using the variational formulation (5.11). Again, let us assume that $\bar{\Omega}$ is a polygon of $\mathbb{R}^2$ and that $\bar{\Omega} = \cup_{i=1}^I \bar{\Omega}_i$ is a triangulation of $\bar{\Omega}$ with triangles $\bar{\Omega}_i$ with diameters $\leqslant h$.

Let $k \geqslant 1$ be an integer. We introduce the following finite-dimensional subspace X_h of the space X:

$$X_h = \{q | q \in X, q|_{\Omega_i} \in (P_{k-1}(\Omega_i))^2, \quad 1 \leqslant i \leqslant I\}. \tag{5.13}$$

We shall study two different choices of the space M_h.

Example 1
Let $m \geqslant 1$ be an integer. We define the finite-dimensional subspace M_h of M:

$$M_h = \left\{ \mu | \mu \in \prod_{i=1}^I H^{1/2}(\partial\Omega_i), \quad \text{there exists a function } v \in C^\circ(\bar{\Omega}) \cap H_0^1(\Omega) \right.$$

$$\left. \text{such that} \quad \mu|_{\partial\Omega_i} = v|_{\partial\Omega_i} \in P_m(\partial\Omega_i) \quad 1 \leqslant i \leqslant I \right\}, \tag{5.14}$$

where $P_m(\partial\Omega_i)$ denotes the space of the restrictions over $\partial\Omega_i$ of all polynomials of degree $\leqslant m$.

Clearly, the V_h-ellipticity of the bilinear form $a(\mathbf{p}, \mathbf{q})$ is always satisfied. On the other hand, condition (2.13) holds if and only if

$$\left. \begin{array}{l} \mu \in P_m(\partial\Omega_i) \\[2mm] \displaystyle\int_{\partial\Omega_i} \mu \mathbf{q} \cdot \mathbf{v}_i \, d\sigma = 0 \quad \text{for all} \quad \mathbf{q} \in (P_{k-1}(\Omega_i))^2 \quad \text{with} \quad \operatorname{div} \mathbf{q} = 0 \end{array} \right\} \Rightarrow \mu = 0. \tag{5.15}$$

One can prove that, for $k = 1, 2$, condition (5.15) (and therefore condition (2.13)) holds if and only if

$$m \leqslant k. \tag{5.16}$$

We conjecture that this property holds for $k > 2$.

Assume condition (5.16). Then, one can check that V_h *is not a subspace of* $V = Q_0(\Omega)$. We get here a *hybrid method of Pian type*: cf. Pian [9].

Example 2
We shall study here a hybrid method *where the space M_h is not a subspace of M*. We set:

$$M_h = \left\{ \mu | \mu \in \prod_{i=1}^{I} S_{m-1}(\partial\Omega_i), \quad \mu|_{\partial\Omega_i} = \mu|_{\partial\Omega_j} \quad \text{on} \quad \partial\Omega_i \cap \partial\Omega_j \right\}. \quad (5.17)$$

In this case, condition (2.13) becomes

$$\left. \begin{array}{l} \mu \in S_{m-1}(\partial\Omega_i), \\[2mm] \int_{\partial\Omega_i} \mu q \cdot v_i \, d\sigma = 0 \quad \text{for all} \quad q \in (P_{k-1}(\Omega_i))^2 \quad \text{with} \quad \text{div } q = 0 \end{array} \right\} \Rightarrow \mu = 0.$$

$$(5.18)$$

Here again, one can prove that, for $k = 1, 2$, condition (5.18) holds if and only if $m \leqslant k$ and we conjecture that this property is still valid for $k > 2$.

Assume now that $m = k$. Then, it is easy to see that V_h *is a subspace of* $V = Q_0(\Omega)$. We get here the so-called *equilibrium method* of Fraejs de Veubeke cf. Fraejs de Veubeke [6], [7].

For a detailed treatment of these dual hybrid methods and the corresponding error analysis, we refer to Thomas [14].

References

[1] Brezzi, F. (1974). On the existence, uniqueness and approximation of saddle point problems arising from Lagrangian multipliers, *R.A.I.R.O.* R.2, 129–151.

[2] Brezzi, F. Sur la méthode des éléments finis hybrides pour le problème biharmonique, (to appear).

[3] Brezzi, F. and Marini, L. D. On the numerical solution of plate bending problems by hybrid methods, (to appear in *R.A.I.R.O. série Mathématiques*).

[4] Ciarlet, P. G. and Raviart, P. A. *La méthode des éléments finis pour les problèmes elliptiques*, (to appear).

[5] Crouzeix, M. and Raviart, P. A. (1973). Conforming and non conforming finite element method for solving the stationary Stokes equations, *R.A.I.R.O.* R.3, 33–76.

[6] Fraejs de Veubeke, B. (1965). Displacement and equilibrium models in the finite element method, *Stress Analysis* (O. C. Zienkiewicz and G. S. Holister, eds.), Ch. 9, 145–197, Wiley.

[7] Fraejs de Veubeke, B. (1973). Diffusive equilibrium models, Lecture notes, University of Calgary.

[8] Irons, B. M. and Razzaque, A. (1972). Experience with the patch test for convergence of finite elements, *The Mathematical Foundations of the Finite Element Method* . . . (A. K. Aziz, ed.), 557–587, Academic Press, New York.

[9] Pian, T. H. H. (1971). Formulations of finite element methods for solid continua, *Recent Advances in Matrix Methods* . . . (R. H. Gallagher, Y. Yamada and J. T. Oden, eds.), 49–83. The University of Alabama Press.

[10] Pian, T. H. H. (1972). Finite element formulation by variational principles with relaxed continuity requirements, *The Mathematical Foundations of the Finite Element Method* . . . (A. K. Aziz, ed.), 671–687, Academic Press.

[11] Pian, T. H. H. and Tong, P. (1969). Basis of finite element methods for solid continua, *Int. J. Numer. Meth. Eng.* 1, 3–28.

[12] Raviart, P. A. and Thomas, J. M. Hybrid finite element methods and 2nd order elliptic problems, (to appear).

[13] Strang, G. and Fix, G. (1973). *An Analysis of the Finite Element Method.* Prentice Hall.

[14] Thomas, J. M., (to appear).

Approximation of Fixed Points of Generalized Contraction Mappings

Ioan A. Rus

Section 1

The metric theory of the fixed point has developed extensively in recent years. A great number of works are concerned with giving some sufficient conditions, which assure the existence and uniqueness of a fixed point for a mapping f, from a metric space (X, d) to itself. Among these many research works we mention: Bianchini–Grandolfi [3], Browder [4], Čirič [5], Furi [6], Gorbunov [7], Iseki [8], Kannon [9], Reich [10], Rus [11], [12], ...

Almost all of these research works are based on the successive approximation method. This also follows from the theorem of Bessaga [2]:

Let X be a nonempty set and $f: X \to X$ with the property that $\cap_{n \in N} f^n(X) = \{a\}$ and let $\alpha \in \,]0, 1[$. Under these conditions, $\exists$ a metric d defined on X with the property that the metric space (X, d) is complete and the mapping f satisfies the condition

$$d(f(x), f(y)) \leqslant \alpha d(x, y).$$

Because such a metric is difficult to work with the following problem arises if (X, d) is a given metric space and $f : X \to X$ a mapping: find conditions on f with respect to the metric d, which assure the existence and uniqueness of a fixed point for f and which can be obtained by the successive approximation method. As a result of this type we mention the following.

Theorem 1 (Furi [6])
Let (X, d) be a complete metric space and $f: X \to X$, such that

$$d(f(x), f(y)) \leqslant \omega(d(x, y)), \quad \forall x, y \in X$$

where $\omega : \mathbb{R}_+ \to \mathbb{R}_+$, $\omega(r) = (1 - \theta(r))r$, satisfies the conditions:

(i) $r_1 \leqslant r_2, \omega(r_1) \leqslant \omega(r_2)$

(ii) $\lim\limits_{n \to \infty} \omega^n(r) = 0, \forall r \geqslant 0$

(*iii*) ω *is continuous*

(*iv*) $0 \leqslant \omega(r) < r$

(*v*) $0 < r_1 \leqslant r_2 \Rightarrow \theta(r_2)/\theta(r_1) \geqslant c > 0$

(*vi*) $F(r) = \displaystyle\int_0^r \frac{ds}{\theta(s)} < +\infty, \ \forall r \geqslant 0.$

Then $\exists \, ! \, x^* \in X : f(x^*) = x^*$ *and for each* $x_0 \in X$, $x_n = f^n(x_0)$ *converges to* x^*. *Moreover*

$$d(x_n, x^*) \leqslant \frac{1}{C} F(d(x_0, f(x_0))).$$

Theorem 2 (Rus [12])
Let (X, d) *be a complete metric space, and* $f:X \to X$, *such that*

(*i*) $\exists \alpha \in R_+, \alpha < 1 : d(f^2(x), f(x)) \leqslant \alpha d(x, f(x)), \ \forall x \in X$

(*ii*) $\exists \chi : R_+^3 \to R_+$, *continuous, such that* $d(f(x), f(y)) \leqslant \chi(d(x, f(x)),$
$d(y, f(y)), d(x, y)), \ \forall x, y \in X,$

and

$$r > \chi(0, r, 0), \quad \text{if} \quad r > 0.$$

Then f *has at least one fixed point. In addition if*

(*iii*) $\exists \psi : R_+^3 \to R_+$, *continuous, such that*

$$d(f(x), f(y)) \leqslant \psi(d(x, f(x)), d(y, f(y)), d(x, y)), \ \forall x, y \in X,$$

and

$$r > \psi(0, 0, r), \quad \text{if} \quad r > 0,$$

then f *has a unique fixed point.*
    Moreover, for each $x_0 \in X$, $x_n = f^n(x_0)$ *converges to the fixed point of* f.
    As a consequence of Theorem 2 we have

Theorem 3 (Reich [10], Rus [12])
Let (X, d) *be a complete metric space, and* $f:X \to X$ *a mapping for which there exist numbers* $\alpha, \beta \in \mathbb{R}_+$, $\alpha + 2\beta < 1$, *such that*

$$d(f(x), f(y)) \leqslant \alpha d(x, y) + \beta[d(x, f(x)) + d(y, f(y))]$$

for all $x, y \in X$.
    Then f *has a unique fixed point. Moreover, for each* $x_0 \in X$, $x_n = f^n(x_0)$ *converges to the fixed point of* f.
    The purpose of the present paper is to give an upper bound for the roundoff error in computing the fixed point by means of Theorems 1 and 3.

Section 2

Let (X, d) be a complete metric space and $f : X \to X$, as in Theorem 1. In addition we suppose

$$\omega(r_1 + r_2) \leqslant \omega(r_1) + \omega(r_2), \quad \forall r_1, r_2 \in \mathbb{R}_+.$$

Let $g : X \to X$ be a mapping which approximates the mapping f. More precisely we assume that

$$d(f(x), g(x)) \leqslant \eta, \quad \text{with given} \quad \eta \in \mathbb{R}, \quad \forall x \in X.$$

Let $x_0 \in X$ and $x_n = f^n(x_0)$, $y_n = g^n(x_0)$. Let x^* be the unique fixed point f. We propose to give an estimate of $d(x^*, y_n)$. We have

Theorem 4
We assume that the conditions given above are satisfied. Then

$$d(y_n, x^*) \leqslant \eta + \omega(\eta) + \ldots + \omega^{n-1}(\eta) + \frac{1}{c} F(d(x_0, x_1)).$$

Proof
We have

$$d(y_n, x^*) \leqslant d(y_n, x_n) + d(x_n, x^*).$$

On the other hand

$$\begin{aligned}
d(y_n, x_n) &\leqslant d(g(y_{n-1}), f(y_{n-1})) + d(f(y_{n-1}), f(x_{n-1})) \\
&\leqslant \eta + \omega(d(y_{n-1}, x_{n-1})) \\
&\leqslant \eta + \omega(\eta + \omega(d(y_{n-2}, x_{n-2}))) \\
&\qquad \cdots \cdots \cdots \cdots \cdots \cdots \cdots \\
&\leqslant \eta + \omega(\eta) + \ldots + \omega^{n-1}(\eta)
\end{aligned}$$

and Theorem 4 follows from Theorem 1.

Section 3

Let (X, d) be a complete metric space and $f : X \to X$, as in Theorem 3. Let $g : X \to X$, be such that

$$d(f(x), g(x)) \leqslant \eta, \quad \forall x \in X.$$

Let $x_0 \in X$, and $x_n = f^n(x_0)$, $y_n = g^n(x_0)$. Let x^* be the unique fixed point of f. We have

Theorem 5
We suppose that the conditions of the Theorem 3 are satisfied. Then

$$d(y_n, x^*) \leqslant \alpha^{n-1}\eta + \alpha^{n-2}F(2) + \ldots + F(n) + \frac{k^n}{1-k} d(x_0, x_1)$$

where

$$F(n) = \eta + \eta\beta \frac{1+\alpha}{1-\alpha-2\beta} + \beta(k^n + k^{n-1})d(x_0, x_1)$$

and

$$k = \frac{\alpha + \beta}{1 - \beta} \, .$$

Proof

We have

$$d(y_n, x^*) \leqslant d(y_n, x_n) + d(x_n, x^*)$$

and

$$d(x_n, x^*) \leqslant \frac{k^n}{1-k} \, d(x_0, x_1).$$

On the other hand (see Bellen-Volčič [1]),

$$d(y_{n-1}, f(y_{n-1})) \leqslant \eta \frac{1+\alpha}{1 -- \alpha - 2\beta} + k^n d(x_0, x_1) \, .$$

Hence we have

$$\begin{aligned}
d(x_n, y_n) &\leqslant d(g(y_{n-1}), f(y_{n-1})) + d(f(y_{n-1}), f(x_{n-1})) \\
&\leqslant \eta + \alpha d(y_{n-1}, x_{n-1}) + \beta[d(y_{n-1}, f(y_{n-1})) + \\
&\quad + d(x_{n-1}, f(x_{n-1}))] \leqslant F(n) + \alpha d(y_{n-1}, x_{n-1})
\end{aligned}$$

$$\dots\dots\dots\dots\dots\dots\dots\dots\dots$$

$$\leqslant \alpha^{n-1}\eta + \alpha^{n-2}F(2) + \dots + F(n)$$

which completes the proof.

References

[1] Bellen, A. and Volčič, A. (1973). Sulla propagazione dell'errore sistematico in alcuni procedimenti iterativi, *Rend. Ist. di Matem. Univ. Trieste* 5, 161-167.

[2] Bessaga, C. (1959). On the converse of the Banach "Fixed point principle", *Colloq. Math.* 7, 41-43.

[3] Bianchini, R. M. and Grandolfi, M. (1968). Trasformazioni di tipo contractivo generalizzato in uno spazio metrico, *Atti Accad. Naz. Lincei* 45, 212-216.

[4] Browder, F. E. (1968). On the convergence of successive approximations for nonlinear functional equations, *Indagationes Math.* 30, 27-35.

[5] Čirič, L. B. (1971). Generalized contraction and fixed-point theorems, *Publ. l'Inst. Math.* 12, 19-26.

[6] Furi, M. (1968). Un teorema di punto fisso per trasformazioni di uno spazio metrico completo in sé, *Atti Accad. Naz. Lincei* 45, 207-211.

[7] Gorbunov, A. D. (1974). Generalization of the method of contracting mapping, *Dokl. Akad. Nauk SSSR* **215**, 1285–1288.

[8] Iseki, K. (1974). Generalizations of Kannan fixed point theorems. Mathematics Seminar Notes, V, January.

[9] Kannon, R. (1968). Some results on fixed points, *Bull. Calcutta Math. Soc.* **60**, 71–76.

[10] Reich, S. (1971). Some remarks concerning contraction mappings, *Canad. Math. Bull.* **14**, 121–124.

[11] Rus, I. A. (1971). Some fixed point theorems in metric spaces, *Rend. Ist. di Matem. Univ. Trieste* **3**, fas. II.

[12] Rus, I. A. (1973). Teoria punctului fix. II: Teoria punctului fix în analiza functională. Cluj.

The Use of Differential Inequalities in the Convergence Analysis of the Numerical Solution of Degenerate Parabolic Equations

E. Schechter

Introduction

Many proofs of the convergence of approximate solutions to the exact one rely on compactness arguments. This is especially the case when the problem is non-linear and smoothness properties of this solution are poor. The disadvantage of such an approach is that only the convergence of a subsequence can be stated, with no indication of its choice.

The aim of this paper is to give an example in which the convergence of the whole sequence can be proved and also to give some information on the rate of convergence. This is done in Section 4 for a second order non-linear degenerate parabolic equation, which is the principal part of a multi-dimensional Boussinesq type equation. It is also a generalization, for degenerate cases, of estimates based on difference inequalities [4]. It seems that the prolongation operator p_h^m used is not the only possible choice; perhaps continuations of the finite-element type [5] of class C^m may lead to better convergence. Finally we mention that, for problems with sufficiently smooth solutions, one can combine the results of Section 3 with classical ways of proving convergence [2].

1. Preliminaries

In this section, we collect some results which will be needed in the paper.

Theorem 1.1
Suppose that:

$$\bar{D}_+ u \leqslant au + b, \quad t \in \,]0, T] \tag{1.1}$$

$$u(0) = u_0 \tag{1.2}$$

where a, b, u_0, T *are constants,* $T \leqslant +\infty.$

Then

$$u(t) \leqslant \frac{b}{a}(e^{at} - 1) + u_0 e^{at}, \quad t \in [0, T].$$ (1.3)

Consider now a function

$$f : \,]0, T] \times G \times \; \mathbb{R}^6 \to \mathbb{R}$$

$T \leqslant +\infty$, $G \subset \mathbb{R}^2$ is a bounded domain and denote

$$L[u] = u_t - f(t, x, y, u, u_x, u_y, u_{xx}, u_{xy}, u_{yy}).$$

Theorem 1.2
Suppose that:

(i) *f is continuous.*

(ii) *f is a locally Lipschitz function with respect to u.*

(iii) *The functions*

$$u, v : \,]0, T[\times G \to \mathbb{R}$$

have continuous derivatives up to the second order in x, y and belong to
$C(\Omega) \cap C^1(\Omega)$

$$\Omega = \,]0, T[\times G.$$

(iv) $L[u] = F(t, x, y), \quad L[v] = G(t, x, y)$
$\quad F, G \in C(\Omega).$

(v) *f is elliptic, i.e.*

$$\sum_{i,j=1}^{2} (r_{ij} - \bar{r}_{ij}) \lambda_i \lambda_j \leqslant 0 \quad \text{for any} \quad \lambda \in \mathbb{R}^2$$

implies

$$f(t, x, y, z, q, r) \leqslant f(t, x, y, z, q, \bar{r}) \quad (t, x) \in \Omega.$$

(vi) $|F(t, x, y) - G(t, x, y)| \leqslant g, \quad (t, x, y) \in \Omega.$

(vii) $|u(0, x, y) - v(0, x, y)| \leqslant \epsilon \quad (x, y) \in G$

(viii) $|u(t, x, y) - v(t, x, y)| \leqslant \epsilon \quad (t, x, y) \in [0, T] \times \partial G.$

Under these assumptions

$$|u(t, x) - v(t, x)| \leqslant \frac{g}{a}(e^{at} - 1) + \epsilon e^{at} \text{ on } \Omega.$$ (1.4)

Here a is the Lipschitz constant corresponding to u and v.

Lemma 1.1
Suppose that:

(i) *The domain G is bounded and satisfies the ordinary cone property.*

(ii) $M \subset C^1(\overline{G})$ and there exists a constant $L > 0$ such that

$$\forall f \in M \quad |f|, |f_x|, |f_y| \leqslant L.$$

Under these assumptions, there is a constant C such that

$$\|f\|_\infty^3 \leqslant C \|f\|_{L_1(G)}, \quad \forall f \in M. \tag{1.5}$$

Proof
Let $f \in M$ and suppose $(x_0, y_0) \in G$ is the maximum point of $|f|$, and let $m_f = f(x_0, y_0)$. Suppose that $m_f \neq 0$. Denote also by S_f the surface represented by $z = f(x, y)$, $(x, y) \in G$. Because of (ii) there exists a vertical (inferior) cone K with vertex angle Γ, such that $\forall P \in S_f$ and $\forall f \in M$, the congruent (by translation) cone K_p does not intersect S_f. Denote by

$K_p(\Gamma, \gamma)$ the cone sector defined by Γ, γ; where γ is the angle defined by (i), and is independent of P. Let $P_0 = (x_0, y_0, m_f)$. Let $k_{p_0}(\Gamma, \gamma)$ denote the projection of $K_p(\Gamma, \gamma)$ on $\{z = 0\}$. If Γ is taken sufficiently small one has

$$k_{p_0}(\Gamma, \gamma) \subset C(x_0, y_0)$$

where $C(x_0, y_0)$ is the circular sector (of angle γ) defined by (i). Hence

$$\int_G |f| \, dx \, dy = \text{mes } K_{p_0}(\Gamma, \gamma) = \text{const} \times m_f \text{ mes } k_{p_0}(\Gamma, \gamma)$$

$$= \text{const} \times (m_f)^3 \text{ ctg } \Gamma/2 = C(m_f)^3.$$

Remarks
If $G \subset \mathbb{R}^n$, then instead of (1.5) we have

$$\|f\|_\infty^{n+1} \leqslant C \|f\|_{L_1(G)}, \quad \forall f \in M \tag{1.6}$$

If in addition $\exists\, \alpha > 0$ such that

$$|f| > \alpha, \quad \forall f \in M, \quad then,$$

$$\|f\|_\infty \leqslant C \|f\|_{L_1(G)}, \quad \forall f \in M. \tag{1.7}$$

2. Continuation of a Discrete Function

We have defined in [7], on the basis of an interpolation operator given by Riabenkii and Filippov [6], the extension operator:

$$p_h^m : c(R_{hk}) \to C^m(0, T \times \mathbb{R}^2)$$

$$m = (m_0, \bar{m}, \bar{m}).$$

Here $R_h = h \ \mathbb{Z}^n, h \in \mathbb{R}_+, R_{hk} = R_k \times R_h; R_k$ being an equidistant mesh $(0 \in R_k)$ placed on $[0, T]$, with the width $k = \lambda h^2, \lambda = $ const. By $c(R_{hk})$ (resp $c(R_h)$) we denote the linear space of real functions defined on R_{hk}(resp R_h). In a similar manner we define:

$$p_{h0}^m : c(R_h) \to C^m(\mathbb{R}^2).$$

We recall some of the properties of p_h^m :

I. If $r_h : C([0, T] \times \mathbb{R}^2) \to c(R_{hk})$ is the restriction operator, then:

$$r_h p_h^m v_h = v_h, \quad \forall v_h \in c(R_{hk}).$$

II. Suppose that $v_h : c(R_{hk}) \to \mathbb{R}$ have, for $h \in]0, h_0]$, their supports in a fixed compact of $\mathbb{R}^3$ then:

(a) If $v_h, h \in]0, h_0]$ are equibounded together with their divided differences up to the first order in t and second order with respect to x, y there exists a constant C independent of h such that:

$$|D^j p_h^m v_h(t, x, y)| \leqslant C, \quad (t, x, y) \in [0, T] \times \mathbb{R}^2$$
$$j_0 \leqslant 1, \quad 0 \leqslant j_1, j_2 \leqslant 2, \quad m_0 \leqslant 1, \quad \bar{m} \leqslant 2.$$

(b) If for $h \in]0, h_0]$ (see Section 3)

$$h^2 \sum_{R_h} \left| \Delta v_h \right|, h^2 \sum_{R_h} \left| \Delta_x^2 v_h \right|, \text{resp } h^2 \sum_{R_h} | \nabla_y^2 v_h |$$

are equibounded for all n, then

$$\int_G \left| \frac{\partial}{\partial t} (p_h^m v_h) \right| dx \, dy, \qquad \int_G \left| \frac{\partial^2}{\partial y^2} (p_h^m v_h) \right| dx \, dy$$

resp.

$$\int_G \left| \frac{\partial^2}{\partial y^2} (p_h^m v_h) \right| dx \, dy$$

are also equibounded, provided that $m_0 \geqslant 1, \bar{m} \geqslant 2$.

Now, let

$$Q_\alpha =]x_{\alpha_1}, x_{\alpha_1} + h] \times]y_{\alpha_2}, y_{\alpha_2} + h], \quad (x_{\alpha_1}, y_{\alpha_2}) \in R_h$$

then

$$\partial G_h = \{(x_{\alpha_1}, y_{\alpha_2}) \in R_n | Q_\alpha \cap \partial G \neq \phi\}$$

III. If $q \leqslant m, s_1 + s_2 \leqslant q, s = s_1 + s_2$

$$\nabla_x^s(t, x) = 0(h^{q - s_1 - s_2}), \quad \nabla_y^s(t, x) = 0(h^{q - s_1 - s_2})$$

for all $(t, x) \in R_k \times \partial_n G$, then,

$$p_h^m v_h|_{[0,T] \times \partial G} = 0(h^q).$$

3. Boundedness of Differences

Consider the Cauchy–Dirichlet problem:

$$u_t = a(u)\,\Delta u \text{ on } \Omega. \tag{3.1}$$

$$u(0, x, y) = \varphi(x, y), \quad (x, y) \in G \tag{3.2}$$

$$u|_{[0,T]\times\partial G} = 0 \tag{3.3}$$

with $a(u) \geqslant 0$.

This is replaced by the difference problem:

$$\Delta v_{ij}^n = a(v_{ij}^{n-1})(\nabla_x^2 v_{i+1,j}^{n-1} + \nabla_y^2 v_{i,j+1}^{n-1}) \tag{3.4}$$

$$(t_n, x_i, y_j) \in R_{hk}', \quad R_{hk}' = R_k \times G_h$$

$$v_{ij}^n = 0 \quad (t_n, x_i, y_j) \in R_k \times \partial G_h \tag{3.5}$$

$$v_{ij}^0 = r_h\varphi \quad (x_i, y_j) \in G_h. \tag{3.6}$$

Here $G_h = \{R_h \cap G\}\backslash\{\partial G_h \cap G\}$. The discrete function v_{ij}^n is extended by zero outside $R_k \times \{G_h \cup \partial G_h\}$. We also use the notation:

$$\Delta v_{ij}^n = \frac{1}{k}(v_{ij}^n - v_{ij}^{n-1}), \quad \nabla_x v_{ij}^n = \frac{1}{h}(v_{ij}^n - v_{i-1,j}^n)$$

$$\nabla_y v_{ij}^n = \frac{1}{h}(v_{ij}^n - v_{i,j-1}^n); \quad \nabla_x^2 = \nabla_x \nabla_x, \quad \nabla_y^2 = \nabla_y \nabla_y.$$

(a) Suppose that: $\varphi|_{\partial G} = 0$ and $|\varphi| \leqslant K$ on $\bar{G}$: $a \leqslant A$ for $|u| \leqslant K$ and $\lambda = k/h^2 = \frac{1}{4}A$. Under these assumptions $|v_{ij}| \leqslant K$ on R_{hk}. Indeed, by (3.4)–(3.6)

$$v_{ij}^1 = (1 - 4\lambda a_{ij}^0)v_{ij}^0 + A\lambda[v_{i+1,j}^0 + v_{i,j+1}^0 + v_{i-1,j}^0 + v_{i,j-1}^0]$$

which implies that:

$$|v_{ij}^1| \leqslant \max \{|v_{ij}^0|, |v_{i+1,j}^0|, |v_{i,j+1}^0|, |v_{i-1,j}^0|, |v_{i,j-1}^0|\}$$

i.e. $|v_{ij}^1| \leqslant K$. The assertion follows by induction. We notice that $\varphi \geqslant 0$ implies $v_{ij}^n \geqslant 0$.

(b) In order to uniformly estimate the first order differences we suppose in addition that a is non-decreasing for $u \geqslant 0$. Also assume that $\varphi \geqslant 0$ and $|\varphi_x|, |\varphi_y| \leqslant K$.

Suppose that $(x_i, y_j) \in G_h$, then

$$\Delta \nabla_x v_{i+1,j}^1 = \frac{1}{h} a(v_{i+1,j}^0)(\nabla_x^2 v_{i+2,j}^0 + \nabla_y^2 v_{i-1,j+1}^0) -$$

$$- a(v_{i,j}^0)(\nabla_x^2 v_{i+1,j}^0 + \nabla_y^2 v_{i,j+1}^0).$$

We shall consider all possible cases.

Denote:

$$\alpha = \nabla_x^2 v_{i+2,j}^0 + \nabla_y^2 v_{i+1,j+1}^0,$$

$$\beta = \nabla_x^2 v_{i+1,j}^0 + \nabla_y^2 v_{i,j+1}^0.$$

(1) $\alpha \geqslant 0, \beta \leqslant 0$ then:

$$\Delta \nabla_x v_{i+1,j}^1 \leqslant \frac{A}{h} [\nabla_x^2 v_{i+2,j}^0 + \nabla_y^2 v_{i+1,j+1}^0 - \nabla_x^2 v_{i+1,j}^0 - \nabla_y^2 v_{ij+1}^0]$$

$$= A [\nabla_x^2 \nabla_x v_{i+2,j}^0 + \nabla_y^2 \nabla_x v_{i+1,j+1}^0].$$

$$\nabla_x v_{i+1,j}^1 \leqslant A\lambda [\nabla_x v_{i+2,j}^0 + \nabla_x v_{i,j}^0 + \nabla_x v_{i,j+1}^0 + \nabla_x v_{i,j-1}^0] +$$

$$+ (1 - 4\lambda) \nabla_x v_{i+1,j}^0,$$

which leads to $|\nabla_x v_{i+1,j}^1| \leqslant K$.

Suppose now $\nabla_x v_{i+1,j}^0 \geqslant 0$.

(2) $\alpha < 0$.

$$\nabla_x v_{i+1,j}^1 = \nabla_x v_{i+1,j}^0 + ka(v_{i,j}^0)(\nabla_x^2 \nabla_x v_{i+2,j}^0 + \nabla_y^2 \nabla_x v_{i+1,j+1}^0)$$

$$+ k(a(v_{i+1,j}^0) - a(v_{ij}^0))\alpha,$$

and the last term on the right can be dropped because a is non-decreasing. By the same argument as for (1) we have $|\nabla x_{i+1,j}^1| \leqslant K$.

(3) $\beta > 0$.

$$\nabla_x v_{i+1,j}^1 = \nabla_x v_{i+1,j}^0 + ka(v_{i+1,j}^0)(\nabla_x^2 \nabla_x v_{i+2,j}^0 + \nabla_y^2 \nabla_x v_{i+1,j+1}^0) -$$

$$- k(a(v_{i+1,j}^0) - a(v_{i,j}^0)).$$

and in the same manner $|\nabla_x v_{i+1,j}^1| \leqslant K$.

(4) If $\nabla_x v_{i+1,j}^0 < 0$ and we are not in case (1), we have either

$$\nabla_x v_{i+1,j}^1 = 4A\lambda\alpha \quad \text{if} \quad \alpha \geqslant 0$$

or

$$\nabla_x v_{i+1,j}^1 = 2\lambda a(v_{ij}^0)(-\beta) \quad \text{if} \quad \alpha \leqslant 0.$$

In both cases we have a combination of ∇_x with the sum of coefficients $4A\lambda < 1$.

Again $|\nabla_x v_{i+j,j}^0| \leqslant K$.

Now let us consider boundary points.

If $(x_{i+1}, y_j) \in \partial G_h$, $(x_i, y_j) \in G_h$

$$\nabla_x v_{i+1,j}^1 = -\frac{v_{ij}^0}{h} - \frac{k}{h} a(v_{ij}^0)\beta$$

and the argument is similar to that of (4). If $(x_{i+1}, y_j) \in G_h$, $(x_i, y_j) \in \partial G_h$, $a(v_{ij}^0) = 0$, $\nabla_x v_{i+1,j}^0 \geqslant 0$ and one can follow (1), (2), (3).

By induction we obtain

$$|\nabla_x v_{i+1,j}^n| \leqslant K \quad \text{on} \quad R_{hk}$$

and analogously

$$|\nabla_y v_{i,j+1}^n| \leqslant K \quad \text{on} \quad R_{hk}.$$

(c) We suppose now that φ has continuous second order derivatives and

$$|\varphi_{xx}|, |\varphi_{yy}| \leqslant K.$$

$$\nabla_x^2 v_{i+1,j}^1 + \nabla_y^2 v_{i,j+1}^1$$
$$= \nabla_x^2 v_{i+1,j}^0 + \nabla_y^2 v_{i,j+1}^0 + a(v_{i+1,j}^0)(\nabla_x^2 v_{i+2,j}^0 + \nabla_y^2 v_{i+1,j+1}^0) -$$
$$- 2a(v_{ij}^0)(\nabla_x^2 v_{i+1,j}^0 + \nabla_y^2 v_{i,j+1}^0) + a(v_{i-1,j}^0)(\nabla_x^2 v_{ij}^0 +$$
$$+ \nabla_y^2 v_{i-1,j+1}^0) + a(v_{i,j+1}^0)(\nabla_x^2 v_{i+1,j+1}^0 + \nabla_y^2 v_{i,j+2}^0) -$$
$$- 2a(v_{ij}^0)(\nabla_x^2 v_{i+1,j}^0 + \nabla_y^2 v_{ij+1}^0) + a(v_{i,j-1}^0)(\nabla_x^2 v_{i+1,j-1}^0 +$$
$$+ \nabla_y^2 v_{i,j}^0).$$

Hence

$$h^2 \sum_{R_h} |\nabla_x^2 v_{i+1,j}^1 + \nabla_y^2 v_{i,j+1}^1| \leqslant h^2 \sum_{R_h} |\nabla_x^2 v_{i+1,j}^0 + \nabla_y^2 v_{i,j+1}^0|$$

$$\leqslant h^2 \left\{ \sum_{R_h} |\nabla_x^2 v_{i+1,j}^0| + \sum_{R_h} |\nabla_y^2 v_{i,j+1}^0| \right\} \leqslant C_1.$$

Since

$$h^2 \sum |\nabla_x^2 v_{i+1,j}^n + \nabla_y^2 v_{i,j+1}^n| = k^2 \sum |\nabla_x^2 v_{i+1,j}^{n-1} + \nabla_y^2 v_{i,j+1}^{n-1}|$$

we deduce that

$$h^2 \sum_{R_h} |\nabla_x^2 v_{i+1,j}^n + \nabla_y^2 v_{i,j+1}^n| \leqslant C_1, \quad \forall t_n \in R_k.$$

(d) $h^2 \sum_{R_h} |(v_{ij}^{n+1} - v_{ij}^n)/k| \leqslant h^2 \sum_{R_h} a(v_{ij})|\nabla_x^2 v_{i+1,j}^n + \nabla_y^2 v_{i,j+1}^n| \leqslant 2AC_1.$

Thus we have proved the following:

Theorem 3.1

Suppose that

(i) *$\varphi:\bar{G} \to \mathbb{R}'$ has continuous second order derivatives and there exists a constant K such that,*

$$|\varphi|, |\varphi_x|, |\varphi_y|, |\varphi_{xx}|, |\varphi_{yy}| \leqslant K.$$

(ii) *$\varphi|_{\partial G} = 0, \quad \varphi(x,y) \geqslant 0, \quad (x,y) \in G.$*

(iii) *a is non-decreasing for positive arguments, and $a(u) \leqslant A$ for $0 \leqslant u \leqslant K.$*

   A being a constant.

(iv) $\lambda = k/h^2 = \frac{1}{4}/A$.

Under these assumptions:

(j) $|v^n_{ij}|,\, |\, \nabla_x v^n_{i+1,j}|,\, |\, \nabla_y v^n_{i,j+1}| \leqslant K$ on R_{hk}.

(jj) $h^2 \sum_{R_h} |\, \nabla^2_x v^n_{i+1,j} + \nabla^2_y v^n_{ij+1}| \leqslant 2C_1$ $\forall t_n \in R_k$.

(jjj) $h^2 \sum_{R_h} |\Delta v^n_{ij}| \leqslant 2AG$ $\forall t_n \in R_k$,

where C_1 is a constant depending on K and G.

Remark
With suitable conditions on a, the theorem may be extended to

$u_t = a(t, x, y, u)\, \Delta u.$

In particular all the above results are valid for

$u_t = a(x, y)\, \Delta u$

with $a \geqslant 0$.

4. Convergence of Approximate Solutions

In order to study the strong convergence of the continued numerical solution, we recall [1] a general formulation of the concepts of consistency and adjustability.

Consider the following differential problem:

$Lu = F(t, x, y)$ on Ω (4.1)

$u/(t = 0) = \varphi(x, y)$ on G (4.2)

$u/[0, T] \times \partial G = \psi(t, x, y).$ (4.3)

We associate the difference problem:

$L_h v_h = F_h(x, y)$ on $R_k \times G'_h$ (4.4)

$v_h(0, x, y) = \varphi_h(x, y)$ on G'_h (4.5)

$v_h(t, x, y) = \psi_h(t, x, y)$ on $R_k \times \partial G_h.$ (4.6)

Definition
(a) The operators $L_h;\, h \in\,]0, h_0]$ are said to be *adjustable* of order $q \geqslant 0$ to L if:

$$\max_{\bar{\Omega}} |p^m_h L_h u_h - L p^m_h u_h| = 0(h^q)$$

for all $u_h \in D(L_h)$ such that $|u_h| \leqslant M$, M being a constant independent of h.

(b) L_h are *consistent* of order q with L, on the set $S \subset D(L)$ if

$$\max_{R_{hk}} |r_h Lu - L_h r_h u| = 0(h^q).$$

Theorem 4.1
Suppose that:

(i) $F = 0, F_h = 0$ and $L_h v_h = 0.$
(ii) L_h *are consistent of order q with L on* $S = p_h v_h.$
Under these conditions

$$\max_{\Omega} |L p_h^m v_h| = 0(h^q) \quad \text{uniformly on } S.$$

This immediately follows from Theorem 3.1 and consistency.
Returning now to problems (3.1)-(3.3) and (3.4)-(3.6), we notice that according to Theorem 3.1,

$$h \int_{\mathbb{R}^2} |D_x^3 p_h^m v_h| \, dx \, dy, h \int_{\mathbb{R}^2} |D_y^3 p_h^m v_h| \, dx \, dy$$

are equibounded. That means that we have consistency and adjustability of order zero, i.e.:

$$\frac{\partial}{\partial t} (p_h^m v_h) = a(p_h^m v_h) \Delta (p_h^m v_h) + R_h(t, x, y) \tag{4.7}$$

where

$$|R_h| = 0(1) \quad \text{uniformly in } v_h.$$

Let now $h_n \to 0$. For simplicity we denote

$$p_{h_n}^m v_{h_n} = v_n.$$

Equation (4.7) now reads:

$$\frac{\partial v_n}{\partial t} = a(v_n) \Delta v_n + R_n(t, x, y), \quad n \in \mathbb{N}. \tag{4.7}$$

Suppose that we had two sequences uniformly converging to different limits (such subsequences exist by Arzela's theorem):

$$u_n \to u, \quad v_n \to v$$

and

$$\max_{\bar{G}} |w_n| = \max_{\bar{G}} |u_n - v_n| > \epsilon > 0, \quad \text{for} \quad n \in \mathbb{N}.$$

Then

$$\int_G \left| \frac{\partial}{\partial t} w_n \right| dx \, dy = \int_G |(a(u_n) - a(v_n)) \Delta u_n| \, dx \, dy +$$
$$+ \int_G a(v_n) |\Delta w_n| \, dx \, dy + \int_G \bar{R}_n \, dx \, dy \tag{4.8}$$

where $\bar{R}_n$ has an obvious meaning.

By Theorem 4.1 we get

$$\int\limits_{G} \left| \frac{\partial}{\partial t} w_n \right| dx\, dy \leqslant M,$$

M being a constant independent of n.

This implies that

$$\bar{D}_+ \| w_n \|_1 \leqslant \frac{M}{\epsilon^3} (\max_{\bar{G}} |w_n|)^3$$

and by Lemma 1.1,

$$\bar{D}_+ \| w_n \|_1 \leqslant C \| w_n \|_1.$$

We also have $\| w_n \| (0) \leqslant C h_n^2$. Then by Theorem 1.1

$$\| w_n \|_1 \leqslant C_1 h_n^2 e^{CT}, \quad \text{a contradiction.}$$

Moreover, the convergence is at least $O[(- \ln h)^{-1}]$.

Indeed, let $\varphi(h_n)$ be such that

$$\max_{\bar{G}} |w_n| \varphi(h_n) < C, \quad \varphi > 0$$

and $-\varphi(h_n) \ln h_n \to +\infty$. Then:

$$\bar{D}_+ \| w_n \|_1 \leqslant - 3M (\max_{\bar{G}} w_n)^3 \ln h_n$$

$$\bar{D}_+ \| w_n \|_1 \leqslant - 3M_1 \| w_n \|_1 \ln h_n$$

with $\| w_n \|_1 (0) \leqslant C h_n^2$. This leads to:

$$\| w_n \|_1 \leqslant c h^2 e^{-M_1 t \ln h} = c h^{2 - t M_1}$$

Taking $t_1 = 1/M_1$, we get for $t \in [0, t_1]$

$$\| w_n \|_1 \leqslant ch, \text{a contradiction, (Lemma 1.1).}$$

Thus we have proved:

Theorem 4.2
Let v_h be the solution of problem (4.4)–(4.6). Suppose the conditions of Theorem 4.1 are fulfilled.

Under these assumptions $p_h^m v_n$ uniformly converges to the solution of (3.1)–(3.3). The convergence is at least of order $O[(-\ln h)^{-1}]$.

An important example of a function a which satisfies these conditions is $a(u) = |u|^p, p \in \mathbb{R}, p \geqslant 1$.

References

[1] Aubin, J. P. (1967). Approximation des espaces des distributions et des opérateurs differentiels, *Bull. Soc. Math. France*, Mem. 12.

[2] Ansorge, J. P. and Hass, R. (1970). Konvergentz von Differenzenverfahren für lineare und nichtlineare Aufangswertaufgabe. *Lecture Notes in Math. No. 159.* Springer Verlag, Berlin–Heidelberg–New-York.

[3] Graveleau, J. L. and Jamet, P. (1971). A finite difference approach to some degenerate nonlinear parabolic equations, *SIAM J. Appl. Math.* 20 (2), 199–223.

[4] Krawczyk, R. (1962). Uber Differenzenverfahren bei parabolischen Differentialgleichungen. *Arch. Rat. Mech. Anal.* 2, 81–121.

[5] Raviart, P. A. (1972). *Methode des Elements Finis,* Univ. Paris, VI.

[6] Rjabenkii, V. S. and Filippow, A. F. (1960). *Uber die Stabilität von Differenzengleichungen.* Deutscher Verlag der Wissenschaften, Berlin.

[7] Schechter, E. (1973). Error estimates and convergence in the numerical solution of non-linear parabolic equations. *Topics in Numerical Analysis,* 265–276. Academic Press, London.

[8] Szarski, J. (1965). *Differential Inequalities.* Polish Scientific Publishers, Warsaw.

Hausdorff Approximation of Functions and Point Sets

Bl. Sendov

The present paper is a review of the results in the field of the best approximation in Hausdorff distance (or H-distance). The H-distance, introduced in 1914 by Hausdorff [1] for measuring the deviation between two subsets of a given metric set, was rarely used in Analysis up to the '60s. We first took interest in the Hausdorff distance and its applications to the Theory of Approximations in connection with Kolmogorov's problem for the ϵ-entropy and the ϵ-capacity of sets in functional spaces [2]. Together with B. Penkov we defined the Hausdorff distance between two bounded functions in a suitable way, which is essentially the Hausdorff distance between two point sets in the plane, corresponding to these functions. To every bounded function f we correlate the minimal, closed and convex with respect to the y-axis, point set $\bar{f}$ in the plane, consisting of the graph of the function f.

The expediency of studying this distance is emphasized by A. N. Kolmogorov in [3]:

"When describing the temporal course of a real process by means of the time function $f(t)$, assuming values of the 'phase set' X chosen in an appropriate way, it turns out to be natural and legal to assume that the function f possesses only discontinuities of the first kind (jumps). For the detailed research of such processes the introduction of a corresponding topology is very useful in the set D of the functions with discontinuities of the first kind.

The topology of uniform convergence, natural when studying continuous processes, turns out to be too strong when studying processes with discontinuities of the first kind. For example, it is desirable for the function sequence

$$f_n(t) = \begin{cases} x_1 & \text{when} \quad t < t_n \\ x_2 & \text{when} \quad t > t_n, \end{cases}$$

175

where $t_n \to t_0$ when $n \to \infty$, to be convergent to the function

$$f(t) = \begin{cases} x_1 & \text{when} \quad t < t_0, \\ x_2 & \text{when} \quad t > t_0, \end{cases}$$

since the function f_n for large n deviates from the function f simply by a small shift at the moment of the jump from the state x_1 to the state x_2. As is well known such a convergence does not exist in the space of uniform convergence (when $x_1 \neq x_2$).

On the other hand, the topology D should not be too weak for it is desirable that the most essential properties of the function $f \in D$ be kept at the boundary transition. For example, it is necessary from the convergence property $f_n \to f$ as $t_n \to t, f_n(t_n + 0) - f_n(t_n - 0) \to C \neq 0$ for it to follow that $f(t + 0) - f(t - 0) = C$."

A distance, generating a topology with the properties pointed by A. N. Kolmogorov, was introduced by Y. V. Prohorov [4]. When defining this distance between two functions, the Hausdorff distance between two point sets corresponding to the given functions is also used. But here to every function f we correlate the smallest closed point set $\hat{f}$ containing the graph of f, except for the isolated points of the graph of f.

In a series of applications it seems natural to add to A. N. Kolmogorov's requirements the following one: the topology in D should be of such a kind that the sequence of continuous functions

$$f_n(t) = \begin{cases} -1 & \text{when} \quad t \leqslant -1/n, \\ nt & \text{when} \quad -1/n \leqslant t \leqslant 1/n, \\ 1 & \text{when} \quad t \geqslant 1/n, \end{cases}$$

should converge to the function

$$f(t) = \begin{cases} -1 & \text{when} \quad t < 0, \\ \alpha & \text{when} \quad t = 0, \\ 1 & \text{when} \quad t > 0, \end{cases}$$

where α is an arbitrary number from the interval $[-1, 1]$.

The distance, considered by us, satisfies the requirements mentioned above. In a certain sense it is of equal behaviour towards the coordinate axes. As is well known, the ϵ-neighbourhood of a given function in the uniform distance is obtained by the variation of the points of its graph only in the direction of the y-axis, but the ϵ-neighbourhood of a given function in the H-distance is obtained by the variation of its graph points in all directions in the plane. Since, in a real computation, we find not only the function value with a certain error, but also the value of the argument at which the function is calculated, with a certain error, it seems more natural in this situation to use the H-distance.

On the other hand the study of functional spaces metrized by the H-distance is connected with some difficulties as these spaces are not Banach spaces. This last circumstance explains some, at first glance, unexpected, results in the theory of Hausdorff approximations. Survey papers on Hausdorff approximation are [5]–[8];

The approximation of point sets in the plane by graphs of polynomials in the Hausdorff distance can be considered as the approximation of set valued functions.

1. The Hausdorff Distance and Best Approximation

Let $d(A, B)$ be an arbitrary Minkowski distance between two points A and B of the Euclidean plane R_2, and let $F \subset R_2$. $G \subset R_2$ be two closed point sets. The Hausdorff distance (or the H-distance) $r(F, G\ d)$ between F and G, generated by d, is defined through

$$r(d; F, G) = \max\left(h(F, G; d), h(G, F; d)\right)$$

$$h(F, G; d) = \max_{A \in F} \min_{B \in G} d(A, B).$$

We shall specify $d(A, B)$ as

$$d_\alpha(A, B) = d_\alpha(A(x_1, y_1), B(x_2, y_2)) = \max\left(\alpha^{-1} | x_1 - x_2|, | y_1 - y_2|\right),$$

where $\alpha > 0$. The H-distance between F and G, generated by d_α we denote by $r(F, G; \alpha)$, and when $\alpha = 1$ we write $r(F, G) = r(F, G; 1)$.

Let $\Delta = [a, b]$ be a closed interval on the real line R_1. Denote by F_Δ the class of all bounded and closed subsets of R_2, which are convex with respect to the y-axis and whose projections on the x-axis coincide with Δ.

Let further f be a bounded real function defined in Δ. The subset of R_2 which is the intersection of all elements $F \subset F_\Delta$ containing the graph of f (considered also as a point set in R_2) is called the complete graph of f and is denoted by $\bar{f}$. The complete graph of a continuous function coincides with its graph.

The Hausdorff distance $r(f, g; \alpha)$ between two bounded real functions defined on Δ is by definition the H-distance between their complete graphs, i.e. $r(f, g; \alpha) = r(\bar{f}, \bar{g}; \alpha)$.

The distance $r(f, g; \alpha)$ between two continuous functions can be considered as a generalization of the uniform distance

$$\rho(f, g) = \max_{x \in \Delta} | f(x) - g(x)|,$$

since

$$\rho(f, g) = r(f, g; 0) = \lim_{\alpha \to +0} r(f, g; \alpha).$$

It is easy to obtain the following relation between $r(f, g; \alpha)$ and $\rho(f, g)$ for every $\alpha > 0$:

$$r(f, g; \alpha) \leqslant \rho(f, g) \leqslant r(f, g; \alpha) + \omega(\alpha r(f, g; \alpha)), \tag{1}$$

where $\omega(\delta)$ is the modulus of continuity of f or of g.

Let H_n be the set of algebraic polynomials of degree $\leqslant n$. If $P(x) \in H_n$, then P will be the graph of $P(x)$ on the interval Δ.

Let F be an arbitrary bounded and closed point set in the plane R_2. We call

$$E_n(F;\alpha) = \inf_{P \in H_n} r(F,P;\alpha)$$

the best approximation of F in H-distance by means of algebraic polynomials in the interval Δ, in analogy to the best uniform approximation

$$E_n(f) = \inf_{P \in H_n} \rho(f,P).$$

$P^*(x) \in H_n$ is a polynomial of best approximation for F in Δ, if $r(F,P^*;\alpha) = E_n(F;\alpha)$.

It is easy to show that for every bounded and closed point set F in R_2 and every interval Δ, there exists a polynomial $P \in H_n$ of best approximation for F, but it may not be unique. If F is the complete graph of a monotone function, then the polynomial of best approximation in H_n for F is unique.

The analogue of the Weierstrass theorem for the H-distance is the following:

Let F be a bounded and closed point set in the plane R_2, then $F \in F_\Delta$ if and only if

$$\lim_{n \to \infty} E_n(F;\alpha) = 0$$

One of the most interesting results in this theory is the following:
If $F \in F_\Delta$, then

$$E_{n,r}(F,\alpha) = 0 \ (\ln n/n). \tag{2}$$

It follows immediately from (1) that if $\omega(f;\delta)$ is the modulus of continuity of the function f, then for every $\alpha \geqslant 0$ the inequalities

$$E_n(f;\alpha) \leqslant E_n(f) \leqslant E_n(f;\alpha) + \omega(f;\alpha E_n(f;\alpha)) \tag{3}$$

hold.

The classic result of Jackson for the best uniform approximation is

$$E_n(f) = 0(\omega(f;n^{-1})). \tag{4}$$

The estimate (2) can not be considered as a generalization of (4), because we can not obtain (4) directly from (3). But for a continuous function f it turns out [10] that an estimate of the form

$$E_n(f;\alpha) = 0(\ln(\alpha n \omega(f;n^{-1}))/\alpha n) \tag{5}$$

can be found (naturally, if $\alpha n \omega(f;n^{-1}) \geqslant e$).

For fixed α the estimate (5) in the set of all continuous functions in the interval Δ is of the same order $O(\ln n/n)$ with respect to n as (2), but from (5) we immediately obtain (4). Actually, since α is an arbitrary positive number, we can take $\alpha = e/n\omega(f; n^{-1})$. Then from (3) and (5) it follows that

$$E_n(f) = O(\omega(f; n^{-1})/e) + O(\omega(f; n^{-1})) = O(\omega(f; n^{-1})).$$

In [9] G. G. Lorentz and K. L. Zeller have studied the order of approximation of monotone functions by means of monotone polynomials in the uniform metric. The following theorem is proved [9]:

There exists a constant c_0 with the following property: if f is an increasing function on Δ, then there exists a sequence of polynomials $p_n \in H_n$, increasing in Δ, such that

$$\rho(f, p_n) \leq c_0 \omega(f; n^{-1}), \quad n = 1, 2, 3, \ldots \tag{6}$$

Considering the H-distance we have the following theorem: There exists a constant c_1 with the property: if f is an increasing function on Δ, then there exists a sequence of polynomials $P_n \in H_n$, increasing in Δ, such that

$$r(f, P_n; \alpha) \leq c_1 \omega(f; m^{-1})/(1 + \alpha m \omega(f; m^{-1})), \tag{7}$$

where

$$m = [n/\ln e(1 + \alpha M m^2)], \quad M = \max_{x \in \Delta} |f(x)|.$$

It is evident that (6) follows from (7) for $\alpha = 0$.

2. Order of Best Hausdorff Polynomial Approximation of Certain Functions

One concrete problem for the uniform approximation by algebraic polynomials, solved by S. N. Bernstein, is the approximation of the function $|x|$. The fundamental importance of this result in the Theory of Approximations is well known.

The problem of approximating $|x|$ in the interval $[-1, 1]$ by algebraic polynomials is equivalent to the problem of the approximation of the semi-circle $y = \sqrt{1 - x^2}$ in the same interval.

Denote by α_n the minimal positive number, for which there exists an algebraic polynomial of degree $\leq n$, whose graph in the interval $[-1, 1]$ lies in the strip between the semicircles $y_1 = \sqrt{1 - x^2} - \alpha_n$ and $y_2 = \sqrt{1 - x^2} + \alpha_n$.

S. N. Bernstein's result [11] can be formulated as follows:
There exist positive constants c' and c'', such that

$$c'/n < \alpha_n < c''/n$$

Denote by β_n the minimal positive number, for which there exists an algebraic polynomial P of degree $\leq n$, whose graph lies in the strip between the semicircles

$$y_1 = \sqrt{(1 - \beta_n)^2 - x^2}; \quad -1 + \beta_n \leq x \leq 1 - \beta_n$$

and

$$y_2 = \sqrt{(1 + \beta_n)^2 - x^2}; \quad -1 - \beta_n \leq x \leq 1 + \beta_n,$$

and such that $P(-1) = P(1) = 0$.

It can be proved [15] that there exist positive constants c_1 and c_2, such that

$$c_1 n^{-2} < \beta_n < c_2 n^{-2}.$$

Evidently α_n is the best uniform approximation, and β_n is the best Hausdorff approximation. The last result specifies that the order of the (in this case quite natural) Hausdorff approximation is n^{-2}, while the order of the corresponding uniform approximation is only n^{-1}.

It is interesting to estimate the best approximation in the H-distance by algebraic polynomials of the functions

$$\varphi_\lambda(x) = |x|^\lambda,$$
$$\psi_\lambda(x) = |x|^\lambda \, \text{sgn} \, x,$$
$$\theta_\lambda(x) = (1 - x^2)^\lambda,$$

for $0 < \lambda < 1$ on the interval $[-1, 1]$. While the best uniform polynomial approximation of φ_λ, ψ_λ and θ_λ depends essentially upon α, the best Hausdorff polynomial approximation can be estimated in order by n^{-1}, for φ_λ and ψ_λ or by n^{-2} for θ_λ, for every $0 < \lambda < 1$. More precisely [12]:

$$2^{-2-1/\lambda} n^{-1} < E_n(\varphi_\lambda; 1) < 2^{1/\lambda} n^{-1},$$
$$2^{-2-1/\lambda(1-\lambda)} n^{-1} < E_n(\psi_\lambda; 1) < 2^{1/\lambda} n^{-1},$$
$$E_n(\theta_\lambda; 1) < 2^{3+1/\lambda} n^{-2}.$$

Let us denote by $AM_{[-1,1]}$ the set of functions f analytic in the circle $|z| < 1$, $|f(z)| \leq M$ for $|z| \leq 1$ and having real values in the interval $[-1, 1]$.

We found [13] the following estimate:
If $f \in AM_{[-1,1]}$, then

$$E_n(f; 1) \leq (8 \ln n/n)^2 + 8Mn^{-2} \qquad (8)$$

in the interval $[-1, 1]$.

The estimate (8) is exact in order, because for the function $\tau(z)$, $\tau(z) = 0$ for $|z| \leq 1$ and $\tau(z) = 1$ for $|z| = 1$, we have $\tau \in AM_{[-1,1]}$, $M = 1$ and

$$E_n(\tau; 1) \geq (\ln n/2n)^2.$$

From (8) and (1) the following corollary can be obtained.

If $f \in AM_{[-1,1]}$ and the modulus of continuity of f is $\omega(f; \delta)$, then for the best uniform polynomial approximation in the interval $[-1, 1]$ the inequation

$$E_n(f) = O(\omega(f; (\ln n/n)^2))$$

holds.

We shall formulate one theorem more for the best Hausdorff polynomial approximation, that shows the specific characteristics of the H-distance.

Denote by Z_1 the class of functions f, for which

$$f(x) = \sum_{\nu=0}^{\infty} a_\nu x^\nu$$

and the following conditions

$$\sum_{\nu=0}^{\infty} \nu a_\nu = f'(1) < \infty,$$

$$\Delta^k a_\nu \geqslant 0; \quad \nu = 0, 1, 2, \ldots; \quad k = 0, 1, 2, \ldots,$$

are valid, where $\Delta^0 a_\nu = a_\nu$, $\Delta^{k+1} a_\nu = \Delta^k a_\nu - \Delta^k a_{\nu+1}$.

The following proposition [14] holds:

If $f \in Z_1$, then for the best approximation in the H-distance by algebraic polynomials in the interval $[0, 1]$ the inequality

$$E_n(f; 1) \leqslant f'(1)(n + 1)^{-2}.$$

holds.

3. Numerical Calculation of Polynomials in the Best Hausdorff Approximation

The approximation of step functions by algebraic polynomials in the H-distance plays an important role in some engineering problems connected with antennae and circuit synthesis.

A. Andreev showed that the second Remez algorithm can be modified in a suitable way for calculating the polynomial of the best Hausdorff approximation. Numerical computer experiments demonstrate the efficient application of this method.

We shall give here two numerical examples, kindly provided by A. Andreev, obtained with the help of his algorithm.

For the function

$$f_1(x) = \begin{cases} -1, & -1 \leqslant x \leqslant -1/3 \\ 1, & -1/3 < x \leqslant 1 \end{cases}$$

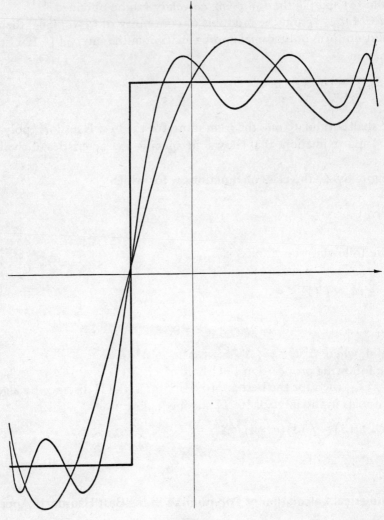

Fig. 1

we have

$$0.2257 \leqslant E_5(f_1; 1) \leqslant 0.2266$$

$$0.1413 \leqslant E_7(f_1; 1) \leqslant 0.1421$$

$$P_5^*(x) \approx 0.22811x^5 + 3.18041x^4 - 1.07950x^3 -$$
$$- 3.96861x^2 + 1.85136x + 1.01437$$

$$P_{10}^*(x) \approx 1.17755x^{10} - 51.87041x^9 + 2.85984x^8 +$$
$$+ 123.85465x^7 - 14.21173x^6 - 100.59484x^5 +$$
$$+ 16.45178x^4 - 30.42548x^3 - 7.39353x^2 -$$
$$- 0.95663x + 1.11606$$

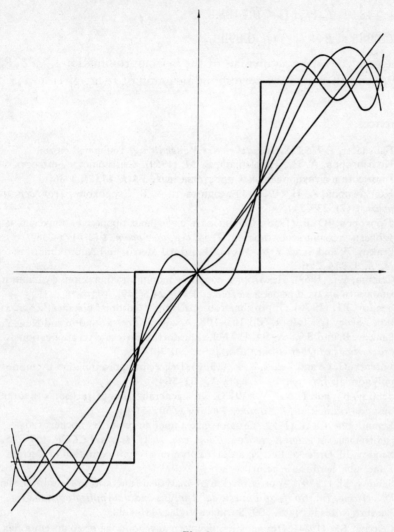

Fig. 2

The graphs of the polynomials of the best approximation P_5^* and P_{10}^*, together with the corresponding graph of f_1, are given in Fig. 1.

For the function

$$f_2(x) = \begin{cases} -1, & -1 \leqslant x \leqslant -1/3 \\ 0, & -1/3 < x \leqslant 1/3, \\ 1, & 1/3 < x \leqslant 1. \end{cases}$$

we have

$$0.2570 \leqslant E_2(f_2; 1) \leqslant 0.2576$$
$$0.1863 \leqslant E_5(f_2; 1) \leqslant 0.1871$$

$$0.1481 \leqslant E_8(f_2; 1) \leqslant 0.1488$$

$$0.0985 \leqslant E_{11}(f_2; 1) \leqslant 0.0995.$$

The graphs of the polynomials of the best approximation P_2^*, P_5^*, P_8^*, and P_{11}^*, together with the corresponding graph of f_2 are given in Fig. 2.

References

[1] Hausdorff, F. (1914). *Grundzüge der Mengenlehre*. Teubner, Leipzig.
[2] Колмогоров, А. Н., Тихомиров, В. М. (1959). ε-энтропия и ε-емкость множеств в функциональных пространствах, *УМН* **14** (2), 3–86.
[3] Колмогоров, А. Н. (1956). О сходимости А. В. Скорохода, *Теор. вер. и ее прим.* **1** (2), 239–247.
[4] Прохоров, Ю. В. (1956). Сходимость случайных процессов и предельные теоремы теории вероятностей, *Теор. вер. и ее прим.* **1** (2), 177–236.
[5] Penkov, B. and Sendov, Bl.(1966). Hausdorff Metrik und Approximation, *Num. Math.* 9, 214–226.
[6] Сендов, Бл. (1969). Некоторые вопросы теории приближений функции и множеств в хаусдорфовой метрике, *УМН* **24** 5(149), 141–178.
[7] Sendov, B1. (1970). Approximation relative to Hausdorff distance, *Approximation Theory* (A. Talbot, ed.) 101–108. Academic Press, London and New York.
[8] Penkov, B. and Sendov, B1. (1971). Hausdorff metric and its applications, *Proceedings of Oberwolfach Conference*, 127–146.
[9] Lorentz, G. G. and Zeller, K. L. (1968). Degree of approximation by monotone polynomials I, *J. Approx. Theory* 1, 501–504.
[10] Sendov, B1. and Popov, V. (1973). On a generalization of Jackson's theorem for best approximation, *J. Approx. Theory* 1, 501–504.
[11] Бернштейн, С. Н. (1952). О наилучшем приближении $|x|$ посредством многочленами данной степени, *Собр. соч. м.* I, *Изд. АН СССР*, 157–205.
[12] Sendov, B1. Order of Best Hausdorff Polynomial Approximation of Certain Functions. Serdica (in print).
[13] Sendov, B1. (1974). Approximation of analytical functions in Hausdorff Metric, *Proceedings Int. Conf. on Functional Analysis and its Applications*, Madras, Texture Notes in Math, 399, Springer-Verlag, 490–500.
[14] Сендов, Бл. (1974). Порядок наилучшего хаусдорфова приближения дяя одного класса аналитиуеских функций, *Доклады БАН*, 27 (12), 1621–1623.
[15] Сендов, Бл. Прибдижение полукруга алгебраическими многочленами Изв, *МИАН Смеклова* (в лечати).

Methods of Simultaneous Iteration for Calculating Eigenvectors of Matrices

G. W. Stewart

1. Introduction

With the generally satisfactory numerical solution of the algebraic eigen-value problem for matrices that can be contained in array form in the high speed storage of a computer [13], numerical analysts have begun examining in detail methods for larger problems [7, 12]. Among these methods, the power method is important because of its computational simplicity. This method finds an eigenvector corresponding to the dominant eigenvalue λ_1 of a matrix A by the following iterative process. Given a nonzero vector q_0, form the sequence $q_0, q_1, q_2, \ldots$ according to the formula

$$Aq_\nu = \rho_{\nu+1} q_{\nu+1}. \tag{1.1}$$

Here $\rho_{\nu+1}$ is a scalar chosen to keep the components of $q_{\nu+1}$ within the range of practical computation (ordinarily it is chosen so that some norm of $q_{\nu+1}$ is unity). It is well known that if λ_1 is larger in magnitude than the remaining eigenvalues of A and if q_0 is not unfortunately chosen, then the vectors q_ν tend in their directions to an eigenvector of A corresponding to λ_1.

The formula (1.1) is particularly nice for computational work with large matrices. It requires only that one be able to form the product of A and a vector, which leaves the elements of A unaltered. Moreover, the conditions required of q_0 for the convergence of the process are very mild; they will be satisfied by almost any randomly chosen vector. In effect, then, the power method provides a simple, always convergent iteration for the dominant eigenvector of a matrix.

However, the method has some drawbacks. In the first place it can calculate only the dominant eigenvector; if other eigenvectors are needed special tricks must be used. Since in many problems it is the dominant eigenvector that is required, this drawback is not as important as it might at first seem. A more serious difficulty is that the process can converge very slowly. If λ_2 is the root next largest in magnitude to λ_1, then the accuracy of q_ν will be proportional to $|\lambda_2/\lambda_1|^\nu$. If $|\lambda_2/\lambda_1|$ is say .99 then the convergence of the method will be intolerably slow.

The purpose of this paper is to describe a technique, now generally known as simultaneous iteration, in which these difficulties are partially overcome by iterating with several vectors at a time. The technique originated under the name *treppeniteration* with Bauer, who used it to calculate factors of polynomials and later generalized it to the eigenvalue problem [1, 2]. The application of the method to real symmetric matrices has been analyzed in several places [4, 8, 10], and Rutishauser [9] has published a program. Some of the practical details of a slight generalization of the method that is suitable for nonsymmetric matrices [2] have been studied by Clint and Jennings [5]. In previously unpublished work, the author has generalized the method in a somewhat different direction.

In Section 2 of this paper the idea of simultaneous iteration will be introduced as a power method for subspaces and the convergence of the method will be analyzed for the special case of a nondefective matrix. In addition the practical problems of iterating with a subspace and of extracting eigenvectors from the iterates will be discussed. In Section 3 Rutishauser's program for symmetric matrices will be examined, and in Section 4 the method proposed by Clint and Jennings. In Section 5 the author's QR variant for nonsymmetric matrices will be described and some new convergence results quoted.

Throughout this paper we shall be concerned with a fixed matrix A of order n whose eigenvalues $\lambda_1, \lambda_2, \ldots, \lambda_n$ have been ordered so that

$$|\lambda_1| \geqslant |\lambda_2| \geqslant \cdots \geqslant |\lambda_n|.$$

The matrix A is not necessarily assumed to have a complete set of eigenvectors; however, when it does they will be denoted by $u_1, u_2, \ldots, u_n$, where u_i corresponds to $\lambda_i (i = 1, 2, \ldots, n)$. For a fixed integer r $(1 \leqslant r < n)$ if $|\lambda_{r+1}| < |\lambda_r|$, then there is a unique invariant subspace $\mathcal{U}_r$ of A satisfying

$$A_r \equiv \{Au: u \in \mathcal{U}_r\} \subset \mathcal{U}_r.$$

We shall be concerned with methods that attempt to approximate $\mathcal{U}_r$, and it is convenient to introduce the notation

$$\Lambda = \text{diag}\,(\lambda_1, \lambda_2, \ldots, \lambda_n),$$
$$\Lambda_1 = \text{diag}\,(\lambda_1, \ldots, \lambda_r), \quad \Lambda_2 = \text{diag}\,(\lambda_{r+1}, \ldots, \lambda_n)$$

and

$$U = (u_1, u_2, \ldots, u_n),$$
$$U_1 = (u_1, \ldots, u_r), \quad U_2 = (u_{r+1}, \ldots, u_n).$$

From the relations $Au_i = \lambda_i u_i$ $(i = 1, 2, \ldots, n)$ it follows that

$$AU_j = U_j \Lambda_j \quad (j = 1, 2). \tag{1.2}$$

2. The Power Method for Subspaces

One way of deriving the simultaneous iteration is to note that the scaling factor $\rho_{\nu+1}$ in (1.1) is arbitrary, which is to say that the iteration is entirely determined by the one-dimensional subspaces spanned by the q_ν. This suggests that the iteration be generalized by starting with an r-dimensional subspace $\mathcal{Q}_0$ and iterating according to the formula

$$\mathcal{Q}_{\nu+1} = A \mathcal{Q}_\nu \quad (\nu = 0, 1, 2, \ldots) \tag{2.1}$$

If $|\lambda_r| > |\lambda_{r+1}|$, it might be expected that the $\mathcal{Q}_\nu$ in some sense converge to the dominant invariant subspace $\mathcal{U}_r$.

For a general matrix A, describing the behaviour of the sequence $\mathcal{Q}_0$, $\mathcal{Q}_1, \ldots$ is a fairly difficult problem (see [6], where the subject is treated in detail). When A is nondefective, the problem is considerably simpler yet still illustrates many of the features of the general case. To analyze this special case, let the columns of the $n \times r$ matrix Q_0 form a basis for $\mathcal{Q}_0 (\mathcal{R}(Q_0) = \mathcal{Q}_0)$. If the sequence $Q_0, Q_1, \ldots$ is defined by

$$Q_{\nu+1} = A Q_\nu \quad (\nu = 0, 1, 2, \ldots) \tag{2.2}$$

then $\mathcal{R}(Q_\nu) = \mathcal{Q}_\nu$.

We can determine if $\mathcal{R}(Q_\nu)$ is near $\mathcal{Q}_r$ by the following device. Set

$$Y_\nu \equiv \begin{pmatrix} Y_1^{(\nu)} \\ Y_2^{(\nu)} \end{pmatrix} = U^{-1} Q_\nu,$$

where Y_1 is $r \times r$. Then

$$Q_\nu = U Y_\nu = U_1 Y_1^{(\nu)} + U_2 Y_2^{(\nu)} \tag{2.3}$$

If $Y_1^{(\nu)}$ is nonsingular,

$$\mathcal{Q}_\nu = \mathcal{R}(Q_\nu) = \mathcal{R}(Q_\nu Y^{(\nu)-1}) = \mathcal{R}(U_1 + U_2 Y_2^{(\nu)} Y_1^{(\nu)-1}) \tag{2.4}$$

Thus if the matrix

$$Z_\nu = Y_2^{(\nu)} Y_1^{(\nu)-1} \tag{2.5}$$

approaches zero, then $\mathcal{R}(Q_\nu)$ will approach $\mathcal{R}(U_1)$, which is $\mathcal{U}_r$.

Now

$$Q_\nu = A^\nu Q_0 = A^\nu U_1 Y_1^{(0)} + A^\nu U_2 Y_2^{(0)}$$
$$= U_1 \Lambda_1^\nu Y_1^{(0)} + U_2 \Lambda_2^\nu Y_2^{(0)},$$

the last equality following from (1.2). It follows that

$$Y_i^{(\nu)} = \Lambda_i^\nu Y_i^{(0)} \quad (i = 1, 2).$$

If $Y_1^{(0)}$ is nonsingular and $|\lambda_r| > |\lambda_{r+1}|$, then Z_ν is well defined and is given by

$$Z_\nu = \Lambda_2^\nu Z_0 \Lambda_1^{-\nu}. \tag{2.6}$$

The (i, j)-element of Z_0 is given by

$$z_{ij}^{(\nu)} = z_{ij}^{(0)} \left(\frac{\lambda_{i+r}}{\lambda_j}\right)^\nu \quad (i = 1, \ldots, n - r; j = 1, 2, \ldots, r). \tag{2.7}$$

Since $|\lambda_{i+r}/\lambda_j| < 1$, it follows that $Z_\nu \to 0$.

Thus we have proved that *if* $Y_1^{(0)}$, *is nonsingular and* $|\lambda_r| > |\lambda_{r+1}|$ *then* $\mathscr{Q}_\nu \to \mathscr{U}_r$.

However, we have proved more than this. It follows from (2.4), (2.5), and (2.6) that if Z_0 is large, the spaces $\mathscr{Q}_\nu$ will require a long time to converge to $\mathscr{U}_r$. Since Z_0 is independent of the choice of basis Q_0 for $\mathscr{Q}_0$, it furnishes a reasonable measure of how good the initial subspace $\mathscr{Q}_0$ is.

A second consequence of the above proof, one that has important practical implications, is that parts of the subspaces $\mathscr{Q}_\nu$ will converge faster than others. For example, it follows from (2.7) that the i-th column of Z_ν vanishes as $|\lambda_{r+1}/\lambda_i|^\nu$. Consequently the i-th column of $U_1 + U_2 Z_\nu$ is $u_i + 0(|\lambda_{r+1}/\lambda_i|^\nu)$. From (2.4) it follows that $\mathscr{Q}_\nu$ contains an approximation to u_i that is accurate to terms of order $|\lambda_{r+1}/\lambda_i|^\nu$.

To see what this means, consider the special case where $\lambda_1 = 1.0$, $\lambda_2 = .99$, and $\lambda_3 = .5$. If we take $r = 1$ (the power method) then the approximations to u_1 will converge as $.99^\nu$, an intolerably slow rate. On the other hand, if $r = 2$, there will be an approximation to u_1 in $\mathscr{Q}_\nu$ that improves as $.5^\nu$, a satisfactory rate. It is an important aspect of methods of simultaneous iteration that one can improve the rate of convergence by working with more vectors.

Turning now to the general problems involved in implementing a subspace power method, we first note that the formula (2.1) is not suitable for implementation on a digital computer. A digital computer cannot byte into a subspace; it requires a numeric representation. This suggests that we proceed as we did in the convergence proof, choosing a basis Q_0 for $\mathscr{Q}_0$ and iterating according to the formula (2.2). However, this iteration is not numerically stable, since the i-th column of Q_ν is given by

$$q_i^{(\nu)} = A^\nu q_i^{(\nu)};$$

i.e., it is just the ν-th iterate of the power method starting with the vector $q_i^{(0)}$. If follows that all the columns of Q_ν may tend to lie along u_1, and any information about the other eigenvectors will be confined to the low order digits of the floating-point representation of Q_ν and will eventually be lost.

The cure for this problem is to observe that the column space of Q_ν is not changed by postmultiplying it by a nonsingular matrix. Thus if we form the sequence Q_ν according to the formula

$$AQ_\nu = Q_{\nu+1}R_{\nu+1}, \qquad (2.8)$$

where the R_ν are nonsingular, we shall still have the relation $\mathscr{R}(Q_\nu) = \mathscr{Q}_\nu$. The R_ν can be chosen to keep the columns of Q_ν strongly independent so that they adequately represent $\mathscr{Q}_\nu$. There are various possible choices for R_ν. Bauer's original *treppeniteration* takes R_ν to be the upper triangular matrix that reduces Q_ν to lower trapezoidal form with zeros above the diagonal. However, this decomposition cannot always be computed stably. A stable alternative choice is to choose R_ν so that Q_ν has orthonormal columns. This variant is mentioned by Wilkinson [13, p. 607], who explores its connections with the QR algorithm, and it will form the basis of the methods discussed in Section 3 and Section 5. A third variant, the basis for Bauer's bi-iteration [2], will be discussed in Section 4.

Whatever the choice of R_ν, the columns of Q_ν need not approach eigenvectors of A, or if they do the approach may be slow. For example, if R_ν is always upper triangular, then the first column of Q_ν is the vector that would have been obtained by starting the power method with $q_1^{(0)}$, and we have seen that the convergence of this method to u_1 can be slow indeed. Thus a second problem that must be faced in implementing a method of simultaneous iteration is to retrieve the eigenvectors of A from the columns of Q_ν. This can be done by computing an approximation to $Y_1^{(\nu)-1}$; for it follows from (2.3) and (2.5) that

$$Q_\nu Y_1^{(\nu)-1} = U_1 + U_2 Z_\nu, \qquad (2.9)$$

and if $|\lambda_r| > |\lambda_{r+1}|$ then $Z_\nu \to 0$.

The usual method for finding an approximation to $Y_1^{(\nu)-1}$ can be motivated heuristically as follows. Let P_ν be a matrix for which $P_\nu^H Q_\nu$ is nonsingular. Then from (2.9),

$$P_\nu^H Q_\nu Y_1^{(\nu)-1} = P_\nu^H U_1 + P_\nu^H U_2 Z_\nu.$$

Also from (2.9) and (1.2),

$$P_\nu^H AQ_\nu Y_1^{(\nu)-1} = P_\nu^H U_1 \Lambda_1 + P_\nu^H U_2 \Lambda_2 Z_\nu.$$

Hence

$$(P_\nu^H AQ_\nu) Y_1^{(\nu)-1} - (P_\nu^H Q_\nu)Y_1^{(\nu)-1}\Lambda_1 = P_\nu^H U_2 (\Lambda_2 Z_\nu - Z_\nu\Lambda_1). \qquad (2.10)$$

Since the right-hand side of (2.10) approaches zero, it is natural to seek an approximation W_ν to $Y_1^{(\nu)-1}$ as the solution to the generalized eigenvalue problem

$$(P_\nu^H AQ_\nu)W_\nu = (P_\nu^H Q_\nu) W_\nu M_\nu, \qquad (2.11)$$

where $M = \text{diag}(\mu_1, \mu_2, \ldots, \mu_r)$, with the μ_i ordered in descending order of magnitude.

It is important to note that the computation of W_ν does not involve a large amount of extra work when $r \ll n$. The calculation of AQ_ν must be done anyway (cf. (2.8)), after which the calculation of $P_\nu^H(AQ_\nu)$ and $P_\nu^H Q_\nu$ involves about the same amount of work as the calculation of $Q_{\nu+1}$ from AQ_ν. The eigenvalue problem (2.11) involves only matrices of order r, which is assumed small. Finally the calculation of the approximate eigenvectors $Q_\nu W_\nu$ is roughly equivalent in work to the formation of $Q_{\nu+1}$.

From previous comments, it may be expected that the i-th column of $Q_\nu W_\nu$ will approximate u_i up to terms of order $|\lambda_{r+1}/\lambda_i|^\nu$. This is usually the case, although the proofs are rather complicated, involving perturbation theory for invariant subspaces. In particular multiple eigenvalues require special treatment.

In summary, a practical implementation of a method will involve three steps. A power step in which AQ_ν is calculated; an orthogonalization step in which $Q_{\nu+1} = AQ_\nu R_{\nu+1}^{-1}$ is formed; and an unscrambling step in which the eigenvalue problem (2.11) is solved and the matrix $Q_\nu W_\nu$ is formed. It is important to realize that the last two steps need not be performed each time. If the columns of Q_ν are sufficiently independent, the orthogonalization can be skipped. Moreover, there is no need to perform an unscrambling step if none of the eigenvectors have converged. These considerations lead to a general program flow that is illustrated in the flowchart below

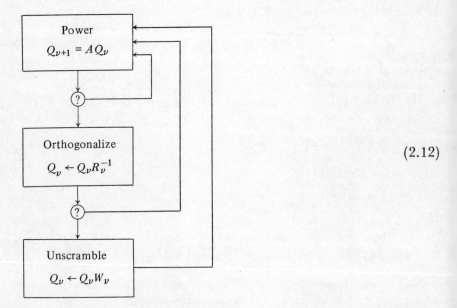

$$(2.12)$$

How the tests, denoted by ⑦ in (2.12), are constructed will depend on the particular variant of the simultaneous iteration.

3. The Orthogonal Variant

When A is real and symmetric, its eigenvalues are real and its eigenvectors can be chosen to form an orthonormal system: $U^T U = I$. This suggests that the column of Q_ν, which will ultimately approximate eigenvectors, should be kept orthonormal. Thus in the orthogonalization step of (2.12), R_ν is chosen to be the (essentially unique) upper triangular matrix that orthogonalizes the columns of Q_ν.

The natural choice of P_ν in the unscrambling step is Q_ν. In this case, since $Q_\nu^T Q_\nu = I$, the eigenvalue problem (2.11) reduces to the symmetric eigenvalue problem

$$Q_\nu^T A Q_\nu W_\nu = W_\nu M_\nu,$$

and W_ν itself will be an orthogonal matrix. The new approximate eigenvectors $Q_\nu' = Q_\nu W_\nu$ are just the Rayleigh-Ritz approximations to the eigenvectors of A in the subspace $\mathscr{Q}_\nu$, and if λ_i is a distinct eigenvalue, then $q_i^{(\nu)'}$ approaches u_i at a rate proportional to $|\lambda_{r+1}/\lambda_i|^\nu$. If $\lambda_i, \lambda_{i+1}, \ldots, \lambda_j$ is a tight cluster of eigenvalues, then the eigenvectors corresponding to them are ill-conditioned. None the less, the space spanned by the columns $q_i^{(\nu)'}, q_{i+1}^{(\nu)'}, \ldots, q_j^{(\nu)'}$ will approach the subspace spanned by $u_i, u_{i+1}, \ldots, u_j$ at a rate proportional to $|\lambda_{r+1}/\lambda_j|^\nu$ [10].

The diagonal elements $\mu_i^{(\nu)}$ of M_ν are the Rayleigh quotients $q_i^{(\nu)'T} A q_i^{(\nu)'}$. Since A is symmetric, they will be about twice as accurate approximations to the eigenvalues λ_i as the $q_i^{(\nu)'}$ are approximations to the eigenvectors u_i [11, p. 311], and indeed the convergence theory shows that the $\mu_i^{(\nu)}$ approach λ_i at a rate proportional to $|\lambda_{r+1}/\lambda_i|^{2\nu}$ [10]. This result holds even if λ_i is a multiple eigenvalue.

Rutishauser [9] has published a program based on the orthogonal variant of simultaneous iteration. The reader is referred to this paper and its predecessor [8] for details; however, some of the general features of the program will now be discussed.

Orthogonalization
After a certain number of power steps the columns of Q_ν must be reorthogonalized. In Rutishauser's program this is done by the Gram-Schmidt method. The usual instabilities associated with the method [13, p. 242] do not arise here, since care is taken that the columns of Q_ν never become too dependent in the power steps.

Convergence and deflation
The natural place to check for convergence is after an unscrambling step. The theory of the method predicts that the first columns of Q_ν will converge first. Once a column has converged it can be held fixed in the power steps.

This deflation procedure can save a considerable amount of work in the later stages of the iteration.

Randomization

The convergence theory of Section 2 indicates that it is important that $Z_0 = Y_2^{(0)} Y_1^{(0)-1}$ not be too large. To avoid this, the program uses two devices. First Q_0 is generated by orthogonalizing the columns of a matrix of pseudo-random numbers. If this matrix is denoted by G and *partitioned* in the form

$$G = \begin{pmatrix} G_1 \\ G_2 \end{pmatrix},$$

where G_1 is $r \times r$, then $Z_0 = G_2 G_1^{-1}$. It has been observed empirically that random matrices are seldom ill-conditioned; consequently it can be expected that G_1^{-1} will not be large and neither will Z_0.

A second safeguard is performed after each of the first three unscrambling steps. The last column of Q_ν is replaced by a random vector which is then orthogonalized against the other columns. This process cannot reduce the accuracy of the previous columns but can improve a deficient Q_ν.

Chebychev steps

Perhaps the most interesting feature of Rutishauser's program is his use of Chebychev steps in the place of the usual power steps. The idea can be motivated by the observation that if the power iteration (2.2) is replaced by an iteration of the form

$$Q_\nu = p_\nu(A)Q_0$$

where p_ν is a polynomial of degree ν, then the expression (2.7) for the elements of Z_ν becomes

$$z_{ij}^{(\nu)} = z_{ij}^{(0)} \frac{p_\nu(\lambda_{i+r})}{p_\nu(\lambda_j)} .$$

If p_ν is chosen to be the Chebychev polynomial on the interval $[-e, e]$, where $|\lambda_r| > e > |\lambda_{r+1}|$, then $|p_\nu(\lambda_{i+r})/p_\nu(\lambda_j)|$ will generally be much smaller than $|\lambda_{i+r}/\lambda_j|^\nu$, and the convergence of the method will be correspondingly improved.

The number e can be taken to be $|\mu_r|$ which is a lower bound for $|\lambda_r|$. It is of course not necessary to compute $p_\nu(A)$. The Chebyshev polynomials can be generated by the three term recurrence

$$p_{\nu+1}(z) = \frac{2}{e} z p_\nu(z) - 2 p_\nu(z) - p_{\nu-1}(z),$$

from which it follows that bases for the subspaces $\mathscr{Q}_\nu$ can be generated according to the recurrence

$$Q_{\nu+1} = \frac{2}{e} AQ_\nu - 2Q_\nu - Q_{\nu-1}.$$

4. Bi-iteration

Bauer's method of bi-iteration [2] is applicable to nonsymmetric matrices and calculates simultaneously a set of r right eigenvectors and a set of r left eigenvectors. This is accomplished by working with a second sequence of subspaces $\mathscr{P}_\nu$ defined by

$$\mathscr{P}_{\nu+1} = A^H \mathscr{P}_\nu \quad (\nu = 0, 1, \ldots).$$

The analysis of Section 2 applies *mutatis mutandi* to show that the spaces $\mathscr{P}_\nu$ contain increasingly good approximations to the left eigenvectors of A, that is to the vectors v_i that satisfy $v_i^H A = \lambda_i v_i^H$.

In practice the method produces two sequences of matrices, $Q_1, Q_2, \ldots$ and and $P_1, P_2, \ldots$, which are bi-orthogonalized at each step; that is the sequences are generated according to the formulas

$$AQ_\nu = Q_{\nu+1} R_{\nu+1}, \quad A^H P_\nu = P_{\nu+1} S_{\nu+1}$$

where $R_{\nu+1}$ and $S_{\nu+1}$ are upper triangular matrices chosen so that

$$P_\nu^H Q_\nu = I.$$

This bi-orthogonalization can be accomplished by a simple algorithm analogous to the Gram-Schmidt algorithm.

If A has distinct eigenvalues and P_0 and Q_0 are not unfortunately chosen, then the columns of P_ν and Q_ν converge to eigenvectors of A. However, the convergence of a particular column may be slow, depending on the ratio of the corresponding eigenvalue to its neighbors. To remedy this, one can take unscrambling steps in which the eigenvalue problem (2.11) is solved for W_ν and the matrix $Q_\nu W_\nu$ is formed. Here the P_ν of this section is used for the P_ν of (2.11), and since $P_\nu^H Q_\nu = I$, the columns of W_ν are just the eigenvectors of the matrix $P_\nu^H A Q_\nu$. Since the left eigenvectors of $P_\nu^H A Q_\nu$ are just the rows of W_ν^{-1}, the columns of P_ν may be unscrambled by forming the matrix $P_\nu W_\nu^{-H}$.

Clint and Jennings [5] have observed that when P_ν and Q_ν have partially converged the matrix $P_\nu^H A Q_\nu$ will have small off-diagonal elements and they propose to use first order perturbation techniques to compute W_ν. While not much work is saved by this device in the most important case where $r \ll n$, the fact that $P_\nu^H A Q_\nu$ has small off-diagonal elements raises a problem that must be faced in all variants of the simultaneous iteration. Specifically, in [3] it is pointed out that the usual QR techniques can fail to produce an accurate system of eigenvectors for a matrix with strong diagonal dominance. In Rutishauser's code this effect is not visible, since he uses a Jacobi routine on $Q_\nu^H A Q$ which is not affected by diagonal dominance. The author believes that the loss of accuracy in solving $P_\nu^H A Q_\nu$ is not too important, since the inaccurate elements of W_ν will be the small ones, which are used only to generate small corrections in columns of Q_ν. However, the subject could stand further analysis and experimentation.

In general there are theoretical reasons for believing that the method of bi-iteration will never be an entirely satisfactory tool for computing dominant invariant subspaces of nonsymmetric matrices. The trouble is that a nonsymmetric matrix may fail to have a complete set of eigenvectors, or the eigenvectors may be nearly linearly dependent. In either case the columns of Q_ν and P_ν will approach dependence and fail numerically to determine the dominant subspace. In the next section we shall consider another variant of simultaneous iteration that does not share this difficulty.

5. Computing Schur Vectors

The success of the orthogonal variant of the simultaneous iteration is largely accounted for by the fact that a symmetric matrix A has a complete set of orthonormal eigenvectors. In other words there is an orthogonal matrix U such that

$$U^T A U = \Lambda = \text{diag}(\lambda_1, \lambda_2, \ldots, \lambda_n).$$

An analogous reduction for nonsymmetric matrices is contained in a theorem of Schur, which states that for any matrix A there is a unitary matrix X such that

$$T \equiv X^H A X \tag{5.1}$$

is upper triangular. The diagonal elements of T are eigenvalues of A, and X can be chosen so that the eigenvalues appear in any order.

When the eigenvalues of A are ordered in decreasing order of magnitude, the leading columns of X form orthogonal bases for the dominant invariant subspaces of A. This suggests that instead of trying to compute eigenvectors of A by simultaneous iteration, we should compute the first r columns of X, which will be called the Schur vectors of A. It should be observed that if $X_r = (x_1, x_2, \ldots, x_r)$ is known, then the first r eigenvectors of A can be obtained by the methods described in Section 2.

The author has analyzed and experimented with methods that compute the Schur vectors. In this section some of the results will be briefly described. More detailed descriptions and proofs will appear in a later publication.

The basic iteration is the same as in the symmetric case; namely given the matrix Q_ν having orthonormal columns, for some k_ν the columns of the matrix $A^{k_\nu} Q_\nu$ are orthogonalized to yield the matrix $Q_{\nu + k_\nu}$. It can be shown that if $|\lambda_{i-1}| > |\lambda_i| > |\lambda_{i+1}|$ and $|\lambda_r| > |\lambda_{r+1}|$, then under mild restrictions on Q_0, the i-th column of Q_ν converges to x_i at a rate proportional to $\max\{|\lambda_i/\lambda_{i-1}|^\nu, |\lambda_{i+1}/\lambda_i|^\nu\}$.

The convergence of this process can still be slow; however, it can be accelerated by a device analogous to the one described in Section 2. Specifically, one forms the matrix

$$B_\nu = Q_\nu^H A Q_\nu$$

and reduces it to Schur form:

$$W_\nu^H B_\nu W_\nu = S_\nu,$$

where W_ν is unitary and S_ν is upper triangular with its diagonal elements decreasing in magnitude. The matrix $Q_\nu' = Q_\nu W_\nu$ then has the property that under the above hypotheses $q_i^{(\nu)'}$ approaches x_i at a rate proportional to $|\lambda_{r+1}/\lambda_i|^\nu$.

A program based on these ideas looks somewhat like Rutishauser's program for symmetric matrices. The orthogonalization step need not be taken at every iteration; only when the columns of Q become too dependent. An estimate of how many steps to take before orthogonalizing can be obtained from the matrix S_ν, which approximates the $r \times r$ leading principal submatrix of T in (5.1). Once a column of Q has converged, it can be held fixed, with a resulting savings in work. Randomization can be performed to insure that an unfortunate choice of Q_0 does not slow convergence; however this process is on a less secure footing in the nonsymmetric case, since it can possibly interfer with vectors that are near convergence.

One feature which distinguishes the iteration for Schur vectors from its symmetric counterpart is the need to deal with the complex eigenvalues of a real matrix. To avoid having to work with complex quantities, one attempts to find the real vectors that reduce A to a quasi-triangular form with 1×1 and 2×2 blocks on the diagonal [11, p. 285]. The 2×2 blocks contain the complex eigenvalues of A, which of course occur in conjugate pairs. It is just this form that is produced by the QR algorithm for real matrices.

References

[1] Bauer, F. L. (1955). Beiträge zur Entwicklung numerische Verfahren für Programmgesteuerte Rechenanlagen I, II. *Bayer. Akad. Wiss. Math.-Natur. Kl. S.-B.*, 275–303; (1956) 163–203.

[2] Bauer, F. L. (1957). Das Verfahren der Treppeniteration und verwandte Verfahren zur Lösung algebraischer Eigenwertprobleme, *Z. Angew. Math. Phys.* 8, 214–235.

[3] Blevins, M. M. and Stewart, G. W. (1974) Calculating the eigenvectors of diagonally dominant matrices, *J. Assoc. Comp. Mach.* 21, 261–271.

[4] Clint, M. and Jennings, A. (1970). The evaluation of eigenvalues and eigenvectors of real symmetric matrices by simultaneous iteration, *Comput. J.* 13, 76–80.

[5] Clint, M. and Jennings, A. (1971). A simultaneous iteration method for the unsymmetric eigenvalue problem, *J. Inst. Math. Appl.* 8, 111–121.

[6] Parlett, B. N. and Poole, W. G. (1973). A geometric theory for the QR, LU, and power iterations, *SIAM J. Numer. Anal.* 10, 389–412.

[7] Ruhe, Axel. (1972). *Iterative eigenvalue algorithms for large symmetric matrices*, Umeå University, Department of Information Processing, Report UMINF-31. 72.

[8] Rutishauser, H. (1969). Computational aspects of F. L. Bauer's simultaneous iteration method, *Numer. Math.* 13, 4–13.

[9] Rutishauser, H. (1970). Simultaneous iteration method for symmetric matrices, *Numer. Math.* 16, 205–223.

[10] Stewart, G. W. (1969). Accelerating the orthogonal iteration for the eigenvalues of a Hermitian matrix, *Numer. Math.* 13, 362–376.

[11] Stewart, G. W. (1973). *Introduction to Matrix Computations.* Academic Press, New York.

[12] Stewart, G. W. (1974). *The numerical treatment of large eigenvalue problems*, in *Information Processing* 74 (IFIP Congress) 666–672.

[13] Wilkinson, J. H. (1964). *The Algebraic Eigenvalue Problem.* Clarendon Press, London.

Biconvergence, Bistability and Consistency of One-Step Methods for the Numerical Solution of Initial Value Problems in Ordinary Differential Equations

Friedrich Stummel

This paper deals with well-known one-step methods for the numerical solution of initial value problems in ordinary differential equations. It establishes new fundamental equivalence theorems for the biconvergence of these methods and associated two-sided estimates for the global discretization error. A one-step method is said to be biconvergent if the convergence of the data of the method is necessary and sufficient for the convergence of the solutions of the one-step method to the solution of the initial value problem. The data of the one-step method are called convergent if the perturbations in the difference equations converge to zero and the initial values converge to the initial value of the solution of the differential equation. The equivalence theorems (13), (20) ensure that consistency is the necessary and sufficient condition for the unique solvability of the difference equations and the biconvergence of the one-step method.

Essential tools in the following treatment are special norms, in particular, a combination of the norms introduced by Bielecki [1] and Spijker (cf. [8]). These norms are specified in Section 1. Then, in Section 2, the basic consistency conditions are defined and characterized. In Section 3, using Banach's fixed point theorem, we can easily establish the unique solvability of the implicit difference equations and the validity of two-sided stability inequalities for Lipschitz continuous, consistent one-step methods. Hence these methods are bistable, that is, both stable and inversely stable in the sense of Stummel [11], [12] and Stummel–Reinhardt [15]. An immediate consequence thereof are the two-sided estimates of the global discretization error in Section 4, Theorem (13), affirming the biconvergence of the solutions of one-step methods in the maximum norm (cf. [12], Section 6). Stetter has remarked in [10], Section 2.2.4, that the validity of such a two-sided error estimate is the unique feature of Spijker's norm.

197

Section 5 shows that an equivalence theorem and corresponding two-sided estimates of the global discretization error hold as well in the norms $\|.\|_{1,p}$ for the solutions of the one-step method and the l^p-norms $\|.\|_p$ for the data space where p may be chosen arbitrarily in $1 \leqslant p \leqslant \infty$. In this way, we obtain not only the convergence of the solutions in the maximum norm but also of the difference quotients of the solutions of one-step methods in the l^p-norm. The equivalence of consistency and convergence of the type considered here for $p = \infty$ has first been stated in Stummel–Hainer [14], Section 11.3. Note that Henrici in [5], Sections 2.2.1, 3.3.1 assumes that the right sides f_ι of the difference equations are continuous jointly as functions of all its arguments and ι whereas our theorems do not make this assumption. Further note that our definition of convergence differs from Dahlquist's in [2] and Henrici's in [6], Section 3.1, because we define the convergence of the method not with respect to all Lipschitz continuous f but only with respect to the given sequence $f, f_\iota, \iota \in I$. The case $p = 1$ corresponds to the choice of norms, for instance, in Stetter [10], Grigorieff [4]. However, one does not find an equivalence theorem in [10], Chapter 2, or biconvergence results and two-sided error estimates in [10], [4] for these norms.

1. Definitions

Let $\mathbb{K}$ be either the real field or the complex field. Let n be a natural number and let $\mathbb{K}^n$ be the associated n-dimensional number space which will be viewed as a normed space with an arbitrary but fixed norm $|.|$. Further let $[a, b]$ be a bounded closed interval of the real axis, let $f : [a, b] \times \mathbb{K}^n \to \mathbb{K}^n$ be a continuous, $\mathbb{K}^n$-valued function of $n + 1$ variables and let w^0 be a vector in $\mathbb{K}^n$. Then we consider the following *initial value problem*

(A) $\quad u(a) = w^0, \quad \dfrac{du}{dt}(t) = f(t, u(t)), \quad t \in [a, b],$

for continuously differentiable $\mathbb{K}^n$-valued solutions u in $[a, b]$. Evidently, this problem is equivalent to the associated integral equation

(C) $\quad u(t) = \displaystyle\int_a^t f(s, u(s))\, ds + w^0, \quad t \in [a, b].$

We assume in the following that this problem has at least one solution u in the space $C[a, b]$ of continuous, $\mathbb{K}^n$-valued functions in $[a, b]$. This integral equation has the form of a *fixed point equation* $u = C(u, w)$ where the mapping C is given by

$$C(v, w)(t) = \int_a^t f(s, v(s))\, ds + w^0, \quad t \in [a, b],$$

for all $v \in C[a, b]$ and $w = (w^0, 0) \in \mathbb{K}^n \times C[a, b]$. In the following, $w^0 \in \mathbb{K}^n$ will denote a fixed vector and $u \in C[a, b]$ a fixed solution of the initial value problem (A) satisfying the initial condition $u(a) = w^0$. Let L, ρ, σ be positive numbers which will later be specified further and let

$$B_\rho(u) = \{v \in C[a, b] \mid \mid v(t) - u(t) \mid e^{-L(t-a)} \leqslant \rho, \quad t \in [a, b]\}.$$

We shall assume that the solution u of (A) or (C) is unique in $B_\rho(u)$.

This initial value problem will now be approximated by one-step difference equations. Let us first state some concepts and notations which are needed for the definition of these methods. Let $I = (1, 2, 3, \ldots)$ be the sequence of natural numbers and let us consider a sequence of finite *partitions* of $[a, b]$,

$$a = t_0^\iota < t_1^\iota < \cdots < t_{N_\iota}^\iota = b, \quad \iota = 1, 2, 3, \ldots .$$

Each partition defines two sets of *net points*,

$$G_\iota = \{t_0^\iota, t_1^\iota, \ldots, t_{N_\iota}^\iota\}, \quad G_\iota' = \{t_1^\iota, \ldots, t_{N_\iota}^\iota\},$$

in $[a, b]$ and associated vector spaces $C(G_\iota)$, $C(G_\iota')$ of $\mathbb{K}^n$-valued functions in G_ι resp. G_ι'. For every net point $t = t_k^\iota \in G_\iota$, we denote by t' the predecessor t_{k-1}^ι in G_ι for $k = 1, \ldots, N_\iota$. The length of the interval $[t', t]$ is assigned to each net point t as *increment* or *step width*

$$\Delta_\iota t = t - t', \quad t \in G_\iota', \quad \iota \in I.$$

We shall assume in the following that the *maximal step widths* h_ι of G_ι form a null sequence,

(1) $\quad h_\iota = \max_{t \in G_\iota'} \Delta_\iota t \to 0 \quad (\iota \to \infty)$.

For every $\iota \in I$ and every function $v_\iota \in C(G_\iota')$, the *backward difference quotient* at the point t or the *forward difference quotient* at the point t' is

$$D_\iota v_\iota(t) = \frac{1}{\Delta_\iota t} (v_\iota(t) - v_\iota(t')), \quad t \in G_\iota'.$$

This paper deals with a class of one-step methods for the approximation of the initial value problem defined by

(A$_\iota$) $\quad u_\iota(a) = w_\iota^0, \quad D_\iota u_\iota(t) = f_\iota(t, u_\iota(t), u_\iota(t')) + w_\iota^1(t), \quad t \in G_\iota', \iota \in I.$

A method of this class is defined by a sequence of functions

$$f_\iota : G_\iota' \times \mathbb{K}^n \times \mathbb{K}^n \to \mathbb{K}^n,$$

of initial values $w_\iota^0 \in \mathbb{K}^n$ and of functions $w_\iota^1 \in C(G_\iota')$. One obtains by summation an equivalent formulation of this problem in the form

$$(C_\iota) \quad u_\iota(t) = \sum_{\substack{s \leqslant t \\ s \in G_\iota'}} f_\iota(s, u_\iota(s), u_\iota(s')) \Delta_\iota s + \hat{w}_\iota(t), \quad t \in G_\iota,$$

where $\hat{w}_\iota$ denotes the function

$$\hat{w}_\iota(t) = w_\iota^0 + \sum_{\substack{s \leqslant t \\ s \in G_\iota'}} w_\iota^1(s) \Delta_\iota s.$$

As usual, the sums on the right sides of these equations are defined by zero for $t = a$. The summed form (C_ι) of the difference equations (A_ι) defines fixed point equations $u_\iota = C_\iota(u_\iota, w_\iota)$ where the mappings C_ι are specified by

$$(2) \quad C_\iota(v_\iota, w_\iota)(t) = \sum_{\substack{s \leqslant t \\ s \in G_\iota'}} f_\iota(s, v_\iota(s) v_\iota(s')) \Delta_\iota s + \hat{w}_\iota(t), \quad t \in G_\iota,$$

for all $v_\iota \in C(G_\iota)$ and all $w_\iota = (w_\iota^0, w_\iota^1) \in \mathbb{K}^n \times C(G_\iota')$, $\iota \in I$.

Example
The simplest examples of one-step methods are the so-called Euler methods. The functions f_ι in the *forward Euler method* are given by

$$f_\iota(t, y, y') = f(t', y'),$$

in the *backward Euler method* by

$$f_\iota(t, y, y') = f(t, y),$$

and in the *modified Euler method* or *trapezoidal rule* by

$$f_\iota(t, y, y') = \tfrac{1}{2}(f(t, y) + f(t', y')),$$

for all $t \in G_\iota', y, y' \in \mathbb{K}^n$ and all $\iota \in I$ (cf. Henrici [5], [6]).

Having made these definitions, we are in a position to introduce the concepts of functional analysis for the treatment of one-step difference approximations. Let E_ι be the vector space $C(G_\iota)$ with the norm

$$\| v_\iota \|_{E_\iota} = \max_{t \in G_\iota} | v_\iota(t) | e^{-L(t - a)}, \quad v_\iota \in C(G_\iota), \quad \iota \in I.$$

These norms are obviously equivalent to the maximum norm on $C(G_\iota)$ and satisfy

$$(3) \quad \| v_\iota \|_{E_\iota} \leqslant \max_{t \in G_\iota} | v_\iota(t) | \leqslant \| v_\iota \|_{E_\iota} e^{L(b - a)}, \quad v_\iota \in C(G_\iota), \quad \iota \in I.$$

Continuous functions v on $[a, b]$ may be restricted to the sets of net points G_ι in $[a, b]$. Thus we obtain functions $v | G_\iota$ or v_{G_ι} in $C(G_\iota)$. In this way, a sequence of restriction operators $r_\iota^E : C[a, b] \to C(G_\iota)$ is defined by

$$r_\iota^E v = v_{G_\iota}, \quad \iota \in I, \quad v \in C[a, b].$$

We denote by $B_\rho^E(u_{G_\iota})$ the closed ball of radius ρ and center u_{G_ι} in the Banach space $C(G_\iota)$, $\|.\|_{E_\iota}$ for each $\iota \in I$.

Further, let F_ι be the vector space $\mathbb{K}^n \times C(G_\iota')$ with the norm

$$\|w_\iota\|_{F_\iota} = \|\hat{w}_\iota\|_{E_\iota} = \max_{t \in G_\iota} \left| w_\iota^0 + \sum_{s \in G_\iota'}^{s \leqslant t} w_\iota^1(s) \, \Delta_\iota s \right| e^{-L(t-a)}$$

for all $w_\iota = (w_\iota^0, w_\iota^1) \in \mathbb{K}^n \times C(G_\iota')$ and all $\iota \in I$. These norms satisfy the relations

(4)
$$\|w_\iota\|_{F_\iota} \leqslant |w_\iota^0| + \max_{t \in G_\iota'} \left| \sum_{s \in G_\iota'}^{s \leqslant t} w_\iota^1(s) \, \Delta_\iota s \right|,$$

$$\max\left(|w_\iota^0|, \max_{t \in G_\iota'} \left| \sum_{s \in G_\iota'}^{s \leqslant t} w_\iota^1(s) \, \Delta_\iota s \right| \right) \leqslant (1 + e^{L(b-a)}) \, \|w_\iota\|_{F_\iota}.$$

Let us define a sequence of restriction operators r_ι^F from

$$\{(w^0, 0)\} \subset \mathbb{K}^n \times C[a, b] \text{ into } \mathbb{K}^n \times C(G_\iota') \text{ by}$$

$$r_\iota^F w = w_{G_\iota} = (w^0, 0_\iota), \quad \iota \in I,$$

where 0_ι denotes the null function in $C(G_\iota')$ and $w = (w^0, 0)$. We write $B_\sigma^F(w_{F_\iota})$ for the closed ball of radius σ and center w_{F_ι} in $\mathbb{K}^n \times C(G_\iota')$ for each $\iota \in I$.

2. Consistency

Consistency is a necessary condition for the discrete convergence of the solutions u_ι of the one-step difference approximations (A_ι) to the solution u of the initial value problem (A). The *local discretization error* or *truncation error* of the sequence (A), (A_ι) at the solution u is defined by

$$d_\iota^A(u) = (d_\iota^0(u), d_\iota^1(u)) \in \mathbb{K}^n \times C(G_\iota') \text{ where}$$

$$d_\iota^0(u) = 0, \quad d_\iota^1(u)(t) = D_\iota u(t) - f_\iota(t, u(t), u(t')), \quad t \in G_\iota', \quad \iota \in I.$$

By means of the local discretization error, the restriction u_{G_ι} of the solution u of (A) can be viewed as a solution of the one-step method (A_ι) using the initial condition $w_\iota^0 = w^0$ and the perturbation $w_\iota^1 = d_\iota^1(u)$ in the difference equations, that is,

$$u(a) = w^0, \quad D_\iota u(t) = f_\iota(t, u(t), u(t')) + d_\iota^1(u)(t), \quad t \in G_\iota', \quad \iota \in I.$$

The one-step method (A_ι) is said to be *consistent* with the initial value problem (A) if and only if the local discretization error is a null sequence,

(5) $\quad \|d_\iota(u)\|_{F_\iota} \to 0 \quad \Leftrightarrow \quad \max_{t \in G_\iota'} \left| \sum_{s \in G_\iota'}^{s \leqslant t} d_\iota^1(u)(s) \, \Delta_\iota s \right| \to 0 \quad (\iota \to \infty),$

The *local discretization error* or *truncation error* of the sequence C, C_ι, $\iota \in I$, at the point (u, w) is defined by the sequence

(6) $d_\iota^C(u, w) = C_\iota(u_{G_\iota}, w_{G_\iota}) - C(u, w)_{G_\iota}$, $\iota \in I$.

The sequence C_ι, $\iota \in I$, is said to be *consistent* with the mapping C at (u, w) if the *consistency condition*

$$\| d_\iota^C(u, w) \|_{E_\iota} \to 0 \quad (\iota \to \infty)$$

is valid. Here u is the solution of the initial value problem (A) such that $u = C(u, w)$ for $w = (w^0, 0)$. Then $w_{G_\iota} = (w^0, 0_\iota)$, $\hat{w}_{G_\iota}(t) = w^0$ for all $t \in G_\iota$, and the representation

(7) $d_\iota^C(u, w)(t) = \sum\limits_{\substack{s \in G_\iota' \\ }}^{s \leqslant t} f_\iota(s, u(s), u(s')) \Delta_\iota s - \int\limits_a^t f(s, u(s))\, ds$

holds for all $t \in G_\iota$, $\iota \in I$. Evidently,

$$\int\limits_a^t f(s, u(s))\, ds = u(t) - u(a) = \sum\limits_{\substack{s \in G_\iota' \\ }}^{s \leqslant t} D_\iota u(s)\, \Delta_\iota s.$$

Consequently, we have the relation

(8) $d_\iota^C(u, w)(t) = - \sum\limits_{\substack{t \in G_\iota' \\ }}^{s \leqslant t} d_\iota^1(u)(s) \Delta_\iota s = -\widehat{d_\iota^A(u)}(t)$, $t \in G_\iota$,

and

$$\| d_\iota^C(u, w) \|_{E_\iota} = \| d_\iota^A(u) \|_{F_\iota}, \quad \iota \in I.$$

The main result of this section is the following consistency theorem.

(9)   *The one-step method* (A_ι) *is consistent with the initial value problem* (A) *at the solution* u *if and only if the sequence of mappings* C_ι, $\iota \in I$, *is consistent with the mapping* C *at* (u, w) *where* $u = C(u, w)$. *These consistenc conditions hold if and only if the following, equivalent conditions are valid:*

(i) $\| d_\iota^A(u) \|_{F_\iota} = \| d_\iota^C(u, w) \|_{E_\iota} \to 0$ $(\iota \to \infty)$;

(ii) $\max\limits_{t \in G_\iota'} \left| \sum\limits_{\substack{s \in G_\iota' \\ }}^{s \leqslant t} d_\iota^1(u)(s) \Delta_\iota s \right| \to 0$ $(\iota \to \infty)$;

(iii) $\max\limits_{t \in G_\iota'} \left| \sum\limits_{\substack{s \in G_\iota' \\ }}^{s \leqslant t} \{f_\iota(s, u(s), u(s')) - f(s, u(s))\} \Delta_\iota s \right| \to 0$ $(\iota \to \infty)$.

Proof

Using the relation (8) between $d_\iota^A(u)$ and $d_\iota^C(u, w)$, one easily sees that (A), (A_ι) is consistent at u if and only if C, (C_ι) is consistent at (u, w), that is, if and only if condition (9 i) holds. The equivalence of (9 i) and (9 ii) follows from (8) and the inequality (4). The function $f(., u(.))$ is uniformly continuous in $[a, b]$. Hence the integral in (7) can be approximated uniformly in G'_ι by Riemannian sums in the form

$$\max_{t \in G'_\iota} \left| \int_a^t f(s, u(s)) \, ds - \sum_{s \in G'_\iota}^{s \leq t} f(s, u(s)) \Delta_\iota s \right| \to 0 \quad (\iota \to \infty).$$

Consequently, the representation (7) implies the equivalence of the consistency conditions (9 i) and (9 iii).

3. Bistability

The one-step difference approximation (A_ι) is said to be *Lipschitz continuous* in the neighbourhood $U = B_\rho(u)$ of the solution u of (A) if there exist constants $L_0 \geq 0, L_1 \geq 0$ such that

$$|f_\iota(t, y, y') - f_\iota(t, z, z')| \leq L_0 |y - z| + L_1 |y' - z'|$$

for all $\iota \in I$ and all $(t, y, y'), (t, z, z') \in U_\iota$ where

$$U_\iota = \{(t, y, y') \in G'_\iota \times \mathbb{K}^n \times \mathbb{K}^n \mid |y - u(t)| \leq \rho \, e^{L(t-a)},$$
$$|y' - u(t')| \leq \rho \, e^{L(t'-a)}\}.$$

Lipschitz continuous one-step methods have the following basic property.

(10) *For every $L > L_0 + L_1$ there exists an index v such that for all $\iota \geq v$ the inequalities $L_0 e^{Lh_\iota} + L_1 \leq L$ and*

(i) $\quad \| C_\iota(u_\iota, w_\iota) - C_\iota(v_\iota, w_\iota) \|_{E_\iota} \leq (1 - e^{-L(b-a)}) \| u_\iota - v_\iota \|_{E_\iota}$

hold and the equalities

(ii) $\quad \| C_\iota(u_\iota, w_\iota) - C_\iota(u_\iota, z_\iota) \|_{E_\iota} = \| w_\iota - z_\iota \|_{F_\iota}$

are valid for all $u_\iota, v_\iota \in B_\rho^E(u_{G_\iota}), w_\iota, z_\iota \in \mathbb{K}^n \times C(G'_\iota)$.

Proof
Note that $(t, u_\iota(t), u_\iota(t')) \in U_\iota$ whenever $\iota \in I$, $u_\iota \in B_\rho^E(u_{G_\iota})$ and $t \in G'_\iota$. Let u_ι, v_ι be an arbitrary pair of elements in $B_\rho^E(u_{G_\iota})$ and let x_ι denote $u_\iota - v_\iota$. From the above Lipschitz condition one obtains the inequality

$$| (C_\iota(u_\iota, w_\iota) - C_\iota(v_\iota, w_\iota))(t) | \leq \sum_{s \in G'_\iota}^{s \leq t} (L_0 |x_\iota(s)| + L_1 |x_\iota(s')|) \Delta_\iota s$$

for all $w_\iota \in F_\iota$. Obviously,

(11) $\quad |x_\iota(t)| \leq \| x_\iota \|_{E_\iota} e^{L(t-a)}, \quad t \in G_\iota, \quad \iota \in I.$

Hence the right-hand side of the above inequality is bounded by

$$\| x_\iota \|_{E_\iota} (L_0 e^{Lh_\iota} + L_1) \sum_{s \in G'_\iota}^{s \leq t} e^{L(s-a)} \Delta_\iota s.$$

As one easily sees,

$$\sum_{\substack{s \in G'_\iota \\ }}^{\substack{s \leqslant t \\ }} e^{L(s'-a)} \Delta_\iota s \leqslant \int_a^t e^{L(s-a)} ds = \frac{1}{L} (e^{L(t-a)} - 1),$$

so that the inequality

$$| (C_\iota(u_\iota, w_\iota) - C_\iota(v_\iota, w_\iota))(t) | e^{-L(t-a)} \leqslant \alpha_\iota \| u_\iota - v_\iota \|, \quad t \in G_\iota,$$

follows where

$$\alpha_\iota = \frac{1}{L} (L_0 e^{Lh_\iota} + L_1)(1 - e^{-L(b-a)}), \quad \iota \in I.$$

Since $h_\iota \to 0 \ (\iota \to \infty)$, for every $L > L_0 + L_1$, there exists an index ν such that $L_0 e^{Lh_\iota} + L_1 \leqslant L$ and hence $\alpha_\iota \leqslant 1 - e^{-L(b-a)}$ for all $\iota \geqslant \nu$. This proves the inequalities for C_ι. Obviously, for every $\iota \in I$, for every $u_\iota \in B_\rho^E(u_{G_\iota})$ and every pair of elements $w_\iota, z_\iota \in F_\iota$, the equations

$$\| C_\iota(u_\iota, w_\iota) - C_\iota(u_\iota, z_\iota) \|_{E_\iota} = \| \hat{y}_\iota \|_{E_\iota} = \| y_\iota \|_{F_\iota}, \quad y_\iota = w_\iota - z_\iota,$$

that is, the equations (10 ii) are valid.

The above theorem enables us to prove the following main bistability theorem which affirms the unique solubility and the bistability of the one-step difference approximations (A_ι).

(12) *Let the one-step method (A_ι) be Lipschitz-continuous and consistent with the initial value problem (A) at the solution u. Then, for every $L > L_0 + L_1$, there exist an index ν and a number $\sigma > 0$ defined by*

(i) $\| d_\iota^A(u) \|_{F_\iota} \leqslant \sigma = \frac{1}{2} \rho e^{-L(b-a)}, \quad L_0 e^{Lh_\iota} + L_1 \leqslant L, \quad \iota \geqslant \nu.$

For every $\iota \geqslant \nu$ and every $w_\iota = (w_\iota^0, w_\iota^1) \in B_\sigma^F(w_{F_\iota})$, the difference equations (A_ι) and the fixed point equations (C_ι) are uniquely soluble in $B_\rho^E(u_{G_\iota})$. For every $w_\iota, z_\iota \in B_\sigma^F(w_{G_\iota})$, the solutions $u_\iota = C_\iota(u_\iota, w_\iota)$, $v_\iota = C_\iota(v_\iota, z_\iota)$ satisfy the following bistability condition, the constant η being $e^{L(b-a)}$,

(ii) $\dfrac{\eta}{2\eta - 1} \| w_\iota - z_\iota \|_{F_\iota} \leqslant \| u_\iota - v_\iota \|_{E_\iota} \leqslant \eta \| w_\iota - z_\iota \|_{F_\iota}.$

Proof
(i) Since the one-step method (A_ι) and the initial value problem (A) are consistent at u, it follows that $\| d_\iota^A(u) \|_{F_\iota} \to 0 \ (\iota \to \infty)$. Hence there exists an index ν such that condition (12 i) holds. Theorem (10) shows that, for each $w_\iota \in \mathbb{K}^n \times C(G'_\iota)$ and each $\iota \geqslant \nu$, the mapping $C_\iota(., w_\iota)$ is a contraction in $B_\rho^E(u_{G_\iota})$, having the Lipschitz constant $\alpha = 1 - e^{-L(b-a)}$. By (6), (8), the solution u of (A) or (C): $u = C(u, w)$ satisfies the condition

$$\| d_\iota^A(u) \|_{F_\iota} = \| d_\iota^C(u, w) \|_{E_\iota} = \| u_{G_\iota} - C_\iota(u_{G_\iota}, w_{G_\iota}) \|_{E_\iota}.$$

Using the assumption (12 i), it follows that

$$\| u_{G_\iota} - C_\iota(u_{G_\iota}, w_\iota) \|_{E_\iota} \leqslant \| u_{G_\iota} - C_\iota(u_{G_\iota}, w_{G_\iota}) \|_{E_\iota} +$$

$$+ \| w_\iota - w_{G_\iota} \|_{F_\iota} \leqslant \| d_\iota^A(u) \|_{F_\iota} + \sigma \leqslant (1 - \alpha) \rho$$

for all $w_\iota \in B_\sigma^F(w_{G_\iota})$, so that $C_\iota(., w_\iota)$ maps the closed ball $B_\rho^E(u_{G_\iota})$ into itself. The Banach fixed point theorem then ensures the existence of a unique solution $u_\iota = C_\iota(u_\iota, w_\iota)$ in $B_\rho^E(u_{G_\iota})$.

(ii) Let $\iota \geqslant v$, let $w_\iota, z_\iota \in B_\sigma^F(w_{G_\iota})$ and let $u_\iota = C_\iota(u_\iota, w_\iota), v_\iota = C_\iota(v_\iota, z_\iota)$ be the corresponding, uniquely determined fixed points of C_ι in $B_\rho^E(u_{G_\iota})$. By Theorem (10), the relation

$$\| u_\iota - v_\iota \|_{E_\iota} \leqslant \| C_\iota(u_\iota, w_\iota) - C_\iota(v_\iota, w_\iota) \|_{E_\iota} + \| C_\iota(v_\iota, w_\iota) - C_\iota(v_\iota, z_\iota) \|_{E_\iota}$$

$$\leqslant \alpha \| u_\iota - v_\iota \|_{E_\iota} + \| w_\iota - z_\iota \|_{F_\iota}$$

and hence the inequality

$$\| u_\iota - v_\iota \|_{E_\iota} \leqslant \frac{1}{1 - \alpha} \| w_\iota - z_\iota \|_{F_\iota}$$

holds which proves the second inequality in (12 ii). Finally, we have

$$\| w_\iota - z_\iota \|_{F_\iota} = \| C_\iota(u_\iota, w_\iota) - C_\iota(u_\iota, z_\iota) \|_{E_\iota} \leqslant \| u_\iota - v_\iota \|_{E_\iota} +$$

$$+ \| C_\iota(v_\iota, z_\iota) - C_\iota(u_\iota, z_\iota) \|_{E_\iota} \leqslant (1 + \alpha) \| u_\iota - v_\iota \|_{E_\iota}.$$

This establishes the first inequality in (12 ii).

4. Biconvergence

We are now in a position to establish the main convergence theorem and the associated error estimates. For Lipschitz continuous one-step methods we obtain the interesting result that our consistency condition is necessary and sufficient for the unique solubility of (A), (A_ι) and the validity of the biconvergence relation.

(13) *Let* (A_ι) *be Lipschitz continuous in the neighbourhood* $B_\rho(u)$ *of the solution* u *of* (A). *Then the consistency of* (A), (A_ι) *at* u, *defined by* (5), *is necessary and sufficient for the unique solubility of the difference equations* (A_ι) *for almost all* ι *and the validity of the biconvergence relation*

(i) $$\max_{t \in G_\iota} | u_\iota(t) - u(t) | \to 0 \quad \Leftrightarrow \quad w_\iota^0 \to w^0, \ \max_{t \in G_\iota'} \left| \sum_{s \in G_\iota'}^{s \leqslant t} w_\iota^1(s) \Delta_\iota s \right| \to 0$$

$$(\iota \to \infty)$$

for all $w_\iota = (w_\iota^0, w_\iota^1) \in B_\sigma^F(w_{G_\iota})$ *and the corresponding solutions* u_ι *of* (A_ι) *in* $B_\rho^E(u_{G_\iota})$. *If* (A), (A_ι) *is consistent, for each* $L > L_0 + L_1$ *there exists an index* v *such that, for all* $\iota \geqslant v$, *the following two-sided estimate of the discretization error holds*

(ii) $$\frac{\eta}{2\eta - 1} \tau_\iota(u_\iota) \leqslant \max_{t \in G_\iota} | u_\iota(t) - u(t) | \ e^{-L(t-a)} \leqslant \eta \tau_\iota(u_\iota)$$

where $\eta = e^{L(b-a)}$ and

$$\tau_{\iota}(u_{\iota}) = \max_{t\in G_{\iota}} \left| w_{\iota}^0 - w^0 + \sum_{s\in G_{\iota}'}^{s\leqslant t} (w_{\iota}^1(s) - d_{\iota}^1(u)(s)) \Delta_{\iota}s \right| e^{-L(t-a)}.$$

Proof
(i) If (A_{ι}) is Lipschitz continuous and (A), (A_{ι}) consistent, Theorem (12) affirms the unique solubility of the difference equations for all $\iota \geqslant \nu$ and the validity of the bistability condition (12 ii). Using the representation (8), we have

$$u_{G_{\iota}} = C_{\iota}(u_{G_{\iota}}, w_{G_{\iota}}) - d_{\iota}^C(u, w) = C_{\iota}(u_{G_{\iota}}, w_{G_{\iota}} + d_{\iota}^A(u)), \quad \iota \in I.$$

It follows that $\| d_{\iota}^A(u) \| \leqslant \sigma$ for $\iota \geqslant \nu$ and thus $w_{G_{\iota}} + d_{\iota}^A(u) \in B_{\sigma}^F(w_{G_{\iota}})$. Hence, on setting $v_{\iota} = u_{G_{\iota}}, z_{\iota} = w_{G_{\iota}} + d_{\iota}^A(u)$ in the bistability condition (12 ii), one obtains the two-sided discretization error estimate in the form

$$\frac{\eta}{2\eta - 1} \tau_{\iota}(u) \leqslant \| u_{\iota} - u_{G_{\iota}} \|_{E_{\iota}} \leqslant \eta \tau_{\iota}(u), \tau_{\iota}(u) = \| w_{\iota} - w_{G_{\iota}} - d_{\iota}^A(u) \|_{F_{\iota}}.$$

(ii) Conversely, let us assume that the difference equations (A_{ι}) are uniquely soluble in $B_{\rho}^E(u_{G_{\iota}})$ for all $w_{\iota} \in B_{\sigma}^F(w_{G_{\iota}})$ and almost all ι, and that the above biconvergence relation (13 i) holds. For $w_{\iota} = w_{G_{\iota}}$, the difference equations (A_{ι}) then have unique solutions $u_{\iota} = C_{\iota}(u_{\iota}, w_{\iota}) \in B_{\rho}^E(u_{G_{\iota}})$ for almost all ι. Since $w_{G_{\iota}} \to w^0(\iota \to \infty)$, the biconvergence relation ensures the convergence of $u_{\iota} \to u$ for $\iota \to \infty$. Using inequality (10 i), we obtain the estimate

$$\| d_{\iota}^A(u) \|_{F_{\iota}} = \| d_{\iota}^C(u, w) \|_{E_{\iota}} \leqslant \| C_{\iota}(u_{G_{\iota}}, w_{G_{\iota}}) - C_{\iota}(u_{\iota}, w_{G_{\iota}}) \|_{E_{\iota}} + $$
$$+ \| u_{\iota} - u_{G_{\iota}} \|_{E_{\iota}} \leqslant (1 + \alpha) \| u_{\iota} - u_{G_{\iota}} \|_{E_{\iota}}.$$

Hence $\| u_{\iota} - u_{G_{\iota}} \|_{E_{\iota}} \to 0$ implies $\| d_{\iota}^A(u) \|_{F_{\iota}} \to 0$ for $\iota \to \infty$ so that (A), (A_{ι}) is consistent.

A-posteriori error estimates are of fundamental interest in numerical analysis. By means of a one-step method, one computes numerical solutions v_{ι} of the difference equations which in general suffer from round-off errors. Further errors arise in the approximate solution of implicit difference equations. Together with the computation of v_{ι}, it is possible to compute an approximation of the local truncation error $d_{\iota}(u_{G_{\iota}})$ (cf. Grigorieff [4], Section 1.3; Lapidus-Seinfeld [7], Section 2.8). Thus one can determine approximately the *defect*

$$A_{\iota}v_{\iota} - A_{\iota}u_{G_{\iota}} = (v_{\iota}(a) - w^0, A_{\iota}^1 v_{\iota} - d_{\iota}^1(u))$$

and the associated error bound

$$(14) \quad \tau_{\iota}(v_{\iota}) = \max_{t\in G_{\iota}} \left| v_{\iota}(a) - w^0 + \sum_{s\in G_{\iota}'}^{s\leqslant t} \{A_{\iota}^1 v_{\iota}(s) - d_{\iota}^1(u)(s)\} \Delta_{\iota}s \right| e^{-L(t-a)}.$$

Using the inequalities (22), these terms establish the two-sided *a-posteriori estimate of the total error* with the constant $\eta = e^{L(b-a)}$ in the form

$$(15) \quad \frac{\eta}{2\eta - 1} \, \tau_\iota(v_\iota) \leqslant \max_{t \in G_\iota} |v_\iota(t) - u(t)| e^{-L(t-a)} \leqslant \eta \tau_\iota(v_\iota).$$

5. Stronger Bistability and Biconvergence results

Bistability and biconvergence of one-step methods can be established as well in stronger norms, including difference quotients of the approximants, than those considered up to now. For this purpose, let us introduce the following norms on $C(G'_\iota)$,

$$\|v_\iota\|_\infty = \max_{t \in G'_\iota} |v_\iota(t)| e^{-L(t'-a)} \quad (p = \infty),$$

$$\|v_\iota\|_p = \left(\sum_{t \in G'_\iota} |v_\iota(t)|^p e^{-Lp(t'-a)} \Delta_\iota s \right)^{1/p} \quad (1 \leqslant p < \infty),$$

for all $v_\iota \in C(G'_\iota)$, $\iota \in I$. In this way, for each p in $1 \leqslant p \leqslant \infty$, we obtain the p-norm

$$\|w_\iota\|_p = \max\{|w^0_\iota|, \|w^1_\iota\|_p\}$$

for all $w_\iota = (w^0_\iota, w^1_\iota)$ in the space $\mathbb{K}^n \times C(G'_\iota)$, $\iota \in I$. Analogously, for each p in $1 \leqslant p \leqslant \infty$, we define the norm

$$\|u_\iota\|_{1,p} = \max \left(\|u_\iota\|_{E_\iota}, \|D_\iota u_\iota\|_p \right)$$

for all u_ι in the space $C(G_\iota)$, $\iota \in I$. Obviously,

$$(16) \quad \|u_\iota\|_{E_\iota} \leqslant \|u_\iota\|_{1,p}, \quad u_\iota \in C(G_\iota), \quad \iota \in I,$$

so that this new norm is stronger than the norm $\|.\|_{E_\iota}$ for $C(G_\iota)$. Concerning the norms for $\mathbb{K}^n \times C(G'_\iota)$, one easily proves the following statement.

(17) *For each p in $1 \leqslant p \leqslant \infty$ and all $w_\iota = (w^0_\iota, w^1_\iota) \in \mathbb{K}^n \times C(G'_\iota)$, the inequality*

$$\|w_\iota\|_{F_\iota} \leqslant |w^0_\iota| + (b - a)^{1/p'} \|w^1_\iota\|_p, \quad \iota \in I,$$

holds where $1/p' = 1 - 1/p$ for $1 < p < \infty$, $(b - a)^{1/p'} = b - a$ for $p = \infty$, and $(b - a)^{1/p'} = 1$ for $p = 1$.

Proof
We have

$$\|w_\iota\|_{F_\iota} = \max_{t \in G'_\iota} |\hat{w}_\iota(t)| \, e^{-L(t-a)}, \quad \hat{w}_\iota(t) = w^0_\iota + \sum_{s \in G'_\iota}^{s \leqslant t} w^1_\iota(s) \, \Delta_\iota s.$$

Hence it follows that

$$|\hat{w}_\iota(t)| e^{-L(t-a)} \leqslant |w^0_\iota| + \sum_{s \in G'_\iota} |w^1_\iota(s)| e^{-L(s'-a)} \Delta_\iota s = |w^0_\iota| + \|w^1_\iota\|_1.$$

Evidently,

$$\| w_\iota^1 \|_1 \leqslant (b - a) \| w_\iota^1 \|_\infty, \quad w_\iota^1 \in C(G_\iota'), \quad \iota \in I,$$

which implies the above inequality for $p = \infty$. Using Hölder's inequality for $1 < p < \infty$, we obtain the estimate

$$\| w_\iota^1 \|_1 \leqslant (b - a)^{1/p'} \| w_\iota^1 \|_p, \quad w_\iota^1 \in C(G_\iota'), \quad \iota \in I,$$

which proves the above inequality in the case $1 < p < \infty$.

Further, we shall need the following important lemma.

(18)   *Let the one-step method* (A_ι) *be Lipschitz continuous. Then, for all* $w_\iota, z_\iota \in \mathbb{K}^n \times C(G_\iota')$ *and each pair of corresponding fixed points* $u_\iota = C_\iota(u_\iota, w_\iota), v_\iota = C_\iota(v_\iota, w_\iota)$ *in* $B_\rho^E(u_{G_\iota})$, *the inequalities*

$$\| D_\iota u_\iota - D_\iota v_\iota - (w_\iota^1 - z_\iota^1) \|_p \leqslant (L_0 e^{L h_\iota} + L_1)(b - a)^{1/p} \| u_\iota - v_\iota \|_{E_\iota}$$

are valid for all $\iota \in I$ *and* $1 \leqslant p \leqslant \infty$ *where* $(b - a)^{1/p} = 1$ *for* $p = \infty$.

Proof

The solutions u_ι, v_ι satisfy the difference equations

$$D_\iota u_\iota(t) - w_\iota^1(t) = f_\iota(t, u_\iota(t), u_\iota(t'))$$
$$D_\iota v_\iota(t) - z_\iota^1(t) = f_\iota(t, v_\iota(t), v_\iota(t')), \quad t \in G_\iota', \quad \iota \in I.$$

For brevity, we set $x_\iota = u_\iota - v_\iota$, $y_\iota = D_\iota u_\iota - D_\iota v_\iota - (w_\iota^1 - z_\iota^1)$. Using the Lipschitz condition for f_ι and the inequality (11), one obtains the estimate

$$| y_\iota(t) | \leqslant | f_\iota(t, u_\iota(t), u_\iota(t')) - f_\iota(t, v_\iota(t), v_\iota(t')) | \leqslant L_0 | x_\iota(t) | + $$
$$+ L_1 | x_\iota(t') | \leqslant (L_0 e^{L h_\iota} + L_1) e^{L(t' - a)} \| x_\iota \|_{E_\iota}.$$

Hence

$$\| y_\iota \|_\infty \leqslant (L_0 e^{L h_\iota} + L_1) \| x_\iota \|_{E_\iota}, \quad \iota \in I, \quad p = \infty,$$

and

$$\| y_\iota \|_p \leqslant (L_0 e^{L h_\iota} + L_1)(b - a)^{1/p} \| x_\iota \|_{E_\iota}, \quad \iota \in I, \quad 1 \leqslant p < \infty.$$

By these means and the bistability theorem (12), we can prove the following stronger bistability theorem.

(19)   *Let the one-step method* (A_ι) *be Lipschitz continuous and consistent with the initial value problem* (A). *Let the constants* $L > L_0 + L_1, \sigma > 0$ *and* v *be specified by the condition*

(i) $\| d_\iota^A(u) \|_{F_\iota} \leqslant \sigma = \frac{1}{2} \rho e^{-L(b - a)}, \quad L_0 e^{L h_\iota} + L_1 \leqslant L, \quad \iota \geqslant v.$

Then, for each p *in* $1 \leqslant p \leqslant \infty$, *each* $\iota \geqslant v$ *and all* $w_\iota = (w_\iota^0, w_\iota^1)$, $z_\iota = (z_\iota^0, z_\iota^1) \in B_\sigma^F(w_{G_\iota})$, *the corresponding, uniquely determined solutions* u_ι *of the difference equations* (A_ι) *or the fixed point equations* $u_\iota = C_\iota(u_\iota, w_\iota), v_\iota = C_\iota(v_\iota, z_\iota)$ *in* $B_\rho^E(u_{G_\iota})$ *satisfy the stronger bistability inequality*

(ii) $\beta \| w_\iota - z_\iota \|_p \leqslant \| u_\iota - v_\iota \|_{1,p} \leqslant \gamma \| w_\iota - z_\iota \|_p$

where

$$\beta = \frac{1}{1 + L(b - a)^{1/p}}, \quad \gamma = 1 + \max\{1, L(b - a)^{1/p}\}\eta(1 + (b - a)^{1/p'}).$$

Proof

The bistability theorem (12) in Section 3 affirms the unique solubility of the difference equations (A_ι) or fixed point equations (C_ι) for all w_ι, $z_\iota \in B_\sigma^F(w_{G_\iota})$ and the condition $L_0 e^{L h_\iota} + L_1 \leqslant L$ for all $\iota \geqslant \nu$. Lemma (18) implies the estimate

$$\| w_\iota^1 - z_\iota^1 \|_p \leqslant \| D_\iota u_\iota - D_\iota v_\iota \|_p + L(b - a)^{1/p} \| u_\iota - v_\iota \|_{E_\iota}$$
$$\leqslant (1 + L(b - a)^{1/p}) \| u_\iota - v_\iota \|_{1,p}.$$

Obviously,

$$|w_\iota^0 - z_\iota^0| = |u_\iota(a) - v_\iota(a)| \leqslant \| u_\iota - v_\iota \|_{E_\iota} \leqslant \| u_\iota - v_\iota \|_{1,p}$$

This proves the first inequality in (19 ii). The above Lemma (18) further yields the estimate

$$\| D_\iota u_\iota - D_\iota v_\iota \|_p \leqslant \| w_\iota^1 - z_\iota^1 \|_p + L(b - a)^{1/p} \| u_\iota - v_\iota \|_{E_\iota}$$

and thus

$$\| u_\iota - v_\iota \|_{1,p} \leqslant \| w_\iota - z_\iota \|_p + \max\{1, L(b - a)^{1/p}\} \| u_\iota - v_\iota \|_{E_\iota}.$$

From the bistability theorem (12) and the inequality (17), it follows that

$$\| u_\iota - v_\iota \|_{E_\iota} \leqslant \eta \| w_\iota - z_\iota \|_{F_\iota} \leqslant \eta(1 + (b - a)^{1/p'}) \| w_\iota - z_\iota \|_p.$$

Hence the second inequality in (19 ii) is valid too.

Remark

The constants γ, β have the values

$$\beta = \frac{1}{1 + L}, \quad \gamma = 1 + \eta(1 + b - a) \max\{1, L\}$$

in the case $p = \infty$, and

$$\beta = \frac{1}{1 + L(b - a)}, \quad \gamma = 1 + 2\eta \max\{1, L(b - a)\}$$

in the case $p = 1$.

The one-step method (A_ι) is said to be *strongly consistent* with the initial value problem (A) at the solution u if and only if the *stronger consistency condition*

$$\| d_\iota^A(u) \|_p \to 0 \quad (\iota \to \infty)$$

holds. Using the inequality (17), the discretization error $d_\iota^A(u)$, $\iota \in I$, of (A), (A_ι) at u satisfies the estimate

$$\| d_\iota^A(u) \|_{F_\iota} \leqslant (b-a)^{1/p'} \| d_\iota^A(u) \|_p, \quad \iota \in I.$$

Hence, the stronger consistency condition implies the consistency of (A), (A_ι) in the sense of Section 2. For $p = \infty$ the stronger consistency condition takes on the form (cf. Henrici [5], Stummel–Hainer [14])

$$\| d_\iota^A(u) \|_\infty \to 0 \quad \Leftrightarrow \quad \max_{t \in G_\iota'} |d_\iota^1(u)(t)| \to 0 \quad (\iota \to \infty).$$

In the case $1 \leqslant p < \infty$, we obtain

$$\| d_\iota^A(u) \|_p \to 0 \quad \Leftrightarrow \quad \sum_{t \in G_\iota'} |d_\iota^1(u)(t)|^p \Delta_\iota t \to 0 \quad (\iota \to \infty),$$

and hence for $p = 1$ a consistency condition in the sense of Grigorieff [4], Stetter [10], Chapter 2.

The stronger bistability theorem immediately implies the stronger biconvergence theorem:

(20)   *Let (A_ι) be Lipschitz continuous in the neighbourhood $B_\rho(u)$ of the solution u of (A) and let $1 \leqslant p \leqslant \infty$. Then the stronger consistency condition $\| d_\iota^A(u) \|_p \to 0$ $(\iota \to \infty)$ is necessary and sufficient for the unique solubility of the difference equations (A_ι) for almost all ι and the validity of the stronger biconvergence relation*

(i) $\displaystyle \max_{t \in G_\iota} |u_\iota(t) - u(t)| \to 0, \quad \left\| D_\iota u_\iota - \frac{du}{dt} \,\Big|\, G_\iota \right\|_p \to 0 \quad \Leftrightarrow \quad w_\iota^0 \to w^0,$

$$\| w_\iota^1 \|_p \to 0 \quad (\iota \to \infty)$$

for all $w_\iota = (w_\iota^0, w_\iota^1) \in B_\sigma^F(w_{G_\iota})$ and the corresponding solutions u_ι of (A_ι) in $B_\rho^E(u_{G_\iota})$. If (A), (A_ι) is strongly consistent, for each $L > L_0 + L_1$ there exists an index ν such that the following two-sided estimate of the discretization error holds,

(ii) $\beta \tau_\iota(u) \leqslant \| u_\iota - u_{G_\iota} \|_{1,p} \leqslant \gamma \tau_\iota(u)$

where

$$\tau_\iota(u) = \| w_\iota - w_{G_\iota} - d_\iota^A(u) \|_p, \quad \iota \geqslant \nu.$$

Proof
(i) Let (A), (A_ι) be strongly consistent. As in the proof of Theorem (13), the unique solubility of the difference equations follows from the bistability theorem (12). The function u_{G_ι} satisfies the fixed point equation

$$u_{G_\iota} = C_\iota(u_{G_\iota}, z_\iota), \quad z_\iota = w_{G_\iota} + d_\iota^A(u), \quad \iota \in I.$$

Since $z_\iota \in B_\sigma^F(w_{G_\iota})$ for $\iota \geqslant \nu$, the stronger bistability inequality for $v_\iota = u_{G_\iota}$ yields the discretization error estimate (20 ii). By the stronger consistency condition, we have $\| d_\iota^A(u) \|_p \to 0$ $(\iota \to \infty)$ so that the inequalities (20 ii) imply the stronger biconvergence relation.

(ii) Conversely, let us assume now that the difference equations (A_ι) or the fixed point equations (C_ι) are uniquely soluble in $B_\rho^E(u_{G_\iota})$ for all $\iota \geq \nu$. Then, in particular, for $w_\iota = w_{G_\iota} = (w_\iota^0, 0_\iota)$ there exist solutions $u_\iota = C_\iota(u_\iota, w_{G_\iota}) \in B_\rho^E(u_{G_\iota})$ for all $\iota \geq \nu$. The biconvergence relation ensures $\| u_\iota - u_{G_\iota} \|_{1,p} \to 0$ $(\iota \to \infty)$ because $\| w_\iota - w_{G_\iota} \|_p \to 0$ for $\iota \to \infty$. Using Lemma (18) for $v_\iota = u_{G_\iota}$, $z_\iota = w_{G_\iota} + d_\iota^A(u)$, we obtain

$$\| d_\iota^A(u) \|_p \leq \{1 + (L_0 e^{Lh_\iota} + L_1)(b - a)^{1/p'}\} \| u_\iota - u_{G_\iota} \|_{1,p} \to 0$$

for $\iota \to \infty$ so that (A), (A_ι) is strongly consistent.

References

[1] Bielecki, A. (1956). Une remarque sur la méthode de Banach-Cacciopoli-Tikhonov dans la théorie des équations differentielles ordinaires, *Bull. Acad. Polon. Sci.*, 4, 261–264.

[2] Dahlquist, G. (1956). Convergence and stability in the numerical integration of ordinary differential equations, *Math. Scand.* 4, 33–53.

[3] Gear, C. W. (1971). *Numerical Initial Value Problems in Ordinary Differential Equations.* Englewood Cliffs: Prentice Hall.

[4] Grigorieff, R. D. (1972). *Numerik Gewöhnlicher Differentialgleichungen I.* Stuttgart: Teubner.

[5] Henrici, P. (1962). *Discrete Variable Methods in Ordinary Differential Equations.* New York-London: Wiley.

[6] Henrici, P. (1963). *Error Propagation for Difference Methods.* New York-London: Wiley.

[7] Lapidus, L. and Seinfeld, J. H. (1971). *Numerical Solution of Ordinary Differential Equations.* New York-London: Academic Press.

[8] Spijker, M. N. (1971). On the structure of error estimates for finite-difference methods, *Numer. Math.* 18, 73–100.

[9] Spijker, M. N. (1972). Equivalence theorems for nonlinear finite-difference methods, *Proc. Conf. "Numerische Lösung nichtlinearer partieller Differential-und Integrodifferentialgleichungen", Oberwolfach 1971. Lecture Notes in Mathematics* 267, 233–264. Berlin-Heidelberg-New York: Springer.

[10] Stetter, H. J. (1973). *Analysis of Discretization Methods for Ordinary Differential Equations.* Berlin-Heidelberg-New York: Springer.

[11] Stummel, F. (1973). Discrete convergence of mappings, *Proc. Conf. Numerical Analysis Dublin, August 1972*, 285–310. New York-London: Academic Press.

[12] Stummel, F. (1973). Approximation methods in analysis. *Lecture Notes, Spring Semester. Aarhus Universitet.*

[13] Stummel, F. (1974). Difference methods for linear initial value problems. To appear in the *Proc. Conf. "Numerische Lösung nichtlinearer partieller Differential-und Integrodifferentialgleichungen", Oberwolfach, December 1973. Lecture Notes in Mathematics*, 395, 123–135. Berlin-Heidelberg-New York: Springer.

[14] Stummel, F. and Hainer, K. (1971). *Praktische Mathematik.* Stuttgart: Teubner.

[15] Stummel, F. and Reinhardt, J. (1973). Discrete convergence of continuous mappings in metric spaces. *Proc. Conf. "Numerische, insbesondere approximations-theoretische Behandlung von Funktionalgleichungen", Oberwolfach December 1972. Lecture Notes in Mathematics*, 333, 218–242. Berlin-Heidelberg-New York: Springer.

On the Error Estimation of Chebyshev Series Approximate Solutions to Boundary Value Problems

M. Urabe

Abstract

In his previous papers [2, 3], the author gave a method of obtaining an error bound of a finite Chebyshev series approximate solution to a boundary value problem. In these papers an error bound is obtained by the use of a certain pair of positive numbers M_1 and M_2, and there these numbers are obtained by means of the numerical integration of the first variation equation of the given differential equation with respect to the approximate solution in question. In the present paper, the author will show that one can obtain these positive numbers algebraically without performing the numerical integration of a differential equation.

1. Introduction

We are concerned with the boundary value problem of the form

$$\frac{dx}{dt} = X(x, t), \quad -1 < t < 1,$$

$$\sum_{i=0}^{N} L_i x(t_i) = l, \quad -1 = t_0 < t_1 < \cdots < t_{N-1} < t_N = 1, \tag{1.1}$$

where x and $X(x, t)$ are real vectors of the same dimension, $X(x, t)$ is defined and twice continuously differentiable with respect to x and t in the region D of the tx-space intercepted by two hyperplanes $t = -1$ and $t = 1$, L_i $(i = 0, 1, 2, \ldots, N)$ are given square matrices, and l is a given vector.

Let $x_m(t)$ be an m-th order Chebyshev series approximate solution to the problem (1.1), that is, a finite Chebyshev series of order m of the form

$$x_m(t) = a_0 + \sqrt{2} \sum_{n=1}^{m} a_n T_n(t) \tag{1.2}$$

satisfying the equations

$$\frac{dx_m(t)}{dt} = P_{m-1}X[x_m(t), t], \quad -1 \leqslant t \leqslant 1,$$

$$\sum_{i=0}^{N} L_i x_m(t_i) = l, \tag{1.3}$$

where $T_n(t)$ is the Chebyshev polynomial of degree n and P_{m-1} is an operator which expresses the truncation of the Chebyshev series of the operand discarding the terms of degree higher than $m - 1$. As shown in [2], the system of equations (1.3) is equivalent to the system of equations

$$F_0(\alpha) := \sum_{i=0}^{N} L_i x_m(t_i) - l = 0,$$

$$F_n(\alpha) := \frac{\sqrt{2}}{\pi e_{n-1}} \int_0^\pi X[x_m(\cos\theta), \cos\theta] \cos(n-1)\theta \, d\theta - \tag{1.4}$$

$$- a'_{n-1}(\alpha) = 0 \quad (n = 1, 2, \ldots, m).$$

In (1.4),

$$\alpha = (a_0^*, a_1^*, a_2^*, \ldots, a_m^*)^*, \tag{1.5}$$

$$e_n = \begin{cases} \sqrt{2} & \text{if } n = 0, \\ 1 & \text{otherwise,} \end{cases} \tag{1.6}$$

and $a'_n(\alpha)$ $(n = 0, 1, 2, \ldots, m - 1)$ denote the coefficients of the Chebyshev series of $dx_m(t)/dt$, that is,

$$a'_n(\alpha) = \frac{2}{e_n} \sum_{p=1}^{[(m-n+1)/2]} (n + 2p - 1) a_{n+2p-1} \tag{1.7}$$

$$(n = 0, 1, 2, \ldots, m - 1),$$

where the symbol * denotes the transpose of vectors and $[(m - n + 1)/2]$ denotes the greatest integer less than or equal to $(m - n + 1)/2$. Since an m-th order Chebyshev series approximate solution $x_m(t)$ to the problem (1.1) is determined by a solution of the system of equations (1.4), in [2] the system of equations (1.4) is called a *determining equation* of an m-th order Chebyshev series approximate solution. In what follows, the system of equations (1.4) will be written in a vector form as follows:

$$F^{(m)}(\alpha) = 0. \tag{1.8}$$

As shown in [3], a convenient method of solving numerically the determining equation (1.8) is the Newton method in which a solution is computed by the iterative process

$$J_m(\alpha_p)h_p + F^{(m)}(\alpha_p) = 0,$$
$$\alpha_{p+1} = \alpha_p + h_p \quad (p = 0, 1, 2, \ldots),$$
(1.9)

starting from an appropriate approximate solution α_0 where $J_m(\alpha)$ denotes the Jacobian matrix of $F^{(m)}(\alpha)$ with respect to α.

After having obtained an m-th order Chebyshev series approximate solution, it is necessary to assure the existence of an exact solution to the given boundary value problem (1.1) and further to get an error estimate for the finite Chebyshev series approximate solution obtained numerically. In [2] and [3], the author gave a method for doing this, making use of an existence theorem for boundary value problems established in [1].

The theorem established in [1] is based on the following proposition established in the same paper.

Proposition
Let

$$\frac{dx}{dt} = A(t)x + \phi(t)$$
(1.10)

be a given linear differential equation where x and $\phi(t)$ are real vectors of the same dimension, $A(t)$ is a continuous square matrix defined on the interval $J = [-1, 1]$, and $\phi(t)$ is a continuous function defined on J.

Let $\Phi(t)$ be the fundamental matrix of the corresponding linear homogeneous differential equation

$$\frac{dy}{dt} = A(t)y$$
(1.11)

such that $\Phi(-1) = E$ (E the unit matrix). If the matrix

$$G := \sum_{i=0}^{N} L_i \Phi(t_i)$$
(1.12)

is non-singular for

$$-1 = t_0 < t_1 < \cdots < t_{N-1} < t_N = 1$$

and given square matrices $L_i (i = 0, 1, 2, \ldots, N)$, then the given differential equation (1.10) has one and only one solution satisfying the boundary condition

$$\sum_{i=0}^{N} L_i x(t_i) = l$$
(1.13)

for any given vector l and such a solution x(t) is given by

$$x(t) = \Phi(t)G^{-1}l + \int_{-1}^{1} H(t,s)\phi(s)\,ds, \tag{1.14}$$

where $H(t,s)$ is the piecewise continuous matrix such that, for $t_{k-1} \leqslant s < t_k$ (k = 1, 2, ..., N),

$$H(t,s) = \begin{cases} \Phi(t)[E - G^{-1}\sum_{i=k}^{N} L_i\Phi(t_i)]\Phi^{-1}(s) & \text{if } s < t, \\ -\Phi(t)G^{-1}\sum_{i=k}^{N} L_i\Phi(t_i)\cdot\Phi^{-1}(s) & \text{if } s \geqslant t. \end{cases} \tag{1.15}$$

For the proof, see Proposition 1 of [1].
If we put

$$\psi(t) = \int_{-1}^{1} H(t,s)\phi(s)\,ds, \tag{1.16}$$

then from the above Proposition it is clear that $x = \psi(t)$ is a solution of (1.10) satisfying the boundary condition

$$\sum_{i=0}^{N} L_i x(t_i) = 0. \tag{1.17}$$

The expression (1.16) defines a linear mapping H in the space of continuous functions defined on the interval J. For fixed L_i and t_i (i = 0, 1, 2, ..., N), the matrix $H(t,s)$ is dependent only on the matrix $\Phi(t)$, that is, it is dependent only on the matrix $A(t)$. Hence, in [1], the mapping H has been called the *H-mapping* corresponding to the matrix $A(t)$. This name will be used also in the present paper and the expression (1.16) will be written sometimes as

$$\psi = H\phi. \tag{1.18}$$

In the present paper, for vectors and matrices, the Euclidean norms and the corresponding ones in the functional analysis sense will be used, and they will be denoted by the symbol $\|\cdot\|$. For vector-valued functions continuous on the interval J, we shall use two kinds of norms denoted by the symbols $\|\cdot\|_n$ and $\|\cdot\|_q$, which are defined by

$$\|f\|_n = \sup_{t\in J} \|f(t)\| = \max_{t\in J} \|f(t)\| \tag{1.19}$$

and

$$\|f\|_q = \left[\frac{1}{\pi}\int_0^{\pi} \|f(\cos\theta)\|^2\,d\theta\right]^{1/2}. \tag{1.20}$$

Once the norms of continuous vector-valued functions have been defined, the norms of H-mappings are defined correspondingly. Thus, for instance,

$$\|H\|_n = \sup_{\phi \in C[J]} \|H\phi\|_n / \|\phi\|_n$$

for a H-mapping H, where $C[J]$ is the set of all vector-valued functions of a certain constant dimension continuous on the interval J. For $\|H\|_n$, by Schwarz's inequality, it is easily seen that

$$\|H\|_n \leqslant [2 \sup_{t \in J} \int_{-1}^{1} \sum_{i,j} H_{ij}^2(t,s)\,ds]^{1/2}, \tag{1.21}$$

where $H_{ij}(t,s)$ are the elements of the matrix $H(t,s)$ of the H-mapping H.

The existence theorem for boundary value problems established in [1] then reads as follows.

Theorem 1.1
Let

$$\frac{dx}{dt} = X(x,t) \tag{1.22}$$

be a given real differential equation, where x and $X(x,t)$ are real vectors of the same dimension, and $X(x,t)$ is continuously differentiable with respect to x in the region D of the tx-space intercepted by the two hyperplanes $t = -1$ and $t = 1$.

Suppose that equation (1.22) has an approximate solution $x = \bar{x}(t)$ lying in D and satisfying the boundary condition

$$\sum_{i=0}^{N} L_i x(t_i) = l \tag{1.23}$$

approximately, where

$$-1 = t_0 < t_1 < t_2 < \cdots < t_{N-1} < t_N = 1,$$

L_i ($i = 0, 1, 2, \ldots, N$) are given square matrices, and l is a given vector. For the approximate solution $\bar{x}(t)$, assume that there are a square matrix $A(t)$ continuous on the interval $J = [-1, 1]$, a positive number δ, and a non-negative number $\kappa < 1$ such that
(i) the matrix

$$G := \sum_{i=0}^{N} L_i \Phi(t_i) \tag{1.24}$$

is non-singular, and
(ii)

$$D_\delta := \{(t,x) \mid \|x - \bar{x}(t)\| \leqslant \delta, \quad t \in J\} \subset D, \tag{1.25}$$

$$\| \Psi(x, t) - A(t) \| \leqslant \kappa/M_1 \qquad (1.26)$$

for any $(t, x) \in D_\delta$ and

$$\frac{M_1 r + M_2 \epsilon}{1 - \kappa} \leqslant \delta. \qquad (1.27)$$

Here $\Phi(t)$ is the fundamental matrix of the linear homogeneous differential equation

$$\frac{dy}{dt} = A(t)y \qquad (1.28)$$

satisfying the initial condition $\Phi(-1) = E$, $\Psi(x, t)$ is the Jacobian matrix of $X(x, t)$ with respect to x, M_1 is a positive number such that

$$\| H \|_n \leqslant M_1 \qquad (1.29)$$

where H is the H-mapping corresponding to $A(t)$, M_2 is a positive number such that

$$\| \Phi(t)G^{-1} \| \leqslant M_2 \qquad (1.30)$$

for any $t \in J$, r is a non-negative number such that

$$\left\| \frac{d\bar{x}(t)}{dt} - X[\bar{x}(t), t] \right\|_n \leqslant r, \qquad (1.31)$$

and ϵ is a non-negative number such that

$$\left\| \sum_{i=0}^{N} L_i \bar{x}(t_i) - l \right\| \leqslant \epsilon. \qquad (1.32)$$

The given differential equation (1.22) then has one and only one solution $x = \hat{x}(t)$ satisfying the boundary condition (1.23) in D_δ, and for $\hat{x}(t)$ it is valid that

$$\| \hat{x} - \bar{x} \|_n \leqslant \frac{M_1 r + M_2 \epsilon}{1 - \kappa}. \qquad (1.33)$$

In Theorem 1.1, suppose that $\bar{x}(t)$ is an m-th order Chebyshev series approximate solution $\bar{x}_m(t)$ obtained numerically to the problem (1.1), that is, a finite Chebyshev series of order m of the form (1.2) satisfying the equations (1.3) approximately. For $\bar{x}(t) = \bar{x}_m(t)$, as shown in [3], one can find easily the value of r satisfying (1.31) by expanding the function $(d\bar{x}_m(t)/dt) - X[\bar{x}_m(t), t]$ in a Chebyshev series. The value of ϵ satisfying (1.32) can be found readily for $\bar{x}(t) = \bar{x}_m(t)$. Taking into account the condition (1.26), let us take $\Psi[\bar{x}_m(t), t]$ for $A(t)$ and denote by $\Phi_m(t), G_m$, and H_m respectively the fundamental matrix of the linear homogeneous differential equation

$$\frac{dy}{dt} = \Psi[\bar{x}_m(t), t]y \qquad (1.34)$$

satisfying the initial condition $\Phi_m(-1) = E$, the matrix $\Sigma_{i=0}^{N} L_i \Phi_m(t_i)$, and the H-mapping corresponding to the matrix $\Psi[\bar{x}_m(t), t]$ if it exists. If one can find the positive numbers M_1 and M_2 satisfying the inequalities (1.29) and (1.30) for $H = H_m$, $\Phi(t) = \Phi_m(t)$, and $G = G_m$, then as shown in [3] one can verify easily the conditions of Thereom 1.1 by checking the existence of the numbers δ and $\kappa < 1$ satisfying the condition (1.25) and the inequalities (1.26) and (1.27). If the conditions of Theorem 1.1 are seen to be all fulfilled, then by Theorem 1.1 one sees that the given boundary value problem (1.1) has a solution and the m-th order Chebyshev series approximate solution $\bar{x}_m(t)$ obtained numerically is within an error bound $(M_1 r + M_2 \epsilon)(1 - \kappa)^{-1}$.

In [3], the author obtains the positive numbers M_1 and M_2 in question by calculating $\Phi_m(t)$, integrating the differential equation (1.34) numerically, and then making use of the formulas (1.15), (1.21), (1.24) and (1.30). In the present paper, the author will show that the positive numbers M_1 and M_2 in question can be obtained algebraically in terms of the norm of $J_m^{-1}(\alpha)$ appearing in Newton's iterative process (1.9) for the determining equation (1.8) and thus without appealing to the numerical integration of the differential equation (1.34).

In the present paper, the author will make use of some properties of Chebyshev series established in [2]. They will be sketched in Section 2. In Section 3, the author will give a method of obtaining algebraically the M_1 in question and, in Section 4, he will give a method of obtaining algebraically the M_2 in question. In Section 5, the author will show that the M_1 and M_2 obtained in Sections 3 and 4 can be, indeed, used effectively for the verification of the conditions of Theorem 1.1, in other words, they are of proper magnitude even if they are larger than those obtained by the method of [3] based on the numerical integration of the differential equation (1.34).

2. Some Properties of Chebyshev Series

Let $f(t)$ be a continuous vector–valued function defined on the interval $J = [-1, 1]$ and let its Chebyshev series be

$$f(t) \sim a_0 + \sqrt{2} \sum_{n=1}^{\infty} a_n T_n(t). \tag{2.1}$$

In (2.1), it is clear that the coefficients $a_0, a_1, a_2, \ldots$ are all vectors of the same dimension as $f(t)$. In the present section some properties of the Chebyshev series (2.1) which will be used in the sequel will be described briefly. For the proof of these properties, see [2].

2.1. Parseval's equality for Chebyshev series
The Chebyshev series (2.1) is clearly equivalent to

$$f(\cos \theta) \sim a_0 + \sqrt{2} \sum_{n=1}^{\infty} a_n \cos n\theta, \tag{2.2}$$

therefore

$$a_n = \frac{\sqrt{2}}{\pi e_n} \int_0^\pi f(\cos \theta) \cos n\theta \, d\theta \quad (n = 0, 1, 2, \ldots), \tag{2.3}$$

where

$$e_n = \begin{cases} \sqrt{2} & \text{if } n = 0, \\ 1 & \text{otherwise.} \end{cases} \tag{2.4}$$

Parseval's equality applied to the Fourier series (2.2) then implies that

$$\|f\|_q^2 = \sum_{n=0}^\infty \|a_n\|^2. \tag{2.5}$$

If $f(t)$ is in particular a finite Chebyshev series of the form

$$f(t) = a_0 + \sqrt{2} \sum_{n=1}^m a_n T_n(t), \tag{2.6}$$

then (2.5) implies that

$$\|f\|_q = \|\alpha\|, \tag{2.7}$$

where $\alpha = (a_0^*, a_1^*, a_2^*, \ldots, a_m^*)^*$ (* denotes the transpose).

2.2. Chebyshev series of derivatives

Suppose that $f(t)$ is continuously differentiable with respect to t on the interval J, and let the Chebyshev series of $\dot{f}(t)$ ($\cdot = d/dt$) be

$$\dot{f}(t) \sim a_0' + \sqrt{2} \sum_{n=1}^\infty a_n' T_n(t). \tag{2.8}$$

Then it is well known that the coefficients $a_0', a_1', a_2', \ldots$ are connected with the coefficients $a_0, a_1, a_2, \ldots$ of the Chebyshev series of $f(t)$ by

$$e_{n-1} a_{n-1}' - a_{n+1}' = 2n a_n \quad (n = 1, 2, 3, \ldots). \tag{2.9}$$

From (2.9) it readily follows that

$$a_n' = \frac{2}{e_n} \sum_{p=1}^\infty (n + 2p - 1) a_{n+2p-1} \quad (n = 0, 1, 2, \ldots). \tag{2.10}$$

2.3. Truncation of Chebyshev series

Let P_m be an operator which expresses the truncation of the Chebyshev series of the operand discarding the terms of degree higher than m. Then for the $f(t)$ of (2.1),

$$P_m f(t) = a_0 + \sqrt{2} \sum_{n=1}^m a_n T_n(t). \tag{2.11}$$

If $f(t)$ is continuously differentiable with respect to t on the interval J, then by the theory of Fourier series we have

$$f(t) = a_0 + \sqrt{2} \sum_{n=1}^{\infty} a_n T_n(t), \qquad (2.12)$$

and hence

$$(I - P_m)f(t) = \sqrt{2} \sum_{n=m+1}^{\infty} a_n T_n(t), \qquad (2.13)$$

where I denotes the identity operator. For $(I - P_m)f(t)$, it is shown in [2] that

$$\| (I - P_m)f \|_n \leqslant \sigma(m) \| (I - P_{m-1})\dot{f} \|_q \leqslant \sigma(m) \| \dot{f} \|_q$$
$$(m = 0, 1, 2, \ldots; P_{-1} \equiv 0) \qquad (2.14)$$

and

$$\| (I - P_m)f \|_q \leqslant \sigma_1(m) \| (I - P_{m-1})\dot{f} \|_q \leqslant \sigma_1(m) \| \dot{f} \|_q$$
$$(m = 0, 1, 2, \ldots; P_{-1} \equiv 0) \qquad (2.15)$$

where

$$\sigma(m) = \sqrt{2} \left[\sum_{n=m+1}^{\infty} n^{-2} \right]^{1/2}, \qquad (2.16)$$

$$\sigma_1(m) = (m + 1)^{-1}. \qquad (2.17)$$

For $\sigma(m)$, it is clearly valid that

$$\frac{\sqrt{2}}{\sqrt{m+1}} < \sigma(m) < \frac{\sqrt{2}}{\sqrt{m}}. \qquad (2.18)$$

2.4. Derivatives of $P_m f(t)$

For a vector-valued function $f(t)$ continuously differentiable on the interval J, put

$$P_m f(t) = f_m(t). \qquad (2.19)$$

Then it is valid that

$$\dot{f}(t) - \dot{f}_m(t) \sim \chi_{m+1}(t) + (I - P_{m+1})\dot{f}(t), \qquad (2.20)$$

where

$$\chi_{m+1}(t) = \begin{cases} \sqrt{2}e_m a'_m [\tfrac{1}{2} + T_2(t) + T_4(t) + \cdots + T_m(t)] + \\ \quad + \sqrt{2}a'_{m+1}[T_1(t) + T_3(t) + \cdots + T_{m+1}(t)] \\ \qquad \text{if } m \text{ is even,} \\ \sqrt{2}a'_{m+1}[\tfrac{1}{2} + T_2(t) + T_4(t) + \cdots + T_{m+1}(t)] + \\ \quad + \sqrt{2}a'_m[T_1(t) + T_3(t) + \cdots + T_m(t)] \\ \qquad \text{if } m \text{ is odd.} \end{cases} \qquad (2.21)$$

Here a'_m and a'_{m+1} are respectively the coefficients of the terms of degree m and $m+1$ in the Chebyshev series (2.8) of $f(t)$.

3. A Method of Obtaining M_1

Let

$$\bar{x}_m(t) = \bar{a}_0 + \sqrt{2} \sum_{n=1}^{m} \bar{a}_n T_n(t) \tag{3.1}$$

be an m-th order Chebyshev series approximate solution obtained numerically to the problem (1.1). Then, since $\bar{x}_m(t)$ satisfies the equations (1.3) approximately, we may suppose that

$$\left\| \frac{d\bar{x}_m(t)}{dt} - P_{m-1} X[\bar{x}_m(t), t] \right\|_n \leqslant r_0 \tag{3.2}$$

for some small positive number r_0.

In (1.1), $X(x, t)$ is supposed to be twice continuously differentiable with respect to x and t in the region D of the tx-space intercepted by the two hyperplanes $t = -1$ and $t = 1$. Therefore let us suppose that on the interval $J = [-1, 1]$,

$$\| X[\bar{x}_m(t), t] \| \leqslant K_0,$$

$$\| \Psi[\bar{x}_m(t), t] \| \leqslant K_1,$$

$$\left\| \frac{\partial X}{\partial t} [\bar{x}_m(t), t] \right\| \leqslant K_2, \tag{3.3}$$

$$\left[\sum_{i,j,l} \left\{ \frac{\partial \Psi_{ij}}{\partial x_l} [\bar{x}_m(t), t] \right\}^2 \right]^{1/2} \leqslant K_3,$$

$$\left[\sum_{i,j} \left\{ \frac{\partial \Psi_{ij}}{\partial t} [\bar{x}_m(t), t] \right\}^2 \right]^{1/2} \leqslant K_4,$$

where $\Psi(x, t)$ is the Jacobian matrix of $X(x, t)$ with respect to x, $\Psi_{ij}(x, t)$ are the elements of the matrix $\Psi(x, t)$, and x_l are the components of the vector x.

First, we shall prove

Lemma 3.1
Let $y = y(t)$ be a solution of the differential equation

$$\frac{dy}{dt} = \Psi[\bar{x}_m(t), t] y + \phi(t), \tag{3.4}$$

where $\phi(t)$ is an arbitrary vector-valued function continuous on J. Then

$$\left\| P_{m-1}\left(\dot{y} - \frac{d}{dt}P_m y\right) \right\|_q \leqslant \frac{1}{\sqrt{2m}}(A_m \| y \|_q + K_1 \| \dot{y} \|_q) +$$

$$+ \frac{\sqrt{m}}{\sqrt{2}} \| (I - P_{m-1})\phi \|_q \quad (\cdot = d/dt) \tag{3.5}$$

for $m \geqslant 1$, where

$$A_m = K_3(K_0 + r_0) + K_4 + \sigma(m-1)K_3[K_1(K_0 + r_0) + K_2]. \tag{3.6}$$

Proof
By (2.20) and (2.21), we have

$$\dot{y}(t) - \frac{d}{dt}P_m y(t) \sim \chi_{m+1}(t) + (I - P_{m+1})\dot{y}(t), \tag{3.7}$$

where

$$\chi_{m+1}(t) = \begin{cases} \sqrt{2}b'_m[\frac{1}{2} + T_2(t) + T_4(t) + \cdots + T_m(t)] + \\ \quad + \sqrt{2}b'_{m+1}[T_1(t) + T_3(t) + \cdots + T_{m+1}(t)] \\ \qquad \text{for even } m \geqslant 2, \\ \sqrt{2}b'_{m+1}[\frac{1}{2} + T_2(t) + T_4(t) + \cdots + T_{m+1}(t)] + \\ \quad + \sqrt{2}b'_m[T_1(t) + T_3(t) + \cdots + T_m(t)] \\ \qquad \text{for odd } m \geqslant 1, \end{cases} \tag{3.8}$$

and b'_m and b'_{m+1} are respectively the coefficients of the terms of degree m and $m+1$ in the Chebyshev series of $\dot{y}(t)$:

$$\dot{y}(t) \sim b'_0 + \sqrt{2} \sum_{n=1}^{\infty} b'_n T_n(t). \tag{3.9}$$

Then by (3.7) and (3.8), we have

$$P_{m-1}\left[\dot{y}(t) - \frac{d}{dt}P_m y(t)\right] = P_{m-1}\chi_{m+1}(t), \tag{3.10}$$

and

$$P_{m-1}\chi_{m+1}(t) = \begin{cases} \sqrt{2}b'_m[\frac{1}{2} + T_2(t) + T_4(t) + \cdots + T_{m-2}(t)] + \\ \quad + \sqrt{2}b'_{m+1}[T_1(t) + T_3(t) + \cdots + T_{m-1}(t)] \\ \qquad \text{for even } m \geqslant 2, \\ \sqrt{2}b'_{m+1}[\frac{1}{2} + T_2(t) + T_4(t) + \cdots + T_{m-1}(t)] + \\ \quad + \sqrt{2}b'_m[T_1(t) + T_3(t) + \cdots + T_{m-2}(t)] \\ \qquad \text{for odd } m \geqslant 1. \end{cases} \tag{3.11}$$

By Parseval's equality, it follows from (3.11) that for even $m > 2$,

$$\| P_{m-1} \chi_{m+1} \|_q^2 = \| b'_m \|^2 \left(\frac{1}{2} + \frac{m-2}{2} \right) + \| b'_{m+1} \|^2 \cdot \frac{m}{2}$$

$$= \frac{m-1}{2} \| b'_m \|^2 + \frac{m}{2} \| b'_{m+1} \|^2,$$

and for odd $m > 1$,

$$\| P_{m-1} \chi_{m+1} \|_q^2 = \| b'_{m+1} \|^2 \left(\frac{1}{2} + \frac{m-1}{2} \right) + \| b'_m \|^2 \cdot \frac{m-1}{2}$$

$$= \frac{m}{2} \| b'_{m+1} \|^2 + \frac{m-1}{2} \| b'_m \|^2.$$

Therefore, whether m is even or not, we have

$$\| P_{m-1} \chi_{m+1} \|_q^2 \leqslant \frac{m}{2} \left(\| b'_m \|^2 + \| b'_{m+1} \|^2 \right) \tag{3.12}$$

for $m \geqslant 1$. Inequality (3.12) then implies that

$$\| P_{m-1} \chi_{m+1} \|_q \leqslant \frac{\sqrt{m}}{\sqrt{2}} \| (I - P_{m-1}) \dot{y} \|_q, \tag{3.13}$$

which by (3.10) implies that

$$\left\| P_{m-1} \left(\dot{y} - \frac{d}{dt} P_m y \right) \right\|_q \leqslant \frac{\sqrt{m}}{\sqrt{2}} \| (I - P_{m-1}) \dot{y} \|_q. \tag{3.14}$$

Now for the solution $y = y(t)$ of (3.4),

$$(I - P_{m-1}) \dot{y} = (I - P_{m-1}) \Psi [\bar{x}_m(t), t] y + (I - P_{m-1}) \phi. \tag{3.15}$$

In order to estimate $\| (I - P_{m-1}) \Psi [\bar{x}_m(t), t] y \|_q$ by the use of (2.15), let us consider

$$\frac{d}{dt} \{ \Psi [\bar{x}_m(t), t] y(t) \}$$

$$= \{ \Psi_x [\bar{x}_m(t), t] \dot{\bar{x}}_m(t) + \frac{\partial \Psi}{\partial t} [\bar{x}_m(t), t] \} y(t) +$$

$$+ \Psi [\bar{x}_m(t), t] \dot{y}(t), \tag{3.16}$$

where $\Psi_x(x, t)$ denotes the tensor of the 3rd order whose components are $(\partial \Psi_{ij}/\partial x_l)(x, t)$. Now, by (3.2), it is valid that

$$\| \dot{\bar{x}}_m - X [\bar{x}_m(t), t] \|_n \leqslant \| \dot{\bar{x}}_m - P_{m-1} X [\bar{x}_m(t), t] \|_n +$$

$$+ \| (I - P_{m-1}) X [\bar{x}_m(t), t] \|_n$$

$$\leqslant r_0 + \| (I - P_{m-1}) X [\bar{x}_m(t), t] \|_n. \tag{3.17}$$

Therefore, by the use of (2.14) and (3.3), we have

$$\| \dot{\bar{x}}_m - X[\bar{x}_m(t), t] \|_n$$

$$\leqslant r_0 + \sigma(m-1) \| \Psi[\bar{x}_m(t), t] \dot{\bar{x}}_m + \frac{\partial X}{\partial t} [\bar{x}_m(t), t] \|_q$$

$$\leqslant r_0 + \sigma(m-1)[K_1 \| \dot{\bar{x}}_m \|_q + K_2]. \tag{3.18}$$

However, if we put

$$\dot{\bar{x}}_m(t) = P_{m-1} X[\bar{x}_m(t), t] + \eta(t),$$

then by Parseval's equality and (3.2), we have

$$\| \dot{\bar{x}}_m \|_q \leqslant \| P_{m-1} X[\bar{x}_m(t), t] \|_q + \| \eta \|_q$$

$$\leqslant \| X[\bar{x}_m(t), t] \|_q + r_0,$$

which by (3.3) implies

$$\| \dot{\bar{x}}_m \|_q \leqslant K_0 + r_0. \tag{3.19}$$

Hence by (3.18) we have

$$\| \dot{\bar{x}}_m - X[\bar{x}_m(t), t] \|_n \leqslant r_0 + \sigma(m-1)[K_1(K_0 + r_0) + K_2].$$

Then by (3.3) we get

$$\| \dot{\bar{x}}_m \|_n \leqslant K_0 + r_0 + \sigma(m-1)[K_1(K_0 + r_0) + K_2]. \tag{3.20}$$

Then from (3.16), by (3.3) and (3.6), we have

$$\left\| \frac{d}{dt} \{ \Psi[\bar{x}_m(t), t] y(t) \} \right\|_q \leqslant A_m \| y \|_q + K_1 \| \dot{y} \|_q. \tag{3.21}$$

By the use of (2.14), from (3.15), we thus obtain

$$\| (I - P_{m-1}) \dot{y} \|_q \leqslant \sigma_1(m-1)[A_m \| y \|_q + K_1 \| \dot{y} \|_q] +$$

$$+ \| (I - P_{m-1}) \phi \|_q. \tag{3.22}$$

Then, since $\sigma_1(m-1) = m^{-1}$ by (2.17), from (3.14) we get the desired inequality (3.5) for $m \geqslant 1$. Q.E.D.

Making use of the above lemma, we shall now prove the theorem on which our method of obtaining M_1 is based.

Let $J_m(\alpha)$ be the Jacobian matrix of $F^{(m)}(\alpha)$ with respect to α, where $F^{(m)}(\alpha)$ is the vector-valued function with components $F_0(\alpha), F_1(\alpha), F_2(\alpha), \ldots, F_m(\alpha)$ given in (1.4), that is, the function on the left-hand side of the determining equation (1.8). Put

$$\rho(m) = \sigma(m) \sum_{i=0}^{N} \| L_i \| + K_1 [(m+1)^{-1} + (2m)^{-1/2}] \quad (m \geqslant 1),$$

$$B_m = 1 + \frac{\pi}{\sqrt{3}} K_1 - K_1(m+1)^{-1}, \tag{3.23}$$

$$\lambda(m) = [1 - K_1(m+1)^{-1}] \sqrt{\frac{m+2}{2}} + \rho(m)$$

and let $\bar{\alpha} = (\bar{a}_0^*, \bar{a}_1^*, \bar{a}_2^*, \ldots, \bar{a}_m^*)^*$ for $\bar{x}_m(t)$ given by (3.1). Then our theorem reads as follows:

Theorem 3.1
Suppose that

$$\| J_m^{-1}(\bar{\alpha}) \| \leqslant M_m', \tag{3.24}$$

and put

$$C_m = 1 - K_1 (m + 1)^{-1} - K_1 M_m' \rho(m). \tag{3.25}$$

Then for any solution $y = y(t)$ of (3.4) satisfying the boundary condition

$$\sum_{i=0}^{N} L_i y(t_i) = 0, \tag{3.26}$$

if any exists, it is valid that

$$\| y \|_n \leqslant \mu_1(m) \| \phi \|_n \tag{3.27}$$

for $m \geqslant$ provided that

$$C_m, \ 1 - K_1(m + 1)^{-1}, \ 1 - \frac{1}{\sqrt{2m}} M_m' A_m B_m C_m^{-1} > 0, \tag{3.28}$$

where

$$\mu_1(m) = \frac{\dfrac{\pi}{\sqrt{3}} + M_m' B_m C_m^{-1} \lambda(m)}{\left[1 - K_1(m + 1)^{-1} \right] \left[1 - \dfrac{1}{\sqrt{2m}} M_m' A_m B_m C_m^{-1} \right]}. \tag{3.29}$$

Proof
For any solution $y = y(t)$ of (3.4), rewriting (3.4) we have

$$\left(\frac{dy}{dt} - \frac{d}{dt} P_m y \right) + \frac{d}{dt} P_m y$$

$$= \Psi [\bar{x}_m(t), t](y - P_m y) + \Psi [\bar{x}_m(t), t] P_m y + \phi(t),$$

that is,

$$\frac{d}{dt} P_m y = \Psi [\bar{x}_m(t), t] P_m y + \{ \Psi [\bar{x}_m(t), t](y - P_m y) -$$

$$- \left(\frac{dy}{dt} - \frac{d}{dt} P_m y \right) + \phi(t) \}. \tag{3.30}$$

Since $d(P_m y)/dt$ is a polynomial of degree $m - 1$ at most for $m \geqslant 1$, from (3.30) we have

$$\frac{d}{dt} P_m y = P_{m-1}\{\Psi[\bar{x}_m(t), t] P_m y\} + \psi(t), \tag{3.31}$$

where

$$\psi(t) = P_{m-1}\{\Psi[\bar{x}_m(t), t](y - P_m y)\} - P_{m-1}\left(\frac{dy}{dt} - \frac{d}{dt} P_m y\right) +$$

$$+ P_{m-1}\phi(t). \tag{3.32}$$

By Lemma 3.1, using (3.3) and (2.15), we then have

$$\| \psi \|_q \leqslant K_1 \sigma_1(m) \|\dot{y}\|_q + \frac{1}{\sqrt{2m}}(A_m \|y\|_q + K_1 \|\dot{y}\|_q) +$$

$$+ \frac{\sqrt{m}}{\sqrt{2}} \|(I - P_{m-1})\phi\|_q + \|P_{m-1}\phi\|_q,$$

from which by Schwarz's inequality and Parseval's equality follows

$$\| \psi \|_q \leqslant \frac{1}{\sqrt{2m}} A_m \|y\|_q + \left(\frac{1}{m+1} + \frac{1}{\sqrt{2m}}\right) K_1 \|\dot{y}\|_q +$$

$$+ \left(\frac{m}{2} + 1\right)^{1/2} [\|(I - P_{m-1})\phi\|_q^2 + \|P_{m-1}\phi\|_q^2]^{1/2}$$

$$= \frac{1}{\sqrt{2m}} A_m \|y\|_q + \left(\frac{1}{m+1} + \frac{1}{\sqrt{2m}}\right) K_1 \|\dot{y}\|_q + \sqrt{\frac{m+2}{2}} \|\phi\|_q. \tag{3.33}$$

Now put

$$c_0 = -\sum_{i=0}^{N} L_i P_m y(t_i). \tag{3.34}$$

Then for a solution of $y = y(t)$ of (3.4) satisfying the boundary condition (3.26), we have

$$c_0 = \sum_{i=0}^{N} L_i (I - P_m) y(t_i), \tag{3.35}$$

from which by (2.14) follows

$$\| c_0 \| \leqslant \sum_{i=0}^{N} \| L_i \| \cdot \sigma(m) \|\dot{y}\|_q. \tag{3.36}$$

Let the Chebyshev series of $P_m y(t)$ and $\psi(t)$ be respectively

$$P_m y(t) = u_0 + \sqrt{2} \sum_{n=1}^{m} u_n T_n(t),$$

$$\psi(t) = c_1 + \sqrt{2} \sum_{n=1}^{m-1} c_{n+1} T_n(t),$$

(3.37)

and let $\xi = (u_0^*, u_1^*, u_2^*, \ldots, u_m^*)^*$ and $\gamma = (c_0^*, c_1^*, c_2^*, \ldots, c_m^*)^*$. Then corresponding to the equalities (3.34) and (3.31), we have

$$J_m(\bar{\alpha})\xi + \gamma = 0,$$

(3.38)

because

$$J_m(\bar{\alpha})\xi = \lim_{h \to 0} \frac{1}{h} [F^{(m)}(\bar{\alpha} + h\xi) - F^{(m)}(\bar{\alpha})]$$

and, by (1.4), the first component $F_0(\bar{\alpha})$ of $F^{(m)}(\bar{\alpha})$ is equal to

$$\sum_{i=0}^{N} L_i \bar{x}_m(t_i) - l$$

and the $(n + 1)$th component $F_n(\bar{\alpha})$ of $F^{(m)}(\bar{\alpha})$ $(n = 1, 2, \ldots, m)$ is the coefficient of the term of degree $n - 1$ in the Chebyshev series of the function

$$X[\bar{x}_m(t), t] - \frac{d\bar{x}_m(t)}{dt}.$$

By (3.24) and Parseval's equality, we then have

$$\begin{aligned}\|P_m y\|_q &= \|\xi\| = \|J_m^{-1}(\bar{\alpha})\gamma\| \\ &\leqslant M_m' \|\gamma\| \\ &= M_m'(\|c_0\|^2 + \|\psi\|_q^2)^{1/2} \\ &\leqslant M_m'(\|c_0\| + \|\psi\|_q).\end{aligned}$$

(3.39)

Then by (3.23), (3.33) and (3.36), we have

$$\|P_m y\|_q \leqslant \frac{M_m' A_m}{\sqrt{2m}} \|y\|_q + M_m' \rho(m) \|\dot{y}\|_q + M_m' \sqrt{\frac{m+2}{2}} \|\phi\|_q.$$

(3.40)

Now, since $y = y(t)$ is a solution of (3.4),

$$\frac{dy}{dt} = \Psi[\bar{x}_m(t), t](y - P_m y) + \Psi[\bar{x}_m(t), t]P_m y + \phi(t),$$

therefore by the use of (2.15) we have

$$\|\dot{y}\|_q \leqslant K_1 \sigma_1(m) \|\dot{y}\|_q + K_1 \|P_m y\|_q + \|\phi\|_q,$$

from which readily follows

$$\|\dot{y}\|_q \leqslant [1 - K_1(m+1)^{-1}]^{-1}(\|\phi\|_q + K_1 \|P_m y\|_q)$$

(3.41)

provided that

$$1 - K_1(m + 1)^{-1} > 0. \tag{3.42}$$

Then substituting (3.41) into (3.40), by (3.23) and (3.25) we get

$$\| P_m y \|_q \leqslant \frac{1 - K_1(m + 1)^{-1}}{\sqrt{2m}} M'_m A_m C_m^{-1} \| y \|_q + M'_m C_m^{-1} \lambda(m) \| \phi \|_q \tag{3.43}$$

provided that

$$C_m > 0. \tag{3.44}$$

Now by (2.14),

$$\| y - P_0 y \|_n \leqslant \sigma(0) \| \dot{y} \|_q. \tag{3.45}$$

Since by (2.16)

$$\sigma(0) = \sqrt{2} \left[\sum_{n=1}^{\infty} n^{-2} \right]^{1/2} = \pi/\sqrt{3}, \tag{3.46}$$

by (3.37) we then have

$$\| y \|_n \leqslant \| P_0 y \|_n + \frac{\pi}{\sqrt{3}} \| \dot{y} \|_q$$

$$= \| u_0 \| + \frac{\pi}{\sqrt{3}} \| \dot{y} \|_q$$

$$\leqslant \| P_m y \|_q + \frac{\pi}{\sqrt{3}} \| \dot{y} \|_q, \quad (m \geqslant 1) \tag{3.47}$$

which by (3.23) and (3.41) implies that

$$\| y \|_n \leqslant [1 - K_1(m + 1)^{-1}]^{-1} \left[B_m \| P_m y \|_q + \frac{\pi}{\sqrt{3}} \| \phi \|_q \right]. \tag{3.48}$$

Then substituting (3.43) into (3.48), we get

$$\| y \|_n \leqslant \frac{1}{\sqrt{2m}} M'_m A_m B_m C_m^{-1} \| y \|_q + [1 - K_1(m + 1)^{-1}]^{-1} \times$$

$$\times \left[\frac{\pi}{\sqrt{3}} + M'_m B_m C_m^{-1} \lambda(m) \right] \| \phi \|_q.$$

Since $\| y \|_q \leqslant \| y \|_n$ and $\| \phi \|_q \leqslant \| \phi \|_n$, we thus have (3.27) provided that

$$1 - \frac{1}{\sqrt{2m}} M'_m A_m B_m C_m^{-1} > 0. \tag{3.49}$$

Q.E.D.

As in [2], let us consider the case where the boundary value problem (1.1) has an isolated solution $x = \hat{x}(t)$ satisfying the internality condition

$$U = \{(t, x) \mid \quad \|x - \hat{x}(t)\| \leqslant \delta_0, \quad t \in J = [-1, 1]\} \subset D \tag{3.50}$$

for some positive number δ_0. By an isolated solution we mean a solution $\hat{x}(t)$ such that the matrix $\Sigma_{i=0}^N L_i \Phi(t_i)$ is non-singular for an arbitrary fundamental matrix $\Phi(t)$ of the first variation equation of the differential equation in (1.1) with respect to $\hat{x}(t)$.

In [2], it is shown that in the above case, for

$$\hat{x}_m(t) = P_m \hat{x}(t) = \hat{a}_0 + \sqrt{2} \sum_{n=1}^m \hat{a}_n T_n(t), \tag{3.51}$$

there is a positive integer m_0 such that for any $m \geqslant m_0$,

$\det J_m(\hat{\alpha}) \neq 0$,

$\|J_m^{-1}(\hat{\alpha})\| \leqslant M'$,   (3.52)

$\|J_m(\alpha_1) - J_m(\alpha_2)\| \leqslant K_3' \sqrt{2m+1}\, \|\alpha_1 - \alpha_2\|$,

where $\hat{\alpha} = (\hat{a}_0^*, \hat{a}_1^*, \hat{a}_2^*, \ldots, \hat{a}_m^*)^*$, $\alpha_1 = (a_0^{(1)*}, a_1^{(1)*}, a_2^{(1)*}, \ldots, a_m^{(1)*})^*$ and $\alpha_2 = (a_0^{(2)*}, a_1^{(2)*}, a_2^{(2)*}, \ldots, a_m^{(2)*})^*$ are vectors such that

$$\theta\left[a_0^{(1)} + \sqrt{2} \sum_{n=1}^m a_n^{(1)} T_n(t)\right] + (1 - \theta)\left[a_0^{(2)} + \sqrt{2} \sum_{n=1}^m a_n^{(2)} T_n(t)\right] \in U$$

on J for any $\theta \in [0, 1]$, M' is a positive number independent of m, and K_3' is the positive number independent of m such that

$$\left[\sum_{i,j,l} \left\{\frac{\partial \Psi_{ij}}{\partial x_l}(x, t)\right\}^2\right]^{1/2} \leqslant K_3' \tag{3.53}$$

for any $(t, x) \in U$. Making use of (3.51), it is shown in [2] that in the case under consideration, the determining equation (1.8) has a solution $\tilde{\alpha}$ such that

$$\|J_m(\tilde{\alpha}) - J_m(\hat{\alpha})\| \leqslant \kappa/M',$$
$$\|\tilde{\alpha} - \hat{\alpha}\| \leqslant \frac{M'K}{1 - \kappa}\, m^{-3/2}, \tag{3.54}$$

where κ is an arbitrary positive number smaller than unity and K is a non-negative number independent of m. The existence of a solution $\tilde{\alpha} = (\tilde{a}_0^*, \tilde{a}_1^*, \tilde{a}_2^*, \ldots, \tilde{a}_m^*)^*$ of the determining equation (1.8) clearly implies the existence of an m-th order Chebyshev series approximate solution

$$\tilde{x}_m(t) = \tilde{a}_0 + \sqrt{2} \sum_{n=1}^m \tilde{a}_n T_n(t) \tag{3.55}$$

to the given boundary value problem (1.1) and the second inequality of (3.54), as shown in [2], implies the uniform convergence of $\tilde{x}_m(t)$ to $\hat{x}(t)$ as $m \to \infty$.

Since the finite Chebyshev series approximate solution (3.1) is the one obtained by the numerical solution of the determining equation (1.8), let us suppose that

$$\|\bar{\alpha} - \tilde{\alpha}\| \leq \omega_m. \tag{3.56}$$

Then, as is seen easily, we have

$$\|\bar{x}_m - \tilde{x}_m\|_n \leq \sqrt{2m+1} \cdot \omega_m. \tag{3.57}$$

Now let us suppose that

$$\omega_m = o(m^{-1/2}) \quad \text{as} \quad m \to \infty. \tag{3.58}$$

Then, since $\tilde{x}_m(t)$ converges to $\hat{x}(t)$ uniformly as $m \to \infty$, by (3.57) we see that $\bar{x}_m(t)$ also converges to $\hat{x}(t)$ uniformly as $m \to \infty$. Then by (3.56) and the last inequality of (3.52), we have

$$\|J_m(\bar{\alpha}) - J_m(\tilde{\alpha})\| \to 0 \quad \text{as} \quad m \to \infty.$$

Then by (3.54) we see that there is a positive number $\kappa' < 1$ such that

$$\|J_m(\bar{\alpha}) - J_m(\hat{\alpha})\| \leq \kappa'/M' \tag{3.59}$$

for any $m \geq m_1 \geq m_0$ provided m_1 is sufficiently large. Then by (3.52), we have

$$J_m(\bar{\alpha}) = J_m(\hat{\alpha})(E + \Lambda_m) \tag{3.60}$$

where

$$\Lambda_m = J_m^{-1}(\hat{\alpha})[J_m(\bar{\alpha}) - J_m(\hat{\alpha})]. \tag{3.61}$$

By (3.52) and (3.59), it is clear that

$$\|\Lambda_m\| \leq \kappa' < 1.$$

Then from (3.60) it readily follows that

$$\|J_m^{-1}(\bar{\alpha})\| \leq \frac{M'}{1 - \kappa'}. \tag{3.62}$$

This means that if the errors in the numerical solution of the determining equation (1.8) are so small that (3.58) is valid, then the positive number M'_m satisfying (3.24) may be replaced by a positive number independent of m. If the errors in the numerical solution of (1.8) are so small that (3.58) is valid, then as shown above, $\bar{x}_m(t)$ converges to $\hat{x}(t)$ uniformly as $m \to \infty$, therefore we may suppose that in (3.3), K_0, K_1, K_2, K_3 and K_4 are all independent of m. The condition (3.28) in Theorem 3.1 is then seen to be always fulfilled provided m is sufficiently large.

We shall now show that a number M_1 satisfying (1.29) for the H-mapping $H = H_m$ corresponding to the matrix $\Psi[\bar{x}_m(t), t]$ is indeed given by the $\mu_1(m)$ defined in (3.29) if $m \geqslant 1$ and that the inequalities (3.24) and (3.28) in Theorem 3.1 are valid. First we shall show that under the conditions (3.24) and (3.28) of Theorem 3.1, there is indeed the H-mapping $H_m (m \geqslant 1)$ corresponding to the matrix $\Psi[\bar{x}_m(t), t]$. Let $\Phi_m(t)$ be the fundamental matrix of the linear homogeneous differential equation

$$\frac{dy}{dt} = \Psi[\bar{x}_m(t), t]y \tag{3.63}$$

satisfying the initial condition $\Phi_m(-1) = E$, and let G_m be the matrix defined by

$$G_m = \sum_{i=0}^{N} L_i \Phi_m(t_i). \tag{3.64}$$

By the Proposition in Section 1, it is then clear that the H-mapping H_m corresponding to the matrix $\Psi[\bar{x}_m(t), t]$ exists if G_m is non-singular. Suppose that G_m is singular. Then there is a non-trival vector c such that

$$G_m c = 0. \tag{3.65}$$

For such c, the function

$$y(t) := \Phi_m(t)c \tag{3.66}$$

satisfies the differential equation (3.63) and moreover the boundary condition

$$\sum_{i=0}^{N} L_i y(t_i) = 0. \tag{3.67}$$

Then by (3.27) in Theorem 3.1, we have

$$y(t) \equiv 0, \tag{3.68}$$

because (3.63) is a special case of the differential equation (3.4) with $\phi(t) \equiv 0$. Since $\det \Phi_m(t)$ never vanishes, (3.68) implies $c = 0$ by (3.66). This is a contradiction. Thus we see that G_m is nonsingular for $m \geqslant 1$. This implies that there is indeed the H-mapping H_m corresponding to the matrix $\Psi[\bar{x}_m(t), t]$ for $m \geqslant 1$. By the Proposition in Section 1, the solution $y = y(t)$ of (3.4) satisfying the boundary condition (3.26) is then given by

$$y = H_m \phi. \tag{3.69}$$

The inequality (3.27) in Theorem 3.1 then implies that

$$\|H_m \phi\|_n \leqslant \mu_1(m) \|\phi\|_n. \tag{3.70}$$

Since $\phi(t)$ is an arbitrary vector-valued function continuous on J, (3.70) implies that

$$\|H_m\|_n \leqslant \mu_1(m). \tag{3.71}$$

By (1.29), this proves that the number M_1 in question can indeed be given by the $\mu_1(m)$ defined in (3.29).

4. A Method of Obtaining M_2

Our method of obtaining M_2 is based on the following theorem.

Theorem 4.1
Suppose that the inequalities (3.24) and (3.28) in Theorem 3.1 are valid. Then for any solution $y = y(t)$ of the linear homogeneous differential equation

$$\frac{dy}{dt} = \Psi[\bar{x}_m(t), t]y \tag{4.1}$$

satisfying the boundary condition

$$\sum_{i=0}^{N} L_i y(t_i) = l, \tag{4.2}$$

if any exists, it is valid that

$$\| y \|_n \leqslant \mu_2(m) \| l \| \tag{4.3}$$

for $m \geqslant 1$, where

$$\mu_2(m) = \left[1 - \frac{1}{\sqrt{2m}} M'_m A_m B_m C_m^{-1} \right]^{-1} M'_m B_m C_m^{-1}. \tag{4.4}$$

Proof
Since (4.1) is a special case of (3.4) with $\phi(t) \equiv 0$, we have (3.31) with

$$\psi(t) = P_{m-1}\{\Psi[\bar{x}_m(t), t](y - P_m y)\} - P_{m-1}\left(\frac{dy}{dt} - \frac{d}{dt} P_m y\right).$$

Hence in the present case, by (3.33) we have

$$\| \psi \|_q \leqslant \frac{1}{\sqrt{2m}} A_m \| y \|_q + \left(\frac{1}{m+1} + \frac{1}{\sqrt{2m}} \right) K_1 \| \dot{y} \|_q. \tag{4.5}$$

Now put

$$c_0 = -\sum_{i=0}^{N} L_i P_m y(t_i). \tag{4.6}$$

Then by (4.2) we have

$$c_0 = -l + \sum_{i=0}^{N} L_i(I - P_m)y(t_i),$$

therefore by the use of (2.14) we have

$$\| c_0 \| \leqslant \| l \| + \sum_{i=0}^{N} \| L_i \| \cdot \sigma(m) \| \dot{y} \|_q. \tag{4.7}$$

Then by (3.39), (4.5) and (4.7), we have

$$\| P_m y \|_q \leqslant M'_m \| l \| + \frac{1}{\sqrt{2m}} M'_m A_m \| y \|_q + M'_m \rho(m) \| \dot{y} \|_q. \tag{4.8}$$

Now in the present case it is clear that (3.41) is valid for $\phi(t) \equiv 0$ provided that (3.42) is valid. Hence substituting (3.41) with $\phi(t) = 0$ into (4.8), we get

$$\| P_m y \|_q \leqslant [1 - K_1 (m+1)^{-1}] M'_m C_m^{-1} \left(\| l \| + \frac{1}{\sqrt{2m}} A_m \| y \|_q \right) \tag{4.9}$$

provided $C_m > 0$. Now in the present case (3.47) is also valid. Hence substituting (3.41) with $\phi(t) \equiv 0$ into (3.47) and making use of (3.23) and (4.9), we get

$$\| y \|_n \leqslant M'_m B_m C_m^{-1} \left[\| l \| + \frac{1}{\sqrt{2m}} A_m \| y \|_q \right]. \tag{4.10}$$

Since $\| y \|_q \leqslant \| y \|_n$, by (4.4) we thus readily get (4.3) provided that (3.49) is valid.   Q.E.D.

As was shown in Section 3, the matrix G_m defined by (3.64) is nonsingular for $m \geqslant 1$ if the conditions (3.24) and (3.28) of Theorem 3.1 are fulfilled. In such a case, by the Proposition in Section 1, the differential equation (4.1) has a unique solution $y = y(t)$ satisfying the boundary condition (4.2) and such a solution is given by

$$y(t) = \Phi_m(t) G_m^{-1} l \tag{4.11}$$

as is seen from (1.14). Then by (4.3) we have

$$\| \Phi_m(t) G_m^{-1} l \| \leqslant \mu_2(m) \| l \| \tag{4.12}$$

for any $t \in J$. Since l is an arbitrary vector, (1.12) clearly implies that

$$\| \Phi_m(t) G_m^{-1} \| \leqslant \mu_2(m) \tag{4.13}$$

for any $t \in J$. By (1.30), this proves that the number M_2 in question can indeed be given by the $\mu_2(m)$ defined in (4.4).

5. The Verification of the Conditions of Theorem 1.1

First we shall prove the following lemma concerning the norms of the derivative of a finite Chebyshev series.

Lemma 5.1
Let

$$f_m(t) = a_0 + \sqrt{2} \sum_{n=1}^{m} a_n T_n(t) \tag{5.1}$$

be an m-th order vector-valued finite Chebyshev series. Then it is valid that

$$\|\dot{f}_m\|_q \leqslant \frac{(m+1)\sqrt{m(2m+1)}}{\sqrt{3}} \|f_m\|_q \tag{5.2}$$

and

$$\|\dot{f}_m\|_n \leqslant \frac{(m+1)\sqrt{m(4m^2-1)}}{\sqrt{3}} \|f_m\|_q, \tag{5.3}$$

where $\cdot = d/dt$.

Proof
For $m \geqslant 2$, put

$$\dot{f}_m(t) = a_0' + \sqrt{2} \sum_{n=1}^{m-1} a_n' T_n(t). \tag{5.4}$$

Then by the use of (2.7) and (2.10), we successively have

$$\|\dot{f}_m\|_q^2 = \sum_{n=0}^{m-1} \|a_n'\|^2$$

$$= \sum_{n=0}^{m-1} \frac{4}{e_n^2} \|(n+1)a_{n+1} + (n+3)a_{n+3} + \cdots\|^2$$

$$\leqslant 4 \sum_{n=0}^{m-1} [(n+1)\|a_{n+1}\| + (n+3)\|a_{n+3}\| + \cdots]^2, \tag{5.5}$$

where a_p is assumed to be zero for $p > m$. By Schwarz's inequality, we then have

$$\|\dot{f}_m\|_q^2 \leqslant 4\|f_m\|_q^2 [(1^2 + 2^2 + 3^2 + \cdots + m^2) + (3^2 + 4^2 + \cdots + m^2) + \cdots + (m-1)^2 + m^2].$$

Since

$$1^2 + 2^2 + 3^2 + \cdots + m^2 = \tfrac{1}{6}m(m+1)(2m+1),$$

we thus have

$$\|\dot{f}_m\|_q^2 \leqslant 4\|f_m\|_q^2 \cdot \tfrac{1}{12}m(m+1)^2(2m+1)$$

$$= \tfrac{1}{3}m(m+1)^2(2m+1)\|f_m\|_q^2,$$

from which readily follows (5.2) for $m \geqslant 2$. For $m = 1$ and $m = 0$, it is easily verified that (5.2) is also valid. Thus (5.2) is valid for all $m \geqslant 0$.

Since

$$\| \dot{f}_m \|_n \leqslant \sqrt{2m-1} \, \| \dot{f}_m \|_q$$

as is seen from Schwarz's inequality, we readily get (5.3) from (5.2).   Q.E.D.

In what follows, we consider the case where the boundary value problem (1.1) has an isolated solution $x = \hat{x}(t)$ satisfying the internality condition (3.50) for some positive number δ_0. In such a case, as is mentioned in Section 3, there is an m-th order Chebyshev series approximate solution $\tilde{x}_m(t)$ of the form (3.55) satisfying the determining equation (1.8) exactly, in other words, satisfying the equations (1.3) exactly. Let $\bar{x}_m(t)$ be an m-th order Chebyshev series approximate solution of the form (3.1) obtained by the numerical solution of the determining equation (1.8). For the coefficients $\bar{\alpha} = (\bar{a}_0^*, \bar{a}_1^*, \bar{a}_2^*, \ldots, \bar{a}_m^*)^*$ and $\tilde{\alpha} = (\tilde{a}_0^*, \tilde{a}_1^*, \tilde{a}_2^*, \ldots, \tilde{a}_m^*)^*$ of $\bar{x}_m(t)$ and $\tilde{x}_m(t)$, let us suppose that (3.56) is valid. Then by Lemma 5.1, we readily see that

$$\| \dot{\tilde{x}}_m - \dot{\bar{x}}_m \|_n \leqslant \frac{(m+1)\sqrt{m(4m^2-1)}}{\sqrt{3}} \, \omega_m. \tag{5.6}$$

In what follows, strengthening (3.58), we assume that

$$\omega_m = o(m^{-7/2}) \quad \text{as} \quad m \to \infty. \tag{5.7}$$

Since (5.7) clearly implies (3.58), we then may suppose as stated in Section 3 that K_0, K_1, K_2, K_3 and K_4 in (3.3) are all independent of m. Moreover, since both of $\tilde{x}_m(t)$ and $\bar{x}_m(t)$ converge uniformly to $\hat{x}(t)$ as $m \to \infty$, we may suppose that if m is sufficiently large, all inequalities in (3.3) are valid also when $\bar{x}_m(t)$ is replaced by $\tilde{x}_m(t)$. Then since $\tilde{x}_m(t)$ satisfies the differential equation in (1.3), for large m we have

$$\| \dot{\tilde{x}}_m \|_q = \| P_{m-1} X[\tilde{x}_m(t), t] \|_q$$
$$\leqslant \| X[\tilde{x}_m(t), t] \|_q$$
$$\leqslant K_0. \tag{5.8}$$

On the other hand, by (2.14) we have

$$\| \dot{\tilde{x}}_m \|_n = \| X[\tilde{x}_m(t), t] - (I - P_{m-1}) X[\tilde{x}_m(t), t] \|_n$$
$$\leqslant K_0 + \sigma(m-1) \left\| \frac{d}{dt} X[\tilde{x}_m(t), t] \right\|_q. \tag{5.9}$$

Then, since

$$\frac{d}{dt} X[\tilde{x}_m(t), t] = \Psi[\tilde{x}_m(t), t] \dot{\tilde{x}}_m(t) + \frac{\partial X}{\partial t}[\tilde{x}_m(t), t], \tag{5.10}$$

by (3.3) and (5.8) we have

$$\| \dot{\tilde{x}}_m \|_n \leqslant K_0 + \sigma(m-1)(K_1 K_0 + K_2) \tag{5.11}$$

for large m.

Making use of (5.11), we shall now prove the following lemmas.

Lemma 5.2
Let K_5 be a non-negative number such that

$$\left\| \frac{\partial^2}{\partial t^2} X[\tilde{x}_m(t), t] \right\| \leq K_5 \tag{5.12}$$

on the interval $J = [-1, 1]$. If m is sufficiently large, then it is valid that

$$\left\| \frac{d^2}{dt^2} X[\tilde{x}_m(t), t] \right\|_q \leq D_m, \tag{5.13}$$

where

$$D_m = \left[1 - K_1 \left\{ \frac{\sqrt{m+1}}{\sqrt{2}(m-1)} + \frac{1}{m+1} \right\} \right]^{-1} \times$$

$$\times \left[K_0^2 K_3 + 2K_0 K_4 + K_1(K_0 K_1 + K_2) + K_5 + \right.$$
$$\left. + \sigma(m-1)K_0 K_3(K_0 K_1 + K_2) \right]. \tag{5.14}$$

Proof
Differentiating both sides of (5.10) with respect to t, we have

$$\frac{d^2}{dt^2} X[\tilde{x}_m(t), t] = \Psi_x[\tilde{x}_m(t), t]\dot{\tilde{x}}_m(t)\dot{\tilde{x}}_m(t) +$$

$$+ 2\frac{\partial \Psi}{\partial t}[\tilde{x}_m(t), t]\dot{\tilde{x}}_m(t) + \Psi[\tilde{x}_m(t), t]\ddot{\tilde{x}}_m(t) +$$

$$+ \frac{\partial^2 X}{\partial t^2}[\tilde{x}_m(t), t]. \tag{5.15}$$

Here $\Psi_x(x, t)$ denotes the tensor whose components are $(\partial \Psi_{ij}/\partial x_l)(x, t)$, where $\Psi_{ij}(x, t)$ are the elements of the matrix $\Psi(x, t)$ and x_l are the components of the vector x. By (3.3), (5.8), (5.11) and (5.12), from (5.15) it follows that

$$\left\| \frac{d^2}{dt^2} X[\tilde{x}_m(t), t] \right\|_q \leq K_3[K_0 + \sigma(m-1)(K_0 K_1 + K_2)]K_0 +$$

$$+ 2K_4 K_0 + K_1\|\ddot{\tilde{x}}_m\|_q + K_5. \tag{5.16}$$

Now, since $\tilde{x}_m(t)$ satisfies the differential equation in (1.3), we have

$$\ddot{\tilde{x}}_m(t) = \frac{d}{dt} P_{m-1} X[\tilde{x}_m(t), t]. \tag{5.17}$$

Since $X[\tilde{x}_m(t), t]$ is twice continuously differentiable with respect to t, by means of (2.20) we can rewrite (5.17) in the following form:

$$\ddot{\tilde{x}}_m(t) = \frac{d}{dt} X[\tilde{x}_m(t), t] - \chi_m(t) - (I - P_m) \frac{d}{dt} X[\tilde{x}_m(t), t], \qquad (5.18)$$

where $\chi_m(t)$ is a function of the form of (2.21). As is seen easily from (2.21), it is valid that

$$\| \chi_m \|_q \leqslant \frac{\sqrt{m+1}}{\sqrt{2}} \| (I - P_{m-2}) \frac{d}{dt} X[\tilde{x}_m(t), t] \|_q \qquad (5.19)$$

for $m \geqslant 2$. Hence by the use of (2.15) and (5.10), from (5.18) we get

$$\| \ddot{\tilde{x}}_m \|_q \leqslant (K_0 K_1 + K_2) + \frac{\sqrt{m+1}}{\sqrt{2}} \sigma_1(m-2) \left\| \frac{d^2}{dt^2} X[\tilde{x}_m(t), t] \right\|_q +$$

$$+ \sigma_1(m) \left\| \frac{d^2}{dt^2} X[\tilde{x}_m(t), t] \right\|_q$$

$$= (K_0 K_1 + K_2) + \left[\frac{\sqrt{m+1}}{\sqrt{2}(m-1)} + \frac{1}{m+1} \right] \cdot \left\| \frac{d^2}{dt^2} X[\tilde{x}_m(t), t] \right\|_q$$

$$(5.20)$$

for $m \geqslant 2$. Then substituting (5.20) into (5.16), we have

$$\left\| \frac{d^2}{dt^2} X[\tilde{x}_m(t), t] \right\|_q \leqslant K_3 K_0^2 + 2K_0 K_4 + K_5 + \sigma(m-1) K_0 K_3 (K_0 K_1 +$$

$$+ K_2) + K_1 (K_0 K_1 + K_2) + K_1 \left[\frac{\sqrt{m+1}}{\sqrt{2}(m-1)} + \frac{1}{m+1} \right] \times$$

$$\times \left\| \frac{d^2}{dt^2} X[\tilde{x}_m(t), t] \right\|_q,$$

from which by (5.14) readily follows (5.13) provided that

$$1 - K_1 \left[\frac{\sqrt{m+1}}{\sqrt{2}(m-1)} + \frac{1}{m+1} \right] > 0. \qquad (5.21)$$

However it is clear that (5.21) is valid if m is sufficiently large. Thus (5.13) is always valid if m is sufficiently large.   Q.E.D.

Lemma 5.3
If m is sufficiently large, then for an m-th order Chebyshev series approximate solution $\bar{x}_m(t)$ of the form (3.1) obtained by the numerical solution

of the determining equation (1.8), it is valid that

$$\| \dot{\tilde{x}}_m - X[\bar{x}_m(t), t] \|_n \leqslant \left[\frac{(m + 1)\sqrt{m(4m^2 - 1)}}{\sqrt{3}} + K_1 \sqrt{2m + 1} \right] \omega_m +$$

$$+ \frac{\sigma(m - 1)}{m - 1} D_m \tag{5.22}$$

and

$$\| \dot{\tilde{x}}_m - P_{m-1} X[\bar{x}_m(t), t] \|_n \leqslant \left[\frac{(m + 1)\sqrt{m(4m^2 - 1)}}{\sqrt{3}} + \right.$$

$$\left. + K_1 \sqrt{2m - 1} \right] \omega_m \tag{5.23}$$

where ω_m is a non-negative number satisfying (3.56).

Proof

Since $\tilde{x}_m(t)$ satisfies the differential equation in (1.3) exactly, we can write the difference $\dot{\tilde{x}}_m(t) - X[\bar{x}_m(t), t]$ as follows:

$$\dot{\tilde{x}}_m(t) - X[\bar{x}_m(t), t] = [\dot{\tilde{x}}_m(t) - \dot{\bar{x}}_m(t)] + \{P_{m-1}X[\tilde{x}_m(t), t] -$$
$$- X[\tilde{x}_m(t), t]\} + \{X[\tilde{x}_m(t), t] - X[\bar{x}_m(t), t]\}.$$

Then by (2.14) and (5.6), we have

$$\| \dot{\tilde{x}}_m - X[\bar{x}_m(t), t] \|_n \leqslant \frac{(m + 1)\sqrt{m(4m^2 - 1)}}{\sqrt{3}} \omega_m +$$

$$+ \sigma(m - 1) \left\| (I - P_{m-2}) \frac{d}{dt} X[\tilde{x}_m(t), t] \right\|_q +$$

$$+ \left\| \int_0^1 \Psi[\bar{x}_m(t) + \theta\{\tilde{x}_m(t) - \bar{x}_m(t)\}, t] d\theta \right\|_n \cdot \| \tilde{x}_m - \bar{x}_m \|_n$$

provided m is sufficiently large. Making use of (2.15) and (3.3), we thus have

$$\| \dot{\tilde{x}}_m - X[\bar{x}_m(t), t] \|_n \leqslant \frac{(m + 1)\sqrt{m(4m^2 - 1)}}{\sqrt{3}} \omega_m +$$

$$+ \sigma(m - 1)\sigma_1(m - 2) \left\| \frac{d^2}{dt^2} X[\tilde{x}_m(t), t] \right\|_q +$$

$$+ K_1 \sqrt{2m + 1} \cdot \omega_m.$$

Then substituting (5.13) into the above inequality, we readily get (5.22).

Next rewrite the difference $\dot{\bar{x}}_m(t) - P_{m-1}X[\bar{x}_m(t), t]$ as follows:

$$\dot{\bar{x}}_m(t) - P_{m-1}X[\bar{x}_m(t), t]$$
$$= [\dot{\bar{x}}_m(t) - \dot{\tilde{x}}_m(t)] + P_{m-1}\{X[\tilde{x}_m(t), t] - X[\bar{x}_m(t), t]\}.$$

Then by the use of (5.6), we successively have

$$\| \dot{\bar{x}}_m(t) - P_{m-1}X[\bar{x}_m(t), t] \|_n$$

$$\leqslant \frac{(m+1)\sqrt{m(4m^2-1)}}{\sqrt{3}} \omega_m +$$

$$+ \sqrt{2m-1}\|P_{m-1}\{X[\tilde{x}_m(t), t] - X[\bar{x}_m(t), t]\}\|_q$$

$$\leqslant \frac{(m+1)\sqrt{m(4m^2-1)}}{\sqrt{3}} \omega_m + \sqrt{2m-1}\|X[\tilde{x}_m(t), t] -$$

$$- X[\bar{x}_m(t), t] \|_q$$

$$\leqslant \frac{(m+1)\sqrt{m(4m^2-1)}}{\sqrt{3}} \omega_m + \sqrt{2m-1}\, K_1\|\tilde{x}_m - \bar{x}_m\|_q$$

$$= \frac{(m+1)\sqrt{m(4m^2-1)}}{\sqrt{3}} \omega_m + \sqrt{2m-1}\, K_1\|\tilde{\alpha} - \bar{\alpha}\|.$$

Hence by (3.56) we readily get (5.23).   Q.E.D.

Comparing (3.2) with (5.23), we see that r_0 may be taken so that

$$r_0 = \left[\frac{(m+1)\sqrt{m(4m^2-1)}}{\sqrt{3}} + K_1\sqrt{2m-1} \right] \omega_m \qquad (5.24)$$

for sufficiently large m. Then by our assumption (5.7) on ω_m, it is clear that

$$r_0 = o(m^{-1}) \quad \text{as} \quad m \to \infty. \qquad (5.25)$$

We shall now show that after obtaining an m-th order Chebyshev series approximate solution $\bar{x}_m(t)$ by the numerical solution of the determining equation (1.8), one can indeed assure the existence of an exact solution to the given boundary value problem (1.1) with twice continuously differentiable $X(x, t)$ by verifying the conditions of Theorem 1.1 by the use of the positive numbers M_1 and M_2 obtained in the way described in Sections 3 and 4, and further one can obtain an error bound for the approximate solution $\bar{x}_m(t)$ obtained, if we are concerned with the case where the given boundary value problem (1.1) has an isolated solution satisfying an internality condition of the form (3.50). Note that in practical problems the existence of an isolated solution satisfying the internality condition is usually not known before we obtain the approximate solution $\bar{x}_m(t)$ in question.

In the verification of the conditions of Theorem 1.1 for $\bar{x}(t) = \bar{x}_m(t)$, by (3.29), (3.71), (4.4), (4.13), (5.7) and (5.22), we may suppose that

$$M_1 = \mu_1(m) = 0(m^{1/2}),$$
$$M_2 = \mu_2(m) = 0(1),$$

$$r = \left[\frac{(m+1)\sqrt{m(4m^2-1)}}{\sqrt{3}} + K_1\sqrt{2m+1}\right]\omega_m + \frac{\sigma(m-1)}{m-1}D_m$$

$$= o(m^{-1})$$

(5.26)

as $m \to \infty$. Since $\bar{x}_m(t)$ satisfies the second equality in (1.3), by (3.56) we have further

$$\left\|\sum_{i=0}^{N} L_i\bar{x}_m(t_i) - l\right\| = \left\|\sum_{i=0}^{N} L_i[\bar{x}_m(t_i) - \tilde{x}_m(t_i)]\right\|$$

$$\leqslant \sum_{i=0}^{N} \|L_i\| \cdot \sqrt{2m+1} \cdot \omega_m.$$

Hence for ϵ in the conditions of Theorem 1.1, by (5.7) we may suppose that

$$\epsilon = \sum_{i=0}^{N} \|L_i\| \cdot \sqrt{2m+1} \cdot \omega_m = o(m^{-3})$$

(5.27)

as $m \to \infty$.

Now in the verification of the conditions of Theorem 1.1 for $\bar{x}(t) = \bar{x}_m(t)$, $\Psi[\bar{x}_m(t), t]$ is taken for $A(t)$. However, by (3.53), we have

$$\|\Psi(x, t) - \Psi[\bar{x}_m(t), t]\| \leqslant K_3'\|\bar{x} - \bar{x}_m(t)\| \leqslant K_3'\delta$$

(5.28)

for any $(t, x) \in D_\delta$ provided that m is sufficiently large. Hence we see that the condition (1.26) of Theorem 1.1 is fulfilled if the inequality

$$K_3'\delta \leqslant \kappa/M_1$$

(5.29)

is valid. As was shown at the end of Section 3, the condition (i) of Theorem 1.1 is fulfilled automatically when $\mu_1(m)$ is obtained by the method described in Section 3. Then in order to verify the conditions of Theorem 1.1, it is sufficient to verify whether or not the following inequalities are valid for sufficiently large m and that there δ can be made smaller than some positive number, say, $\delta_0/2$:

$$\frac{M_1 r + M_2 \epsilon}{1 - \kappa} \leqslant \delta \leqslant \frac{\kappa}{K_3' M_1},$$

(5.30)

where κ is a certain positive number smaller than unity. The inequalities (5.30), however, can be rewritten as follows:

$$M_1^2 r + M_1 M_2 \epsilon \leqslant \kappa(1 - \kappa)/K_3' \tag{5.31}$$

and

$$\frac{M_1 r + M_2 \epsilon}{1 - \kappa} \leqslant \delta \leqslant \frac{\kappa}{K_3' M_1} . \tag{5.32}$$

The inequality (5.31), however, is valid for an arbitrary positive number κ smaller than unity if m is sufficiently large, because by (5.26) and (5.27),

$$M_1^2 r + M_1 M_2 \epsilon = o(1) \quad \text{as} \quad m \to \infty. \tag{5.33}$$

The inequality (5.32) is then valid clearly if one chooses δ so that

$$\delta = \frac{M_1 r + M_2 \epsilon}{1 - \kappa} . \tag{5.34}$$

For such δ, it is valid that

$$\delta < \delta_0/2 \tag{5.35}$$

for sufficiently large m, because by (5.26) and (5.27),

$$M_1 r + M_2 \epsilon = o(n^{-1/2}) \quad \text{as} \quad m \to \infty. \tag{5.36}$$

The above results show that in the case under consideration, the conditions of Theorem 1.1 can all be verified by the use of the positive numbers M_1 and M_2 obtained in the way described in Sections 3 and 4, and hence that one can thereby assure the existence of an exact solution to the given boundary value problem (1.1) and one can obtain an error bound to the approximate solution $\bar{x}_m(t)$ obtained numerically.

References

[1] Urabe, M. (1966). An existence theorem for multi-point boundary value problems, *Funkcial. Ekvac.* 9, 43–60.
[2] Urabe, M. (1967). Numerical solution of multi-point boundary value problems in Chebyshev series—Theory of the method, *Numer. Math.* 9, 341–366.
[3] Urabe, M. (1969). Numerical solution of boundary value problems in Chebyshev series—A method of computation and error estimation, *Conference on the Numerical Solution of Differential Equations*, Lecture Notes in Mathematics, Springer, Berlin, 109, 40–86.

Numerical Solution of Evolution Problems in Banach Spaces

Emil Vitásek and Jiří Taufer†

1. Introduction

In [1] we have introduced the class of so-called overimplicit multistep methods for solving ordinary differential equations and we have shown that in this class there exist A-stable (in Dahlquist's sense) methods of arbitrarily high orders. The aim of this paper is the study of the applicability of these methods to the approximate solution of evolution problems in Banach spaces, especially of abstract parabolic equations. Throughout the whole paper we will deal only with so-called selfstarting methods. These methods form a certain subclass of the overimplicit methods introduced in [1] and we now remind ourselves of their definition. The methods will be formulated for the approximate solution of the differential equation

$$x' = f(t, x), \quad t \in \langle a, b \rangle \tag{1.1}$$

with the initial condition

$$x(a) = \eta. \tag{1.2}$$

The right-hand term of this differential equation is assumed to be defined, continuous and to satisfy a Lipschitz condition with respect to x in the strip $a \leqslant t \leqslant b$, $-\infty < x < \infty$, so that the solution of the problem (1.1), (1.2) exists and is unique in the whole interval $\langle a, b \rangle$. Put $t_i = a + ih$, $i = 0, 1, \ldots$ where $h > 0$ is a constant and denote the approximate solution at the point t_i by x_i. One step of a selfstarting overimplicit method consists in computing the values $x_{n+1}, \ldots, x_{n+k}$ of the approximate solution at the points $t_{n+1}, \ldots, t_{n+k}$—assuming x_n to be known—simultaneously from the system

$$\begin{bmatrix} x_{n+1} \\ \vdots \\ x_{n+k} \end{bmatrix} = \begin{bmatrix} x_n \\ \vdots \\ x_n \end{bmatrix} + hC \begin{bmatrix} f_{n+1} \\ \vdots \\ f_{n+k} \end{bmatrix} + hf_n d, \tag{1.3}$$

† Institute of Mathematics, Academy of Sciences, Prague, Czechoslovakia.

where $f_i = f(t_i, x_i)$ is a square matrix of order k and d is a k-dimensional vector.

The fact that the function $f(t, x)$ satisfies a Lipschitz condition guarantees the existence and the uniqueness of the solution of (1.3) for any sufficiently small h, so that one step of the method is well-defined. In order to describe the whole method it is necessary, moreover, to indicate how to continue in the following step, i.e., how to choose the new initial value. The method will be practicable obviously only if the new initial value is chosen from the values $x_{n+1}, \ldots, x_{n+k}$. In order to specify it exactly let us take an integer $s, 1 \leqslant s \leqslant k$, and let the new initial value be x_{n+s}. Hence, the overimplicit selfstarting method is characterized not only by the matrix C and the vector d but also by the parameter s. Let us note that if $s < k$ then the values $x_{n+s+1}, \ldots, x_{n+k}$ which have been already computed are recomputed from a different system and, consequently, generally different values are obtained for them. The fact that these different values are denoted by the same symbols will not lead, obviously, to misunderstanding.

Define now the order of the method (1.3) in the usual way. We introduce the vector of the local discretization error of the method by the formula

$$L(x(t), h) = \begin{bmatrix} x(t+h) \\ \vdots \\ x(t+kh) \end{bmatrix} - \begin{bmatrix} x(t) \\ \vdots \\ x(t) \end{bmatrix} - hC \begin{bmatrix} x'(t+h) \\ \vdots \\ x'(t+kh) \end{bmatrix} - hx'(t)d \quad (1.4)$$

and we say that the method is of order p if the components of L are (for any sufficiently smooth function) of order h^{p+1}.

After formulating this definition the following two theorems can be easily proved.

Theorem 1.1
A selfstarting overimplicit method of order at least 1 is convergent.

Theorem 1.2
Let the solution of the problem (1.1), (1.2) $x(t)$ have continuous derivatives up to the order $p + 1$. Then the approximate solution computed by a selfstarting overimplicit method of order p converges to the exact solution with the rate h^p.

Since it may be expected that the accumulation of local discretization errors will be mostly affected by the errors in $x_0, x_s, \ldots, x_{rs}, \ldots$ and by the s-th component of the vector L, it seems reasonable to define the order of the method (1.3) with respect to s. Here we will require only that

$$L_i(x(t), h) = O(h^p), \quad i = 1, \ldots, k \quad (1.5)$$

and

$$L_s(x(t), h) = O(h^{p+1}) \tag{1.6}$$

hold.

Now it can be proved that the assumptions of the convergence theorems formulated above can be weakened in such a way that the words "order of the method" are replaced by "order with respect to s".

2. Application to Differential Equations in Banach Spaces

In this section, we will investigate the applicability of the methods just described to the approximate solution of the differential equation (1.1) with the initial condition (1.2) where $f(t, x)$ is a function of the real variable t and of x from a Banach space B. We assume also that the values of $f(t, x)$ belong to B. Naturally, the solution is then sought in B, too. If it is assumed that $f(t, x)$ satisfies a Lipschitz condition with respect to x then it is rather obvious that no new problems arise and that in fact all assertions formulated in the preceding section remain true. Our task will be therefore to study the possibilities of the use of selfstarting methods for the approximate solution of equation (1.1) in a Banach space, without assuming the boundedness of the operator in the right-hand term of this differential equation. Such a situation is much more complicated than the above one and for that reason we restrict ourselves to the case of linear equations, i.e. equations such that $f(t, x) = Ax + f(t)$ where A is a linear (generally unbounded) operator in B and, moreover, we will assume that A is independent of t. Thus, in what follows, we will deal with the differential equation

$$\frac{dx}{dt} = Ax + f(t), \quad t \in \langle 0, T \rangle \tag{2.1}$$

with the initial condition

$$x(0) = \eta. \tag{2.2}$$

In all that follows the operator A is assumed to be closed and its domain of definition $\mathscr{D}(A)$ is assumed to be dense in B. Moreover, we suppose that A is the generator of a strongly continuous semi-group $U(t)$, i.e., we assume that there exist real constants M and ω such that

$$\| (\lambda I - A)^{-m} \| \leqslant \frac{M}{(\mathrm{Re}\,\lambda - \omega)^n} \tag{2.3}$$

for $n = 1, 2, \ldots$ and for all complex λ for which $\mathrm{Re}\,\lambda > \omega$. Finally, we assume that the function $f(t)$ is continuous in $\langle 0, T \rangle$.

Any function $x(t)$ which belongs to $\mathscr{D}(A)$ for $t \in \langle 0, T \rangle$ and which satisfies the initial condition (2.2) and the equation (2.1) for all $t \in \langle 0, T \rangle$ (i.e., also for $t = 0$) will be called the classical solution of (2.1), (2.2).

The function $x(t)$, continuous in $\langle 0, T \rangle$ and satisfying (2.2), which belongs to $\mathscr{D}(A)$ for $t > 0$ and satisfies (2.1) for $t > 0$, will be called the weak solution of (2.1), (2.2).

Finally, the function of the form

$$x(t) = U(t)\eta + \int\limits_0^t U(t - s)f(s)\, ds \qquad (2.4)$$

will be called the generalized solution of (2.1), (2.2).

It is obvious that our assumptions guarantee the existence of the generalized solution of (2.1), (2.2) at least.

Apply now the method (1.3) to the problem (2.1), (2.2). We obtain the system

$$\begin{bmatrix} x_{n+1} \\ \vdots \\ x_{n+k} \end{bmatrix} = \begin{bmatrix} x_n \\ \vdots \\ x_n \end{bmatrix} + hC \otimes A \begin{bmatrix} x_{n+1} \\ \vdots \\ x_{n+k} \end{bmatrix} + hC \begin{bmatrix} f_{n+1} \\ \vdots \\ f_{n+k} \end{bmatrix} + hD \otimes A \begin{bmatrix} x_n \\ \vdots \\ x_n \end{bmatrix} +$$

$$+ h \begin{bmatrix} d_1 f_n \\ \vdots \\ d_k f_n \end{bmatrix} \quad n = 0, s, 2s, \ldots \qquad (2.5)$$

where D is the diagonal matrix of order k having the components of the vector d in the main diagonal and the symbol $\Gamma \otimes A$ denotes the so-called tensor product of the matrix Γ (of order k) with the operator A, i.e., the operator which maps the set $\mathscr{D}(A) \times \ldots \times \mathscr{D}(A) \subset B_k = B \times \ldots \times B$ into the space B_k and which is defined by

$$\Gamma \otimes A = \begin{bmatrix} \gamma_{11}A & \gamma_{12}A & \cdots & \gamma_{1k}A \\ \vdots & \vdots & & \vdots \\ \gamma_{k1}A & \gamma_{k2}A & \cdots & \gamma_{kk}A \end{bmatrix} \qquad (2.6)$$

Now we cannot conclude as simply as in the preceding that (2.5) has—at least for sufficiently small h—a solution, since the operator $C \otimes A$ is generally unbounded and the Banach fixed point theorem cannot be applied. However, the following lemma can be proved.

Lemma 2.1
Let the matrix C be regular. Then the spectrum $\sigma(C \otimes A)$ of the operator $C \otimes A$ consists of the numbers of the form $\lambda_C \lambda_A$ where $\lambda_C \in \sigma(C)$ and $\lambda_A \in \sigma(A)$.

This lemma easily yields the necessary and sufficient condition for the existence of the bounded operator $(I - hC \otimes A)^{-1}$.

Lemma 2.2

Let C be regular. Then the operator inverse to the operator $I - hC \otimes A$ exists and is bounded for all $h \in (0, h_0)$ if and only if none of the half-lines $1/(h\lambda_C)$, $h \in (0, h_0)$, $\lambda_C \in \sigma(C)$ has common points with the spectrum of the operator A.

Since our operator A is the generator of a semi-group, its spectrum lies in the half-plane $\operatorname{Re} \lambda \leqslant \omega$ and from Lemma 2.2 it follows immediately that the condition

$$\operatorname{Re} \lambda_C > 0, \quad \lambda_C \in \sigma(C) \tag{2.7}$$

is sufficient for the existence of the bounded operator $(I - hC \otimes A)^{-1}$ for any sufficiently small h.

The operator $(I - hC \otimes A)$ maps $\mathscr{D}(A) \times \ldots \times \mathscr{D}(A)$ on B_k and therefore, the inverse operator maps B_k into B_k and, consequently, it is possible to write it in the form:

$$(I - hC \otimes A)^{-1} = \begin{bmatrix} M_{11} & M_{12} & \cdots & M_{1k} \\ \vdots & \vdots & & \\ M_{k1} & M_{k2} & \cdots & M_{kk} \end{bmatrix} \tag{2.8}$$

where M_{ij} are bounded operators from B into B.

Further, put

$$(I - zC)^{-1} = Q^{-1}(z) \begin{bmatrix} p_{11}(z) & \cdots & p_{1k}(z) \\ \vdots & & \\ p_{k1}(z) & \cdots & p_{kk}(z) \end{bmatrix} \tag{2.9}$$

where

$$Q(z) = \det(I - zC). \tag{2.10}$$

The elements of the inverse matrix to $I - zC$ are thus denoted by $p_{ij}(z)Q^{-1}(z)$. It is well-known from algebra that the $p_{ij}(z)$ are polynomials in z of degree at most $k - 1$. The following lemma enables us to write the method (2.5) in a more concise form.

Lemma 2.3

Let C be regular. Then the operator $(I - hC \otimes A)^{-1}$ exists and is bounded if and only if the operator $Q^{-1}(hA)$ exists and is bounded. Moreover, if one of the above operators exists and is bounded then the operators M_{ij} from (2.8) are given by

$$M_{ij} = p_{ij}(hA)Q^{-1}(hA). \tag{2.11}$$

Thus, let us suppose in what follows that, for a given h, the bounded operator $(I - hC \otimes A)^{-1}$ exists. If we put

$$P_s(z) = \sum_{j=1}^{k} p_{ij}(z)(1 + d_j z) \tag{2.12}$$

and

$$R_s(z) = P_s(z)Q^{-1}(z) \tag{2.13}$$

then Lemma 2.3 implies that one step of the method (2.5) can be rewritten in the form

$$x_{(r+1)s} = R_s(hA)x_{rs} + q_r^{(s)}, \quad r = 0, 1, \ldots, \quad x_0 = \eta \tag{2.14}$$

where

$$q_r^{(s)} = h \sum_{i,j=1}^{k} c_{ij}M_{si}f_{rs+j} + h \sum_{i=1}^{k} M_{si}d_i f_{rs}. \tag{2.15}$$

We are now ready to formulate the basic theorem concerning the convergence of overimplicit selfstarting methods applied to the equation (2.1).

Theorem 2.1
Let there be given a selfstarting overimplicit method of order at least 1 with respect to s and let the corresponding matrix C be regular. Further, let the operator A in (2.1) be the generator of a continuous semi-group. Finally, let f(t) in (2.1) be continuous. Then the method is, for the equation (2.1), convergent, i.e. we have

$$\lim_{\substack{h \to 0 \\ t=rsh}} x_{rs} = x(t) \tag{2.16}$$

where x(t) is the generalized solution of (2.1), if and only if positive constants h_0, M_1 and ω_1 exist such that

$$\| R_s^r(hA) \| \leqslant M_1 e^{\omega_1 rh} \tag{2.17}$$

for $r = 1, 2, \ldots$ and for any $h < h_0$.

Let us note that the assertion of this theorem is very similar to the assertions which have been proved in [2] for the case when $f(t)$ in (2.1) is identically equal to zero. From Theorem 2.1, it may be also concluded that only the validity of (2.17) and the existence of bounded operators M_{sj} suffice for the convergence so that, from this point of view, we need not study the existence of the bounded operator $(I - hC \otimes A)^{-1}$. But the study of this question is important from the practical point of view since in practical computation, we will use the formula (2.5) and not (2.14). In this connection, it is necessary to realize that, strictly speaking, our method in the form (2.5) is applicable only in the case when $x_{rs} \in \mathcal{D}(A)$ for all r. In the contrary case the right-hand term of (2.5) may have no sense. The method in the form (2.14) is naturally well-defined in both these cases. Both forms of our method give identical results if they are practicable.

Theorem 2.1 answers only the question about the convergence of the selfstarting overimplicit method. It is rather clear that if we have more information about the solution of the original problem we will be able to say something also about the rate of the convergence. The following theorem can be now proved in a relatively simple way.

Theorem 2.2

Let there be given a selfstarting overimplicit method of order $p(p \geqslant 1)$
with respect to s and let C be regular. Further, let (2.17) hold. Finally, let
the solution of (2.1), (2.2) belong to $\mathscr{D}(A)$ for $t \in \langle 0, T \rangle$ and let it have
continuous derivatives up to the order $p + 1$ in $\langle 0, T \rangle$. Then the approxi-
mate solution computed by the given method converges to the exact one
with the rate h^p.

To conclude this section we show how the above results can be strength-
ened in the case of a homogeneous equation, i.e., in the case that $f(t)$ in
(2.1) is identically equal to zero. In this case we deal in fact with the
approximation of a continuous semi-group $U(t)$ by the sequence of
operators $R_s^r((t/rs)A)$ where $R_s(z)$ in the rational function given by (2.13).
If the selfstarting overimplicit method, which defines the function $R_s(z)$,
is of order p with respect to s then, obviously,

$$R_s\left(\frac{1}{s}z\right) = e^z + o(z^{l+1}) \tag{2.18}$$

for some $l \geqslant p$ where, actually, $l > p$ may occur (for example, in the case
of the trapezoidal rule, $p = 2$; the corresponding function $R(z)$ is the Padé
approximation of the exponential function and, consequently, $l = 3$). Thus,
it can be expected that in this special case the rate of convergence will be
given by the number l rather than by p. The following theorem states that
this is really true.

Theorem 2.3

Let there be given a selfstarting overimplicit method of order p with
respect to s with regular C and let the corresponding function $R_s(z)$ satisfy
(2.17) and (2.18). Further, let $U(t)$ be a continuous semi-group generated
by the operator A. Then there exists a constant K such that

$$\left\|\left(U(t) - R_s^r\left(\frac{t}{rs}A\right)\right)\eta\right\| \leqslant Kt\|A^{l+1}\eta\|\, e^{\omega_1 t}\left(\frac{t}{rs}\right)^l \tag{2.19}$$

holds for any $\eta \in \mathscr{D}(A^{l+1})$.

Note that the fact that $R_s(z)$ arose from a selfstarting overimplicit
method is not substantial here. The theorem remains true for any rational
function $R(z)$ satisfying (2.18) and such that (2.17) holds.

3. The Existence of Methods Satisfying the Assumptions of the Convergence Theorem

In the preceding section we have seen that, in the case of equation (2.1)
with an unbounded operator, the fulfillment of (2.17) determines the convergence of the selfstarting overimplicit method. For example, in the case

of the very simple explicit Euler method the corresponding operator $R_1(hA)$ is $I + hA$ and (2.17) cannot be satisfied since this operator is even unbounded for any $h > 0$. On the other hand, for the implicit Euler method,

$$R_1(hA) = (I - hA)^{-1} = \frac{1}{h}\left(\frac{1}{h}I - A\right)^{-1}$$

and from (2.3) it follows immediately that (2.17) is satisfied.

From these very simple examples, one can see that the inequality (2.17) represents probably a very restrictive requirement on the class of methods under consideration.

Let us study this question in the case that the given space B is a Hilbert space and that the given operator A is selfadjoint, in more detail. In this case, $R_s^r(hA)$—if it exists—is also selfadjoint and we have

$$\| R_s^r(hA) \| = \rho^r(R_s(hA)) \tag{3.1}$$

where $\rho(C)$ denotes the spectral radius of the operator C. Since we have assumed that A is selfadjoint the spectrum of hA is real and from this and from the spectral mapping theorem, we get

$$\rho(R_s(hA)) \leqslant \sup_{\lambda \leqslant \omega} |R_s(h\lambda)| \tag{3.2}$$

since the operator A as the generator of a semi-group has its spectrum in the half-plane Re $\lambda \leqslant \omega$. From (3.1) and (3.2) we obtain easily

Theorem 3.1
Let B be a Hilbert space and A be selfadjoint. Let $R_s(z)$ be the rational function given by a selfstarting overimplicit method of order p with respect to s and let C be regular. Further, let

$$|R_s(h\lambda)| \leqslant 1 \tag{3.3}$$

for $\lambda < 0$ and any sufficiently small h. Then (2.17) holds.

Hence, we see that if A is a selfadjoint operator in a Hilbert space then the $A(0)$-stability, in the Widlund sense, of the selfstarting method under consideration is sufficient for the convergence.

As an example of such a class we introduce the so-called selfstarting methods of Adams type which are based on interpolation and are defined by the formula

$$x_{n+i} = x_n + h \sum_{j=0}^{k} \gamma_{ij} f(t_{n+j}, x_{n+j}), \quad i = 1, \ldots, k \tag{3.4}$$

where

$$\gamma_{ij} = \int_0^i l_j(t)\, dt \tag{3.5}$$

and $l_j(t)$ is the elementary Lagrange interpolating polynomial for the points $t = 0, \ldots, k$, i.e., the polynomial of degree k with the values 0 at the points $t = 0, \ldots, k, t \neq j$ and 1 at the point $t = j$. These formulae are of order $k + 1$ and, for k even, of order $k + 2$ with respect to k. It can be easily verified that

$$R_k(z) = Q(-z)Q^{-1}(z) \tag{3.6}$$

and that the polynomial $Q(z)$ has no real negative zeros. But this implies immediately the validity of (3.3). Hence, we have constructed a class of convergent methods containing methods of arbitrarily high order (naturally in the case of a Hilbert space).

It seems almost obvious that in the case of a general operator, A-stability in the sense of Dahlquist will play the same role as was played by $A(0)$-stability in this special case. But definite results in this direction have not yet been fully developed.

In the preceding pages, we constructed some approximate methods for the solution of differential equations in abstract spaces. Naturally, this paper does not exhaust by far all aspects of the problems under consideration. For example, differential equations with a time-dependent operator have not been mentioned. The reason for this is that the situation is incomparably more complicated in this case. The substance of the difficulties which we meet here consists in the fact that the operator in the time-independent case, which corresponds to the operator $(I - hC \otimes A)^{-1}$, need not exist even under very reasonable assumptions about the given method. We know very little about these problems at present. We are able only, on the basis of the selfstarting methods of Adams type, to construct concrete formulae which converge even in the case of time-dependent problems.

References

[1] Práger, M., Taufer, J. and Vitásek, E. (1973). Overimplicit multistep methods, *Aplikace matematiky* 18, 399–421.
[2] Kato, T. (1966). *Perturbation Theory for Linear Operators.* Springer-Verlag, Berlin-Heidelberg–New York.

Unconditionally Stable Finite Element Schemes for Parabolic Equations

Miloš Zlámal

Abstract

The initial-boundary value problem for a linear parabolic equation in an infinite cylinder under the Dirichlet boundary condition is solved by applying the finite element discretization in the space dimension and a class of implicit Runge–Kutta methods in time. Bounds for the discretization error in the L_2-norm are given.

1. Introduction

For simplicity, we restrict ourselves to the initial-boundary value problem

$$\frac{\partial u}{\partial t} = Lu \quad \text{for} \quad (x, t) \in \Omega \times (0, \infty),$$

$$u = 0 \quad \text{on} \quad \Gamma \times < 0, \infty), \quad u(x, 0) = g(x) \text{ in } \Omega. \tag{1}$$

Here

$$Lu = \sum_{i,j=1}^{N} \frac{\partial}{\partial x_i} \left(a_{ij}(x) \frac{\partial u}{\partial x_j} \right) - a(x)u,$$

$$a_{ij}(x) = a_{ji}(x), \quad \sum_{i,j=1}^{N} a_{ij}(x) \xi_i \xi_j \geqslant \alpha \sum_{i=1}^{N} \xi_i^2 (\alpha > 0), \quad a(x) \geqslant 0 \tag{2}$$

and $x = (x_1, \ldots, x_N)$ is a point of a bounded omain Ω in Euclidean N-space R^N with a smooth boundary Γ. Other boundary conditions as well as the equation with a nonhomogeneous term are easy to treat in the same way.

The finite element method was applied by the engineers for the solution of heat conduction problems a number of years ago (see, e.g., Wilson and Nickell [1]). Their idea is that in the space dimension a finite element discretization is used whereas in time a finite difference method is applied. In this paper the finite element discretization is considered in finite element spaces V_h^p which are finite-dimensional subspaces of H_0^1 (by H^m we denote the Sobolev spaces $W_2^{(m)}(\Omega)$ $(m = 0, 1, \ldots)$ with the usual norm

$$\| v \|_m = \left(\sum_{|i| \leqslant m} \| D^j v \|_0^2 \right)^{1/2};$$

the scalar product in $H^0 = L_2$ is denoted by $(., .)_0$, H_0^1 is the closure of $\mathscr{D}(\Omega)$, the set of infinitely differentiable functions with compact support in Ω, in the norm $\| \cdot \|_1$) and which have the following approximation property: to any $u \in H^{p+1} \cap H_0^1$ there exists $\hat{u} \in V_h^p$ such that

$$\| u - \hat{u} \|_j \leqslant Ch^{p+1-j} \| u \|_{p+1}, \quad j = 0, 1, \tag{3}$$

C being a constant independent of the small positive parameter h and of the function u. Finite element spaces constructed first for special domains and later for arbitrary curved domains possess this property. The parameter h is, in general, the maximum diameter of all elements.

The weak form of (1) is to find, for every $t > 0$, the function $u \in H_0^1$ such that, besides the initial condition, it satisfies

$$(\dot{u}, \varphi)_0 + a(u, \varphi) = 0 \quad \forall \varphi \in H_0^1; \tag{4}$$

here

$$a(u, v) = \int_\Omega \left[\sum_{i,j=1}^N a_{ij}(x) \frac{\partial u}{\partial x_i} \frac{\partial v}{\partial x_j} + a(x)uv \right] dx.$$

A well-known approach for getting an approximate solution of the problem (4) consists in first applying the Galerkin principle to (4): The approximation $U(x, t)$ of $u(x, t)$ is that function from V_h^p which satisfies

$$(\dot{U}, \varphi)_0 + a(U, \varphi) = 0 \quad \forall \varphi \in V_h^p, \quad t > 0,$$
$$U(x, 0) = \hat{g}(x). \tag{5}$$

The function $\hat{g}(x) \in V_h^p$ is chosen according to (3). The existence and uniqueness of $U(x, t)$ follows from the following fact: Set $U(x, t) = \sum_{j=1}^l \alpha_j(t) v_j(x) = (\boldsymbol{\alpha})^T \mathbf{v}$ where $\{v_j(x)\}_{j=1}^l$ form a basis of V_h^p and put $v_k (k = 1, \ldots, l)$ into (5). We easily see that (5) is equivalent to the system of ordinary differential equations

$$\dot{\boldsymbol{\alpha}} = -A\boldsymbol{\alpha}, \quad \boldsymbol{\alpha}(0) = \hat{\mathbf{g}}. \tag{6}$$

Here $\boldsymbol{\alpha}(t)$ is the vector $(\alpha_1(t), \ldots, \alpha_l(t))^T$, $\mathbf{v}(x)$ is the vector $(v_1(x), \ldots, v_l(x))^T$, $A = M^{-1}K$, M is the mass matrix $\{(v_j, v_h)_0\}_{j, h=1}^l$, K is the stiffness matrix $\{a(v_j, v_h)\}_{j, h=1}^l$ and $\hat{\mathbf{g}}$ is the vector $(\hat{g}_1, \ldots, \hat{g}_l)^T$ defined by $\hat{g}(x) = \sum_{j=1}^l \hat{g}_j v_j(x)$. The matrices M and K are positive definite, hence A is regular.

To get a computable approximate solution of (1) we have to discretize (5) in time. Consider the time levels $t = nk, n = 1, 2, \ldots, k$ being the time increment. Denote by U^n the approximate value of $U(x, nk)$. A well-known scheme is the Crank–Nicolson–Galerkin scheme

$$(M + \tfrac{1}{2}kK)\boldsymbol{\alpha}^{n+1} = (M - \tfrac{1}{2}kK)\boldsymbol{\alpha}^n \tag{7}$$

first introduced in [1]. Douglas and Dupont [2] proved (for the nonlinear case) that it is of the second order of accuracy with respect to k. Recently some papers have appeared in which schemes of higher order of accuracy constructed (Hlaváček [3], Bramble and Thomée [4], Zlámal [5]). The main advantage of such schemes in practical computations will be that we shall be able to achieve the needed accuracy with a large time increment, i.e. using a small number of time steps.

Any discretization of (5) represents an approximate method for the solution of the initial value problem (6) (e.g., we get the Crank–Nicolson–Galerkin scheme (7) if we apply the trapezoidal method). Hence a natural approach for the construction of higher accuracy methods consists in analyzing the possibilities of approximate methods for the solution of systems (6) and in choosing suitable methods for this purpose.

One could think of choosing explicit methods. However, in contrast to finite difference schemes such a choice does not have the advantage of producing an explicit scheme. The matrix A is of the form $M^{-1}K$ and M is not a diagonal matrix. This is the reason why, even in the case of explicit methods, we have to solve a system of linear equations at every time step. An important property of the matrix $-A$ which we have to take into consideration is that its eigenvalues are negative and unbounded. If the spaces V_h^p have a certain additional property (which the finite element subspaces used in applications possess) the largest eigenvalue $\lambda_{\max}$ of A is $0(h^{-2})$. Most of the classical methods for solving ordinary differential equations require, for reasons of stability, that $|\lambda k|$ be bounded by a small number, of the order 1 to 10, for all eigenvalues λ. Such methods will provide conditionally stable schemes. The relation $\lambda_{\max} = 0(h^{-2})$ leads to a restriction of the form $k < c_1 h^2$ where c_1 is a constant (usually less than one). As we have to solve a system of equations at every time step these schemes have no practical significance. We arrive at the conclusion that we must restrict ourselves to the methods which do not require the boundedness of $|\lambda k|$, i.e. to methods with a region of absolute stability containing the interval $(-\infty, 0)$. These methods will provide unconditionally stable schemes.

One class of such methods are A-stable methods introduced by Dahlquist [6]. Here we use a family of implicit Runge–Kutta methods introduced by Butcher [7]. These are r-stage methods of order $2r$ and Ehle [8] has shown that they are A-stable. The general r-stage implicit Runge–Kutta method for the system $\dot{\alpha} = f(t, \alpha)$ is defined by

$$\alpha^{n+1} = \alpha^n + k \sum_{i=1}^{r} c_i \beta^i$$

$$\beta^i = f\left(t_n + a_i k, \alpha^n + k \sum_{j=1}^{r} b_{ij} \beta^j\right), \quad a_i = \sum_{j=1}^{N} b_{ij}$$

(8)

The constants b_{ij}, c_i in Butcher's schemes are obtained by means of Gaussian quadrature formulas. For the system (6) we get

$$M\boldsymbol{\beta}^i + k \sum_{j=1}^{r} b_{ij}K\boldsymbol{\beta}^j = -K\boldsymbol{\alpha}^n, \quad i = 1, \ldots, r, \tag{9}$$

and this system together with (8) defines the approximate solution $U^n = (\boldsymbol{\alpha}^n)^T \mathbf{v}(x)$. Alternatively, U^n can be defined in a variational form:

$$(\beta^i, \varphi)_0 + k \sum_{j=1}^{r} b_{ij}a(\beta^j, \varphi) = -a(U^n, \varphi) \quad \forall \varphi \in V_h^p, \quad i = 1, \ldots, r$$

$$\tag{10}$$

$$(U^{n+1}, \varphi)_0 = (U^n, \varphi)_0 + k \sum_{i=1}^{r} c_i(\beta^i, \varphi)_0 \quad \forall \varphi \in V_h^p;$$

here $\beta^i(x) = (\boldsymbol{\beta}^i)^T \mathbf{v}(x)$.

II. Preliminaries

For simplicity, we assume that

$$a_{ij}(x), a(x), g(x) \in C^\infty(\bar{\Omega}), \quad \Gamma \in C^\infty. \tag{11}$$

We first state some facts about the solution $u(x, t)$ of (1). It is of the form $\sum_{i=1}^{\infty} g_i e^{-\lambda_i t} \psi_i(x)$ where λ_i and $\psi_i(x)$ are (positive) eigenvalues and eigenfunctions, respectively, of the problem $-L\psi = \lambda\psi$, $\psi/\Gamma = 0$ and g_i are the Fourier coefficients of the initial value $g(x)$. Ladyženskaja [9, Chapter III, Section 17] showed that if $g \in H^m$ and

$$g/\Gamma = Lg/\Gamma = \cdots = L^{[m-1/2]}g/\Gamma = 0, \tag{12}$$

then $u(x, t) \in H^m$ for $t \geq 0$. Conversely, if $u(x, t) \in H^m$ for $t \geq 0$, then $g \in H^m$ and (12) is satisfied. The proof is based on two inequalities. The first holds for any series $\sum_{i=1}^{\infty} g_i \psi_i(x)$:

$$\left\| \sum_{i=1}^{\infty} g_i \psi_i(x) \right\|_m^2 \leq C \sum_{i=1}^{\infty} \lambda_i^2 g_i^2 \tag{13}$$

(in the sequel, C is a generic constant, not necessarily the same in any two places, which does not depend on h, k, n, l, g). Concerning the other, we need the following consequence: if $g \in H^m$ and (12) is satisfied, then it follows that

$$\sum_{i=1}^{\infty} \lambda_i^m g_i^2 \leq C \|g\|_m^2.$$

An important property of Butcher's formulas is that if we apply them to the scalar equation $\dot{y} = -\lambda y$ we get $y^{n+1} = R_r(\lambda k)y^n$ where $R_r(\tau) = N_r(\tau)/D_r(\tau)$ is the Padé diagonal approximation of the function $e^{-\tau}$. We shall make use of the following properties of $R_r(\tau)$ (see e.g., Varga [10]): for $\tau \geqq 0$ we have

$$D_r(\tau) \geqq 1, \quad |R_r(\tau)| \leqq 1, \quad |e^{-\tau} - R_r(\tau)| \leqq C\tau^{2r+1} \tag{15}$$

(the last inequality is usually stated for τ sufficiently small; however, as $|e^{-\tau} - R_r(\tau)| \leqq 2$ it must hold for all $\tau \geqq 0$).

III. Convergence

Theorem
Let $g(x)$ satisfy (12) *with* $m = \max (p + 1, 4r)$ *and let u^n denote the value $u(x, nk)$. Then, for arbitrary h and k, the discretization error is bounded by*

$$\sup_{1 \leqq n < \infty} \| u^n - U^n \|_{L_2} \leqq \| u^0 - U^0 \|_{L_2} + C(h^{p+1} + k^{2r}) \lg \frac{1}{k} \| g \|_m. \tag{16}$$

Proof
(a) If we apply (8) to the scalar equation $\dot{y} = -\lambda y$ we get $y^{n+1} = y^n + k \times \Sigma_{i=1}^r c_i \beta^i$ where the β^i form the solution of the system $\beta^i + k\lambda \Sigma_{j=1}^r b_{ij}\beta^i = -\lambda y^n, i = 1, \ldots, r$. Let $x_\nu^i(\tau)$ $(\nu = 1, \ldots, r)$ be the solutions of

$$x_\nu^i(\tau) + \tau \sum_{j=1}^r b_{ij} x_\nu^j(\tau) = \begin{cases} 1 & i = \nu \\ 0 & i \neq \nu \end{cases} \quad i = 1, \ldots, r. \tag{17}$$

Evidently, the $x_\nu^i(\tau)$ are of the form

$$x_\nu^i(\tau) = \frac{Q_{i\nu}(\tau)}{Q(\tau)} \tag{18}$$

where $Q_{i\nu}(\tau)$ and $Q(\tau)$ are polynomials of degree $r - 1$ and r, respectively. We can express the β^i by means of x_ν^i: $\beta^i = \Sigma_{\nu=1}^r x_\nu^i(k\lambda) (-\lambda y^n)$. Hence $y^{n+1} = [1 - \Sigma_{i=1}^r \Sigma_{\nu=1}^r c_i k\lambda x_\nu^i(k\lambda)]y^n$. On the other hand, we have mentioned above that $y^{n+1} = R_r(k\lambda)y^n$. As a consequence we have

$$Q(\tau) = D_r(\tau), \quad 1 - \sum_{i=1}^r \sum_{\nu=1}^r c_i \tau x_\nu^i(\tau) = R_r(\tau). \tag{19}$$

(b) We write u^n in the form $u^n = \xi^n + \eta^n$ with $\eta^n \in V_h^p$ being the Ritz approximation of u^n, i.e. the orthogonal projection of u^n onto V_h^p with the energy norm $[a(.,.)]^{1/2}$ (several authors have used this decomposition; we learned it from [4]). Hence

$$a(\eta^n, \varphi) = a(u^n, \varphi) = (-Lu^n, \varphi)_0 \quad \forall \varphi \in V_h^p$$

and with respect to (3) we find that $\| \xi^n \|_0 \leqq Ch^{p+1} \| Lu^n \|_{p-1} \leqq Ch^{p+1} \| u^n \|_{p+1}$. By means of (13) and (14), we immediately obtain $\| \xi^n \|_0 \leqq Ch^{p+1} \| g \|_{p+1}$. Therefore $\| u^n - U^n \|_0 \leqq \| \eta^n - U^n \| + Ch^{p+1} \| g \|_{p+1}$ and it is sufficient to estimate $\epsilon^n = \eta^n - U^n \in V_h^p$.

(c) U^{n+1} is defined by (10). For η^{n+1} we write the identity

$$(\eta^{n+1}, \varphi)_0 = (\eta^n_{\cdot}, \varphi)_0 + k \sum_{i=1}^r c_i(z^i, \varphi)_0 + (\pi^n - \omega^n, \varphi)_0$$

$$\forall \varphi \in H^1_0, \tag{20}$$

$$\pi^n = u^{n+1} - u^n - k \sum_{i=1}^r c_i z^i, \quad \omega^n = \xi^{n+1} - \xi^n,$$

and we choose for z^i the solution of the elliptic system

$$(z^i, \varphi)_0 + k \sum_{j=1}^n b_{ij} a(z^j, \varphi) = -a(u^n, \varphi) \quad \forall \varphi \in H^1_0, \quad i = 1, \ldots, r \tag{21}$$

(we prove later that the z^i exist and belong to H^{p+1}). We decompose z^i in the same way as u^n: $z^i = \lambda^i + \mathcal{H}^i$, $a(\mathcal{H}^i, \varphi) = a(z^i, \varphi) \; \forall \varphi \in V^p_h$. Then we have

$$(\mathcal{H}^i, \varphi)_0 + k \sum_{j=1}^n b_{ij} a(\mathcal{H}^i, \varphi) = -a(\eta^n, \varphi) - (\lambda^i, \varphi)_0 \quad \forall \varphi \in V^p_h. \tag{22}$$

Setting $\delta^i = \mathcal{H}^i - \beta^i$ we get from (10) and (20)

$$(\epsilon^{n+1}, \varphi)_0 = (\epsilon^n, \varphi)_0 + k \sum_{i=1}^r c_i(\delta^i, \varphi)_0 + (\pi^n - \omega^n + k \sum_{i=1}^r c_i \lambda^i, \varphi)_0 \tag{23}$$

$$\forall \varphi \in V^p_h$$

We write this equation in a matrix form. For this purpose, let $\mathbf{w}$ be the vector $\mathbf{w} = M^{-1/2}\mathbf{v}$ ($\mathbf{v}$ is the basis vector: see Introduction) and let us define the vectors $\boldsymbol{\epsilon}^n$, $\boldsymbol{\delta}^i$ by $\epsilon^n = (\boldsymbol{\epsilon}^n)^T \mathbf{w}$, $\delta^i = (\boldsymbol{\delta}^i)^T \mathbf{w}$. Since $(\mathbf{v}, \mathbf{v}^T)_0 = M$ and $a(\mathbf{v}, \mathbf{v}) = K$, we have $(\mathbf{w}, \mathbf{w}^T)_0 = I$ and $a(\mathbf{w}, \mathbf{w}^T) = M^{-1/2}KM^{-1/2}$. The matrix $S = M^{-1/2}KM^{-1/2}$ is symmetric and positive definite. Putting the components w_i ($i = 1, \ldots, l$) of the vector $\mathbf{w}$ for φ into (23), we get

$$\boldsymbol{\epsilon}^{n+1} = \boldsymbol{\epsilon}^n + k \sum_{i=1}^r c_i \boldsymbol{\delta}^i + \boldsymbol{\pi}^n - \boldsymbol{\omega}^n + k \sum_{i=1}^r c_i \boldsymbol{\lambda}^i \tag{24}$$

with

$$\boldsymbol{\pi}^n = (\pi^n, \mathbf{w})_0, \quad \boldsymbol{\omega}^n = (\omega^n, \mathbf{w})_0, \quad \boldsymbol{\lambda}^i = (\lambda^i, \mathbf{w})_0.$$

From (10) and (22) it follows that

$$(\delta^i, \varphi)_0 + k \sum_{j=1}^r b_{ij} a(\delta^j, \varphi) = -a(\epsilon^n, \varphi) - (\lambda^i, \varphi)_0 \quad \forall \varphi \in V^p_h,$$

hence

$$\boldsymbol{\delta}^i + \sum_{j=1}^r b_{ij} k S \boldsymbol{\delta}^j = -S\boldsymbol{\epsilon}^n - \boldsymbol{\lambda}^i, \quad i = 1, \ldots, r.$$

By (17), we easily verify that

$$\delta^i = - \sum_{\nu=1}^{r} x_\nu^i(kS) S \epsilon^n - \sum_{\nu=1}^{r} x_\nu^i(kS) \lambda^i.$$

Setting this formula into (24) and taking into account the second identity in (19) we obtain

$$\epsilon^{n+1} = R_r(kS)\epsilon^n + f^n,$$

$$f^n = \pi^n - \omega^n + k \sum_{i=1}^{r} c_i \left[I - \sum_{\nu=1}^{r} x_\nu^i(kS) \right] \lambda^i. \tag{25}$$

Denote by $\|\cdot\|$ the Euclidean norm of a vector or a matrix. From the definition of the vector ϵ^n it follows that $\|\epsilon^n\|_0 = \|\epsilon^n\|$ and from (25) and the second property (15) (notice that S is positive definite) we easily obtain $\|\epsilon^n\|_0 \leq \|\epsilon^0\|_0 + \Sigma_{j=0}^{n-1} \|f^j\|$; recalling (b) we get

$$\|u^n - U^n\|_0 \leq \|u^0 - U^0\|_0 + Ch^{p+1}\|g\|_{p+1} + \sum_{j=0}^{n-1} \|f^j\|. \tag{26}$$

(d) Let $f \in L_2(\Omega)$ and let $\hat{f}$ be the orthogonal projection of f onto V_h^p with the norm $\|\cdot\|_0$. Then $\hat{f} = (\hat{\mathbf{f}})^T w$ where $\hat{\mathbf{f}} = (f, w)_0$ and $\|\hat{\mathbf{f}}\| = \|\hat{f}\|_0 \leq \|f\|_0$. Therefore $\|\pi^n\|_0$, $\|\omega^n\|_0$ and $\|\lambda^i\|_0$ are bounds for $\|\pi^n\|$, $\|\omega^n\|$ and $\|\lambda^i\|$, respectively. We begin with the estimation of $\|\lambda^i\|_0$.

Since $\lambda^i = z^i - \mathcal{H}^i$, where $\mathcal{H}^i$ are the Ritz approximations of z^i, we have $\|\lambda^i\|_0 \leq Ch^{p+1}\|z^i\|_{p+1}$ provided z^i belong to H^{p+1}. Denote by z_s^i the Fourier coefficients of z^i and assume z^i has the form $z^i = \Sigma_{s=1}^{\infty} z_s^i \psi_i(x)$. From (21) it follows that

$$z_s^i + k\lambda_s \sum_{j=1}^{r} b_{ij} z_s^j = -\lambda_s u_s^n, \quad i = 1, \ldots, r, \tag{27}$$

where $u_s^n = e^{-nk\lambda_s} g_s$ are the Fourier coefficients of u^n. Hence $z_s^i = -\Sigma_{\nu=1}^{r} x_\nu^i(k\lambda_s)\lambda_s u_s^n$. By (18), (19) and (15), $x_\nu^i(\tau)$ are bounded for $\tau \in \langle 0, \infty)$, thus $(z_s^i)^2 \leq C\lambda_s^2 e^{-2nk\lambda_s} g_s^2$. By (14), it follows for $n \geq 1$ that

$$\sum_{s=1}^{\infty} \lambda_s^{p+1}(z_s^i)^2 \leq C \sum_{s=1}^{\infty} \lambda_s^2 e^{-2nk\lambda_s}\lambda_s^{p+1} g_s^2$$

$$\leq Ck^{-2}e^{-nk\lambda_1} \sum_{s=1}^{\infty} (k\lambda_s e^{-1/2nk\lambda_s})^2 \lambda_s^{p+1} g_s^2$$

$$\leq Cn^{-2} e^{-nk\lambda_1} k^{-2} \|g\|_{p+1}^2.$$

Hence by (13), $z^i \in H^{p+1}$ and $\|z^i\|_{p+1} \leq Cn^{-1} e^{-1/2nk\lambda_1} k^{-1}\|g\|_{p+1}$; consequently $\|\lambda^i\|_0 \leq Cn^{-1}e^{-1/2nk\lambda_1} k^{-1} h^{p+1}\|g\|_{p+1}$ and

$$k\left\| \sum_{i=1}^{r} c_i \left[I - \sum_{\nu=1}^{r} x_\nu^i(kS) \right] \lambda^i \right\| \leq Cn^{-1} e^{-1/2nk\lambda_1} h^{p+1}\|g\|_{p+1}, \quad n \geq 1. \tag{28}$$

Also, we easily find that

$$k\left\|\sum_{i=1}^{r} c_i[I - x_\nu^i(kS)]\boldsymbol{\lambda}^i\right\| \leq Ch^{p+1}\|g\|_{p+1}, \quad n = 0. \tag{29}$$

Let us estimate $\|\pi^n\|_0$. Set $\pi^n = \sum_{s=1}^{\infty} \pi_s^n \psi_s(x)$. We have $\pi_s^n = u_s^{n+1} - u_s^n - k\sum_{i=1}^{r} c_i z_s^i$. Consider the initial-value problem $\dot{y} = -\lambda_s y$, $y(t_n) = u_s^n$ and solve it by the Runge–Kutta method (8). We get $y^{n+1} = u_s^n + k\sum_{i=1}^{r} c_i \beta^i$ where $\beta^i + k\lambda_s \sum_{j=1}^{r} b_{ij}\beta^j = -\lambda_s u_s^n$. From (27) it follows that $\beta^i = z_s^i$, hence $\pi_s^n = u_s^{n+1} - y^{n+1} = u_s^{n+1} - R_r(k\lambda_s)y^n = u_s^{n+1} - R_r(k\lambda_s)u_s^n = e^{-nk\lambda_s}[e^{-k\lambda_s} - R_r(k\lambda_s)]g_s$. By the last inequality in (15), we obtain $(\pi_s^n)^2 \leq Ce^{-nk\lambda_1}e^{-nk\lambda_s}k^{4r+2}\lambda_s^{4r+2}g_s^2$ for $n \geq 1$. By (14),

$$\|\pi^n\|_0 \leq Cn^{-1}e^{-1/2nk\lambda_1}k^{2r}\|g\|_{4r}, \quad n \geq 1. \tag{30}$$

In a similar way, we prove that

$$\|\pi^0\|_0 \leq Ck^{2r}\|g\|_{4r}. \tag{31}$$

As far as $\omega^n = \xi^{n+1} - \xi^n$ is concerned we have $\|\omega^n\|_0 \leq Ch^{p+1}\|u^{n+1} - u^n\|_{p+1}$. We leave out the details and state the final bounds:

$$\begin{aligned}\|\omega^n\|_0 &\leq Cn^{-1}e^{-1/2nk\lambda_1}h^{p+1}\|g\|_{p+1}, \quad n \geq 1 \\ \|\omega^0\|_0 &\leq Ch^{p+1}\|g\|_{p+1}.\end{aligned} \tag{32}$$

(e) From (28), (29), (30), (31) and (32) it follows that

$$\|\mathbf{f}^0\| \leq C[h^{p+1} + k^{2r}]\|g\|_m$$

$$\|\mathbf{f}^n\| \leq C[h^{p+1} + k^{2r}]n^{-1}e^{-1/2nk\lambda_1}\|g\|_m, \quad n \geq 1$$

where $m = \max(p + 1, 4r)$. Therefore

$$\sum_{n=0}^{\infty}\|\mathbf{f}^n\| \leq C[h^{p+1} + k^{2r}]\|g\|_m\left[1 + \sum_{n=1}^{\infty} n^{-1}e^{-1/2nk\lambda_1}\right]. \tag{33}$$

Now

$$\sum_{n=1}^{\infty} n^{-1}e^{-1/2nk\lambda_1} = \sum_{n=1}^{\infty} n^{-1}(e^{-1/2k\lambda_1})^n = -\lg(1 - e^{-1/2k\lambda_1}) = 0\left(\lg\frac{1}{k}\right)$$

which together with (33) and (26) proves (16).

Remarks
1) The approach of this paper can be used for deriving error bounds of other methods, e.g. the Calahan method applied in [5].
2) The implicit Runge–Kutta methods (8) suffer from the disadvantage that we have to solve at every step a system of rl equations. This disadvantage disappears if we use linear multistep methods for the solution of the system (6). It is true that the order of an A-stable linear multistep method cannot exceed two. However, A-stability is a stronger property than we

need for our purpose. As all eigenvalues of the matrix $-A$ are negative A_0-stability is sufficient. Such methods are analyzed and error bounds given in [11].

Added: Unconditionally stable schemes of high order are also constructed and error bounds derived in Thomée [12].

Added in proof: The Theorem is generalized in the thesis of M. Crouzeix: Sur l'approximation des équations différentielles opérationnelles linéaires par des méthodes de Runge–Kutta (Université Paris VI, to appear).

References

[1] Wilson, E. L. and Nickell, R. E. (1966). Application of finite element method to heat conduction analysis, *Nuclear Eng. Design* 4, 276–286.
[2] Douglas, J. Jr. and Dupont, T. (1970). Galerkin methods for parabolic equations, *SIAM J. Numer. Anal.* 7, 576–626.
[3] Hlaváček, I. (1972, 1973). On a semi-variational method for parabolic equations. I, II, *Aplikace Mat.* 17, 327–351; 18, 43–64.
[4] Bramble, J. H. and Thomée, V. (1974). Discrete time Galerkin methods for a parabolic boundary value problem, *Ann. Mat. Pura Appl.*, to appear.
[5] Zlámal, M. (1974). Finite element methods for parabolic equations, *Math. Comp.* 28, 393–404.
[6] Dahlquist, G. (1963). A special stability problem for linear multistep methods, *BIT* 3, 27–43.
[7] Butcher, J. C. (1964). Implicit Runge–Kutta processes, *Math. Comp.* 18, 50–64.
[8] Ehle, B. L. (1969). On Padé approximations to the exponential function and *A*-stable methods for the numerical solution of initial value problems, *Univ. of Waterloo Dept. Appl. Analysis and Comp. Science, Research Rep. No. CSSR 2010.*
[9] Ladyženskaja, O. A., Solonnikov, V. A. and Ural'ceva, N. N. (1968). Linear and quasilinear equations of parabolic type, *Translations Math. Monographs* 23, Amer. Math. Soc., Providence R. I.
[10] Varga, R. S. (1961). On higher order stable implicit methods for solving parabolic partial differential equations, *J. Math. Phys.* 40, 220–231.
[11] Zlámal, M. (1975). Finite element multistep discretizations of parabolic boundary value problems, Math. Comp. 29.
[12] Thomée, V. Some convergence results for Galerkin methods for parabolic boundary value problems, to appear.

Author Index

Numbers followed by asterisks refer to pages on which references are listed at the end of chapters.

A

Abramowitz, M., 65, 75*
Alford, R. M., 63 76*
Alterman, Z. S., 60, 61, 62, 63, 76*
Ansorge, J. P., 163 (2), 173*
Ansorge, R., 18 (2, 9), 25*
Atkinson, F. V., 1 (1), 2 (1), 4 (1), 9 (1), 10 (1), 11 (1), 14 (1), 14*
Aubin, J. P., 170 (1), 172*

B

Barnhill, R. E., 91 (1), 106*
Bauer, F. L., 186 (1), 195*
Bellen, A., 160 (1), 160*
Bellman, R., 95 (2), 97 (2), 106*
Bernstein, S. N., 179 (11), 184*
Bessaga, C., 157 (2), 160*
Bianchini, R. M., 157 (3), 160*
Bielecki, A., 197 (1), 211*
Birkhoff, G., 91 (3), 94 (15), 106*
Blevins, M, M., 193 (3), 195*
Bond, T. J., 117 (6), 120*
Boore, D. M., 60, 63, 76*
Bramble, J. H., 18 (8) 25*, 29 (1, 2), 40*, 89 (5), 93 (4), 106*, 255 (4), 257 (4), 261*
Brezzi, F., 141 (1), 142 (1, 2, 3), 144 (1), 154*
Browder, F. E., 157 (4), 160*
Butcher, J. C., 255 (7), 261*

C

Chen, T. C., 51 (1), 53 (1), 56*

Ciarlet

Ciarlet, P. G., 29 (3), 40*, 89 (6), 106 (6), 106* 141 (4), 154*
Čirič, L. B., 157 (5), 160*
Clement, Ph., 29 (4), 40*
Clint, M., 186 (4), 196*
Coffman, D., 1 (2), 14*
Collatz, L., 1 (3), 14* 18 (3, 4), 25*, 90 (7, 8), 100 (8), 106*
Coppel, W., 85 (1), 86*
Crouzeix, M., 150 (5), 154*
Cryer, C. W., 79 (2), 86*

D

Dahlquist, G., 198 (2), 211*, 255 (6), 261*
Davis, P. J., 18 (7), 25*, 103 (9), 106*
De Boor, Carl, 122 (1), 125 (1), 132*
Descloux, J., 28 (5), 29 (6), 40*, 89 (19), 90 (19), 93 (19) 107*
Dieudonné, J., 97 (10), 106*
Dobrin, M. B., 57, 76*
Douglas, J. Jr., 255 (2), 261*
Dupois, G., 29 (7), 40*
Dupont, T., 255 (2), 261*

E

Ehle, B. L., 255 (8), 261*

F

Fichera, G., 30 (8), 40*
Filippow, A. F., 165 (6), 173*
Fix, G. J., 89 (31), 104 (31), 107*, 141 (13), 155*
Fix, G. J., 89 (31), 104 (31), 107*

263